Bootleg Skies

A novel by

Paul Berge

PERFORMANCE PUBLICATIONS

FIRST EDITION

SECOND RELEASE
Ahquabi House Publishing
1184 Hwy. 349
INDIANOLA, IA 50125
Ph. 515-961-0654

Chapter
I

1929

Jake Hollow woke without opening his eyes. Wherever he was, it was quiet, and Jake rarely stayed where it was quiet. He knew if he opened his eyes something would hurt, like waking after a night of drinking and realizing in that second before complete consciousness that a hangover was at hand.

This was no hangover. Fear, vague and cold, kept him from trying to push through the fog. To do so would mean confronting the source of the fear, and he didn't want to know what that was.

A storm.

He remembered a storm, and an airplane, his airplane, which meant he must have been a pilot. This was better, worth remembering, except the storm. Something about that, and the pain he knew awaited him the moment his eyes cracked. His thoughts were jumbled, and an eye tried to pry itself open, only to be stabbed by a bright light, and it closed.

An airplane, he thought. I have an airplane...and there was a storm...

The sky, no longer gray, was boiling itself into a sickly green. Jake Hollow was barely ahead of the cold front, and below, the fields of young corn were, as yet, still. Behind his biplane, however, he could see the trees beginning to nod in the wind, turning silver undersides of their leaves toward the vanishing sun. It wasn't the first time he had ridden the crest of a thunderstorm, but it didn't make it any easier. Jake had fueled his Standard J-1 biplane at a filling station in western Iowa and was hoping to make it to Boone–about eighty miles northeast. He could have easily remained clear of the front had he been able to resist the station owner's offer to trade the gas for a ride over his brother's farm, "...just west of town."

"It'll have to be quick," Jake said. He spoke with the easy confidence of one in control of himself and the world around

1

him. "I don't like playing with thunderstorms, and that one's moving right along." He felt the last five dollars in his pocket, knowing he would be broke again after paying for the fuel. Twenty-two gallons at twenty cents per left little for room and board.

"Oh, it isn't far," the station owner, a short round man with a sweaty forehead and hairy ears, said. "Won't take but a minute."

"Okay. Help me push it away from the pumps. No, don't push there, push on the wing itself. That strut's been patched twice too often."

The twelve-foot tall biplane jerked awkwardly as the two men pushed it onto the dirt road in front of the station. The tailskid dug itself into the many ruts. The station owner, named David, was a blotter of sweat by the time the biplane was pointed down the highway.

Jake knew his next paycheck would come on Saturday night after he flew over the graduating class of Boone High School. With the fuel paid for in advance, from someone else's pocket, he could look forward to his first night in a dry bed since leaving Amarillo. He had left with enough money to reach Lincoln, Nebraska for an appointment at the Standard Aircraft Factory, but the urge to pick up extra change by hopping rides along the way had depleted his reserves. Business had been slower than anticipated. Barnstorming was, anymore, a poor way to earn a living. If a pilot could earn back his expenses he was making it. Jake was just shy of making it.

An offer of confirmed cash in Boone for the graduation airshow had set him on his present course.

"Hop in the front seat," Jake said. "Put your seat belt on and don't touch anything."

"Wait, I've got to tell Mother where I'm going..." David waddled off.

"No, there isn't time!"

"Don't be silly; just be a minute," the fat man called as he made for the house. "Mother? Where are you?"

A screen door slammed, and he disappeared inside. Jake kicked at the dusty road and watched the sky churning to the west. David's flat midwestern voice carried through the open windows into the street.

"...Oh sure it would be safe—government would never let planes fly around if they weren't safe. Anyhow, this fellow shot down twelve Germans in the war–he told me so...what's that? Yes, he does look a little young, but...No, he won't do any loop-de-loops...We're just going up over Eddie's farm. What's that? Of course you have to wear a parachute; that's the law."

"We're ready to go!" David called from the porch. He lead a gray woman, shorter and rounder than himself, down the steps. It was a painfully slow process with Mother stopping twice to whisper in her son's ear while pointing suspiciously at Jake.

"I don't think we can fit the two of you in there." Jake's voice was politely strained. The sky to the west rumbled.

"Fine. Take me first, then come back and give Mother a ride. You got plenty of time before this rain hits."

"But..."

"You just sit here, Mother," David shouted. "And watch the gas pumps."

"You be careful," Mother called as any mother would.

Six landings later, with half the town's eighteen inhabitants looking on, Jake unloaded the Curry twins—Pat and Cathy. Jake worked his way up the side of the fuselage with the gasoline hose to begin refueling. David's brother, Eddie, had seen the biplane circle his farm and, after the second pass, decided it was time to drive into town and see just what was going on. Eddie hated to drive the Ford alone, so Barbara and the girls had to come along, and, of course, they, too, wanted rides with the fellow who had shot down twenty Germans.

The glass dome atop the fuel pump had drained, and David worked the long handle back and forth to refill it. Suddenly, the trees on either side of the road shook from the storm's initial blast.

3

"Great," Jake muttered. The first drops struck the hot engine with a frying pan hiss.

"Well, we're gettin' home before this thing gets too bad," Eddie announced, and he gathered stray daughters and a wife into the truck. "Thanks for the ride!" The truck bounced down the road, the girls knocking about loosely in the bed. The Curry twins ran home with the first serious clap of thunder. Lightning tracked spider-like across the approaching gray wall.

"Never mind the rest!" Jake hollered. "Help me push it back."

Leaning against the wing, they rolled the flying machine onto the street. Jake ran around to the cockpit, reached inside and snapped on the magneto switch.

"Please start, please, please, please start," he muttered and ran to the propeller, took hold of one tip and gave it a swing.

The OX-5 motor fired sending greasy smoke across the fuselage and into the summer air. Jake was around the wing and into the cockpit before David could speak. A quick wave, a glance around the nose, and the throttle came forward. A racket never heard before in the town echoed between the clapboard houses. The biplane rolled a short distance, its tail rose, and the whole plane lifted from the road sending a cloud of rain-pelted dust over David and the pumps. Mother, from her seat on the porch, waved vacantly at the noise.

Jake turned northeast, toward Boone, keeping ahead of the squall line. The storm, however, refused to track straight west to east and, instead, moved from southwest to northeast. This put Jake's course roughly parallel to the storm's leading edge with the two lines slowly converging.

In the rough air two-hundred feet above the ground, the biplane bounced and pitched. Jake fought to keep upright. "Can't say I like this," he said through clenched teeth.

He strained to see through the wall of rain where Boone should have been. The town's airport was not scheduled for opening dedication until August, and the school's vice principal had told him over the phone, "...it would be best not to land at the airport until it's officially dedicated." He suggested Jake use the gravel road beside the football field. The school had no football team, but it did have a football field.

The football field in Boone might as well have been in Burma for all Jake's chances of making it through the squall. It became apparent that the Boone High School, class of 1929, was in serious jeopardy of losing its commencement aerial display when the Standard bored into the rear of the thunderstorm, and Jake stared at a dark mass of cumulus cloud spitting lightning.

The decision whether to press on and challenge the tempest or turn for safer air was made without the slightest hesitation. With a shuddering gasp and burp the OX-5 motor fell silent, its thick wooden propeller stuck helplessly still in the wind.

"OH BEANS!" Jake mouthed the words, while he quickly checked throttle, magneto and fuel valve, finding all in proper order. A glance at the fuel gauge resolved all confusion–he was out of gas. He had never finished refueling at the last stop and had been too concerned with the storm to consider the consequences. He pushed the stick forward to hold airspeed and turned into the wind for the short glide onto a rough corn field.

The biplane, with its dead propeller, slid across a row of hedge trees, rose abruptly on a vagabond updraft only to drop heavily onto both wheels when the air quit lifting. Jake's teeth snapped together with a sharp crack. He held the joystick against his stomach, keeping the nose high, but the impact was too much. The right wheel broke free of its axle, and the shock, not yet dissipated, continued up the landing gear, splintering the strut. What energy still remained from the impact then tore the strut from the fuselage.

Jake still held the joystick in his lap when the airplane, wounded but not dead, rose painfully into the air before it ran out of airspeed and crashed heavily onto the one remaining gear leg collapsing it with ease into the plane's belly. Jake's mouth slammed shut a second time, and he bit hard into his tongue. A warm salt taste flowed. The abrupt stop in the corn field sent his forehead smashing into the instrument panel, shattering the altimeter's glass face.

Slowly, Jake leaned back in his seat and listened to the hiss of rain on the still hot engine. His mouth was thick with blood and a quickly swelling tongue. Blood ran into his left eye from the half-moon gash on his forehead where it had hit the altimeter. The joystick was still clasped tightly against his stomach, and he pushed it away. He released his seat belt.

The engine ticked and hissed. Jake slid from the open cockpit onto the lower wing and then to the muddy field. The broad wing looked out of place resting on the smashed cornstalks and mud. He spit blood, being careful not to hit the wing's fabric. The absurdity of the precaution made him laugh painfully. Red and white flashes of light swam across his blurred vision. At last, the sky opened up, dumping sheets of straight falling rain onto the wreckage. Tilting his face up, he let the warm water rinse the blood clear.

Across the field, a flimsy truck with a flat bed bounced wildly through the corn rows toward him. The driver clung tenaciously to the wheel that kicked with every rut, while a single windshield wiper slapped with ineffective vigor against the rain. The truck stopped several feet from the biplane's nose, and the driver stared through the rain-distorted glass. Jake leaned casually against the propeller, undamaged in the landing, and gave a nonchalant wave.

"Mind if I park here a while?" Red and white flashes criss-crossed his brain, his knees buckled, and his face hit the mud.

Chapter
II

1979

Terry Marcin, 39, was an aircraft mechanic for "Wendall Air Service", Boone Municipal Airport, Boone, Iowa. He had taken the job three years earlier when he and his cat, Eloise, arrived in his Plymouth Valiant with a quarter tank of gas and nine dollars in his wallet. It had been close to nightfall in late November. The heater had been blowing cold damp air across his knees since he had left his wife on Thanksgiving Day after an animated discussion about the proper squeezing of toothpaste tubes had ended with a fully-dressed turkey being kicked across a snow-covered lawn. This proved to be the denouement to an already strained relationship.

Terry had left the Chicago suburbs, heading west, with no destination in mind. Economics, however, dictated he go no further than central Iowa. The fact that there was an airport in Boone offered a vague opportunity for shelter. Harboring no desire to spend another cramped night in the Valiant with Eloise, he accepted an offer of four dollars per hour and free use of a small room in the main hangar.

He was proclaimed chief mechanic and charged with repairing leaky faucets, mowing around hangars and in between, working on airplanes. Wendall's personal plane, a twin-engine Baron, took most of Terry's time, and when he wasn't inside the cowling, he was under the belly with a rag and can of wax.

After nine months in Boone, Terry had settled all accounts with his ex-wife through the services of an attorney named-Marshall Scranton. Terry was awarded custody of Eloise and the Plymouth, while Mary Rose, his ex, kept the house, furniture and related wares. Mr. Marshall Scranton assured Terry this was an equitable settlement, as a court fight could only prove ugly with the incident involving the turkey on the lawn certain to emerge, underscoring Mary Rose's allegations of mental cruelty on Terry's part. Terry signed the papers and kept the cat.

Three years of monkish seclusion in the rear of Wendall's hangar had allowed Terry to rebuild a 1946 Aeronca Champ he had found at a farm sale while looking for furniture. No one at the sale had recognized the odd pile of tubes and ribs as being an airplane. Now, three years later, he was still without table or chair, but in the corner of the hangar, outside his apartment door, sat the Champ.

Its wings were a smooth white, its fuselage the lightest shade of yellow he could mix. The top of the cowling was black to eliminate glare into the cockpit, and thin black lines outlined various portions of the struts and landing gear with subtle economy.

He was proud of his work, and public commentary ranged from the polite: "Isn't that nice," to, "Of course, you know that's not the way they came from the factory." Terry ignored the critics and managed to enjoy the machine.

Boone was a business airport in the center of the agri-industry. The corporate flight world moved bodies worth money, and hardly a nod was afforded Terry or the curious taildragger.

At sunset, when the men in the twins crawled into their stationwagons and headed for home, Terry would roll the Champ from the hangar and do the one thing he truly loved—fly.

It was early June after the office had closed, and the sun was still two hours from the horizon. Terry pulled the Aeronca Champ from the hangar and onto the deserted ramp. The air was warm and fragrant with the odors of spring—plowed fields, new flowers, livestock. His preflight inspection was a loving ritual allowing him to run his fingers along the smooth finish picking off dried bugs and admiring his workmanship.

The surface was like wet glass. Each tape over the ribs had been painstakingly stretched, doped, and the crimped edges pressed down with his fingertips, then redoped, and pressed again until they blended in with the mirror glaze of the

8

surrounding fabric. He had done it alone, the way he did most things. Solitude had become his companion.

Happiness was never an issue, he often mused. The occasional visit to a woman he had met at the library kept him in touch with the human world of emotional need. He never considered himself lonely—alone maybe—but never lonely. Lonely persons, he reasoned, were miserable. They tended to sit apart at weddings and cry, or sigh after brushing their teeth at night.

"Terry!"

He turned. Wendall Grieve, his employer, was marching toward him. The man was in his late seventies, stood over six-feet tall, smoked unfiltered Camel cigarettes, and often visited the airport after hours. He wore a baseball cap low across his eyes. Several times Terry had accompanied him to Little League games, thinking, at first, Wendall was merely showing support for the team he sponsored. It soon became apparent, however, Wendall Grieve loved to sit through any game provided it was inside a ball park. He hated televised baseball.

"Yes, Mr. Grieve," Terry answered. "How are you tonight?"

"Damn near dead!" He coughed a rattling mass of smokey glop from his lungs and spat on the ramp. Terry was used to the routine including the gloomy reference to death. It had been going on since he had moved in.

"Maybe if you quit smoking."

"There's nothing wrong with my smoking; soothes the throat, you know."

"No, I didn't."

"How's that room of yours?"

"Fine."

"Say, this your plane?" Wendall acted as though he were seeing it for the first time—a sign he was about to ask a favor. Often, Terry felt himself talking with a big kid when talking to Grieve. The ubiquitous baseball cap added to this impression.

9

"Yes," Terry answered. "Same one I've had all along." Instantly, he felt he had been too rude, but he tired of the indirect approach Wendall always employed.

"Really? Well, maybe you're right; I guess this is your plane..."

"Thank-you," Terry answered. "Is there something you wanted?"

"Me? Oh, no, not really. You about to go up?" Terry was close enough to the old man to smell the strong tobacco breath and watch his thin lips move in that nervous smile he had over yellow stained teeth. A grain truck with Wendall's name on the side in large print rumbled past on the gravel road. He owned a good portion of the town's wealth, and made certain everyone knew it. His name appeared everywhere: WENDALL ELEVATORS, WENDALL TRUCKING, WENDALL FORKLIFT, TIRE, AND BOILER, and the list went on even to include a pediatric wing at the Methodist Hospital—"The Grieve Ward".

Terry stared at him and asked, "Do you want to go up with me?"

"Oh, you probably got somewhere you want to go alone; I wouldn't want to impose. You still seeing that librarian, Mrs. Waters?"

"It's Miss Waters, yes, but not tonight."

Mr. Grieve was already climbing into the front seat of the Champ and said, "If it's no trouble!"

"No trouble. Switch OFF, brakes ON, and throttle BACK!" Terry called to his boss who knew the routine.

"Brakes are on, throttle's back, and the switch is off!" he answered. Terry pulled the propeller through until gas dripped from the carburetor onto the pavement.

"Switch ON!"

"It's HOT!"

They climbed toward the southwest, over the railroad tracks and along the river, muddy with runoff from surrounding fields. Wendall stared directly ahead without saying a word. Terry knew where the old man wanted to go, and he knew after they passed over the river Wendall would say, "Mind if we swing about ten miles south, so I can look at a piece of property?" He always made the same request, and Terry always obliged.

Wendall directed him again to a section of farmland about fifteen miles away. He never told Terry what exactly he wanted to see, or why his own plane, the Baron, could not serve. He would sit quietly in front of Terry with the window slid back and his hand cupped in the slipstream to direct the fresh air in a violent rush against his face. When they arrived over the parcel he had Terry circle several times, quite low, around an old barn isolated square in the middle of a field.

The barn was surrounded on all four sides by cultivated land, the crop rows sometimes running straight to the walls. No road led to the building. Terry had asked once as they flew past, "What's inside?" Wendall had only answered, "Oh, nothing...just junk." But still they would circle, with Wendall staring at the plain barn with its two eye-like windows high above a tightly sealed double door.

The previous year the field had been planted in corn, but this year it was alfalfa. Terry noted that the hay appeared ready for its first cutting, and said, "Looks like they'll be baling soon. I think we could land there once its cut, if you want to come back."

"No!" Wendall yelled in a screech over the engine noise. "There's nothing down there but junk!"

"Oh."

"Let's go home." Wendall sat quietly for the trip back, and left Terry alone at the airport to put the plane away.

"Eloise, I'm home! What's for dinner?" Terry threw his jacket onto a single bed along the wall. The cat jumped down from the refrigerator and brushed against his leg. He picked her up and rubbed his nose against hers.

"Did you miss me?" The animal stared blankly into his face and gently bit his nose.

"Well, let's see what we have for dinner tonight." He opened a small can and waved it under his nose as if testing a

11

rare wine. "Oh my! This evening we are serving the house specialty—smashed up fish guts and non-specific animal byproducts, yummy!"

Eloise rubbed his leg with extra vigor to show her impatience and gratitude. He slid the bowl under her nose, and ran his hand along her spine until she arched her back in ecstasy. While she purred at his feet he looked inside the refrigerator for himself. The bulb was still burned out—had been for two years—but he could make out two cans of beer, a bottle of thousand island dressing, and something gray wrapped in cellophane. He tried the freezer compartment, above the top shelf, but the glacial formations surrounding the door only allowed a brief peek inside where a three year old potpie scarcely showed through the permafrost. He slammed the door.

"Eloise, don't wait up for me!" The cat blinked as Terry left.

The Boone Municipal Library closed at 6 p.m. on Friday nights. Terry sat in the parking lot at 5:55 and watched two teenage girls and a short boy with glasses and scarlet acne leave the building. They each carried great loads of books, and the girls seemed determined to outdo each other laughing at their escort's wit. The skinnier of the two girls covered her mouth in a self-conscious gesture when she laughed. Braces, Terry mused.

The lights inside the library began to switch off. Terry strode casually up the eight steps to the door and slipped in past the circulation desk and disappeared in the stacks between TRAVEL and BIOGRAPHY.

"We're closing in five minutes." Miss Waters moved entrenched readers into the night. One man slept heavily near the magazine rack and woke after several firm prods.

"What's matter?" His voice was thick with sleep and confusion.

"We're closing, Mr. Glemsen. Time to go home; I'll put the magazine away for you." She ushered him through the door. "Remember, we don't open until 10:30 tomorrow.

Goodnight." Her voice trailed off as the sound of Mr. Glemsen's footsteps diminished down the eight steps. The door closed, and the bolt clicked. "Phwew..."

"Do you have a book on sexual misconduct in small town America?" Terry called from behind the shelves.

"Just when you think all the perverts are gone, something pops out of the Biographies!" She laughed.

"Now you're just trying to sweet talk me."

"And then I'll have my way with you."

"Are all you librarians the same?"

"Underpaid?"

"Over-anxious!" He slipped from behind the shelf and took her hand. She was a full head shorter than him, and weighed no more than a hundred pounds. Her hair was light brown and cut short. She was a year older than Terry, and had only just begun to gray, something Terry found attractive.

She smiled and reached to kiss him when someone called from behind, "Excuse me." It was a thin, embarrassed voice from a small woman near the desk. She looked seventy and flashed a quick friendly smile. "I don't wish to interrupt..."

"Mrs. Harmon, oh!" Meredith Waters broke from Terry's grasp and ran to the desk. "I thought everyone was gone...we close at six..." She took the book from Mrs. Harmon and ran it through a noisy machine on the desk before inserting a book mark and handing it back. Terry slipped behind a shelf and cackled audibly.

"I'm sorry if I interrupted you and your friend," Mrs. Harmon whispered and gave her a knowing wink.

"No, ah, no, not at all. Thank-you for coming." Meredith felt her face glow warm with embarrassment. "Here let me get the door for you." She fumbled with the bolt and sent Mrs. Harmon away. "Oh, God!" She locked the door again. "Is there anyone else hidden away in here?" she called.

"Only your friend." Terry peeked from behind the shelf.

Chapter
III

1929
Lincoln, Nebraska

A man in his late twenties stood beside the Packard touring car, holding a stone. He glanced over his shoulder like a pitcher checking first base, wound his arms over his head, brought his left arm back and threw the stone with all his strength at the wooden hangar wall fifty feet away.

"That's my fastball!" He spoke to a short fleshy man with thinning hair who sat on the car's running board drinking from a brown pint bottle. "They never could hit that one; went by too fast." He picked another stone off the ground and pushed his fedora back on his head. His suit jacket was draped over the car's front fender. He wore a gray vest over clean white shirt, his starched collar dug into his thick neck.

"Curve ball—nobody ever hit this one either," he boasted easily and wound up again for another pitch. He sent the rock crashing into the building within inches of the last one. Chips of wood flew with the impact.

The man on the running board nodded appreciation and said, "Henry, how come you didn't stay pro?" His voice was flat, midwest flat, but with the clipped phrasing of a city dweller, more Chicago than farmer.

"I would have, Art, but after the '19 series they started looking real careful at anybody, like me, picking up a little on the side making book, or just playin' the ponies, for that matter. They said I was too friendly with the wrong element." He laughed and loosed another stone at the wall. It flew wide and hit a drain spout, denting it. Someone turned the corner and yelled, "Hey, stupid, find some place else to do that!"

Henry picked a stone and, without the windup, threw at the complainer hitting him square in the gut. He fell backward, doubled up more in surprise than pain, and stumbled back into the hangar.

15

"That's my change-up!"

Art laughed between swallows on the bottle while Henry wiped dust from his hands and grinned. A Standard LS-5 biplane took off from the grass field behind the hangar. Its water-cooled engine shook the warm air. Henry turned toward the noise and watched the biplane bank overhead, its shadow racing over the Packard.

"Wow! Look at that, Art! Will you look at her climb! That's beautiful; is that what we're gettin', Art?"

"That's it, Henry. A Standard LS-5. Wingspan, 44 feet, 7 inches. Stands 10 feet 7 inches tall; gross weight, 2,922 pounds. Carries 43 gallons of fuel, one pilot and 720 pounds of cargo, passengers, Canadian whiskey or whatever." Henry laughed at the last remark. "And it does all this while cruising at eighty-six miles per hour."

"Art, how come you're so good with numbers?"

"Accountant's curse." He took another swallow from the bottle. Despite the alcohol, his voice was clear, although slow. He seemed tired.

"Think I could learn to fly that?" Henry asked.

"I don't see why not, you're pretty good with the car."

"Think Doc will let me fly it?"

"Ha, now that's a different question. We've already hired that other pilot, and he comes with an airplane. Of course, his won't carry the load the new one will. Who knows, Henry, maybe we'll need two pilots."

"I know I could fly it, Doc'll let me. Hell, that other guy hasn't even shown up yet." Henry threw another stone at the hangar. "There can't be any mystery to it." He tossed a stone. "If that long drink of water—what's his name?"

"Hollow."

"Yeah, him. If he can fly, so can I." He overturned a stone with his toe. "What the hell kind of dumb-shit name is Hollow, anyhow?"

"Pilot's name, I guess," Art said and took a swallow.

"If you ask me..."

"No one did."

16

"I think it was a bonehead mistake giving him any money. He's a bum. He won't show."

"Wasn't your money."

"Okay, Doc's money, whatever." He turned and plunged both hands inside his pockets and glanced at the sky. "I don't figure what Doc saw in him."

Art shifted. "He's like that sometimes. He works on impulse." Henry's face was a question. "Doc," Art continued, "gets a wild hair up his ass now and again. An idea pops, and he lunges for it. It's a gift, really.

"So I suppose when he saw this pilot hopping rides at the fair grounds in Amarillo, his brain just started working the angles. You got to admire the man. Not many can spot opportunity and, at the same time, understand how to exploit it. Like I said, it's a gift. Besides, we wasn't doing so hot in Texas, I gotta admit that. Yea, Doc can find the angles, all right. It's a goddamn gift."

"Do you think like that, Art?"

"No, I try to think as little as possible."

"Me, too," Henry said, then added, "Naw, that ain't really true. I like to think. I just, kinda, get all caught up sometimes. Like right now, I think I could fly better than some yahoo bum pilot from Texas."

"I don't think he's from Texas," Art said. "Can't remember him saying much about where he was from. He sounded a little like he was from the east coast, up north, maybe. Maybe not." He shrugged and took a swallow.

Henry snatched a rock from the dust and wound his arm back as though ready to throw. Instead, he lightly tossed the stone from one hand to the other. "This Hollow's too skinny to be a pilot."

"Less weight in the pilot means more payload. Besides, he wasn't that thin. Looked like a runner, you know, one of those Olympic runners, all muscle and no fat, like a mountain lion, maybe."

"He's still a skinny shit," Henry said. "Hey, here comes Doc."

From behind the hangar walked a thin man in a white linen suit with white Panama hat, gray high top shoes and white spats. He walked quickly and smoked a cigarette with choppy puffs. Henry lit a cigarette and called to him.

"Where's our boy?"

"There's been a slight change in plans..." Doc answered and stepped into a puddle, mud splashed on his white pants. "Oh, will you look at that, now!"

Henry and Art laughed. Doc walked up and scraped his shoe on the running board. Art took a final deep swallow from the bottle, then tossed it feebly across the road. Henry scooped up a large rock, and before the bottle came to rest, he splattered it with a well aimed throw.

"My fast ball!"

"Put the damn rocks down," Doc said. He turned to the accountant. "And how much have you had so far this morning, Art? I swear you've been at it since sun-up."

Doc's accent was hard to place. It was a cultured tone with a trace of deep south, but he spoke too quickly for a southerner. He had once told Henry he had spent fifteen years working for some missionary group in Honduras.

"So where's our pilot?" Art asked.

"He's been detained. He'll be arriving a day or two late— by train."

"Train?" Henry spoke up. "Where's the airplane?"

"Apparently he has had some mechanical difficulties, and the craft is no longer serviceable ."

"What?" the other two asked.

"This is actually working to our advantage as the sales manager has provided us with an excellent discount on a new machine."

"Can we afford it?" Art, the accountant, asked.

"The salesman listened to reason," Doc said smiling. He opened his suit jacket revealing the black handle of a small pistol. "He agreed to some favorable terms; seems business is slow of late, and he's had a few unexpected cancellations." A brief smile crept across his thin lips. Henry enjoyed this and snatched a small round stone off the ground. In a continuous motion he whipped it toward the hangar. It sailed high in the air and crashed through a window.

"Now, I told you to stop that," Doc chided without conviction. "I'm hungry; let's get out of here. I saw a cafe back in town." He climbed into the rear seat of the Packard. Henry helped Art to his feet and slid him into the seat beside Doc and closed the door. He took the driver's seat, and in a rattle of flying gravel and dust the Packard sped away. Two faces watched their departure from the corner of the hangar.

Clara made change from the cash register for the man and woman on the other side of the counter. An oscillating fan stirred the greasy air in the warm cafe, and Clara continually wiped at the sweat on her upper lip.

"Thank-you, Thomas. Now, be sure and come back soon."

"We will, Clara. You know I can't stay away from you," Thomas said. He pulled a soiled white hat on his bald head.

"You're going to make him fat," his wife chirped at Clara while clutching her husband's arm.

"Well, he's a valued customer, Mary," Clara said. "You both are," she quickly added.

Mary continued her chirping, "I don't know what you'd do if we was ever to leave town. Every night he's saying, 'got to go down to Clara's for a taste of her lemonade supreme.'" She pulled her husband away from the counter saying, "Personally, I don't see where it's any better than mine!" Clara smiled at Thomas who winked and left.

Three men, Art, Henry, and Doc, sat at a table across the room from the cash register eating lunch. Henry shoveled dripping forkfuls of roast pork into his mouth while Art picked at a cheese sandwich and talked with Doc.

"We have exactly four thousand, six hundred and thirty-two dollars and seventy-five cents left; less ninety cents for lunch gives us..." He scribbled the figures on the table cloth as he spoke.

Doc interrupted. "It gives us plenty."

"How do you figure? We have to have two grand, up front, when we get to Canada. That plus travel expenses cuts our

19

reserves in half, which was barely enough when we had that jerk lined up to fly his plane for the first load across the border."

"We don't need him," Henry blurted through his food. "I bet I can fly."

"You miss the point," Art said. "We still don't have enough to purchase a new airplane. That by itself is two thousand, four hundred and ninety-five bucks!"

"I told you we got a deal." Doc's accent now sounded more New York than southern, and Art knew it could change just as easily back to magnolias and cotton. "The sales manager will be a temporary partner until after our first trip, then we buy him out for an extra five hundred. He went for the idea."

"You mean the factory's going to let us simply take one of their brand new airplanes on a promise?"

"More or less. Mr. Howard, the sales manager, has a fair amount of leeway with established customers..."

"But we're not established!"

"No, but the Alaska Mining Company is, and they order parts regularly by wire, and the funds arrive later."

Art grinned.

"We," Doc said, "will be the Alaska Mining Company long enough to run up to Canada, fit the storage tank in, pick up a load, run it down to Minnesota..."

"Get the cash, and disassociate ourselves with the mining industry," Art finished and nodded his head.

"We in the mining business?" Henry asked without concern.

"Yes, Henry," Doc said. "And I think it's about time we tested Clara's lemonade supreme."

Doc waved toward Clara asking her to come to the table. She waddled over wiping her hands on a damp rag.

"What can I get for you boys?"

"I would like a glass of your lemonade supreme." Doc said and smiled politely. She hesitated briefly while she sized up the strangers.

"I'm sorry, but we're out of lemonade."

"But it's your lemonade supreme I want." Doc was cool and stared into her eyes. Henry wiped his mouth on a napkin and burped. She turned to leave.

"We are out of lemonade."

"Arthur, I think this woman is mistaken. I believe I see some lemonade behind the counter. Why don't you go take a look for her." Art and Henry both stood—Henry to block Clara from leaving while Art walked around the counter.

"You might as well take a seat," Doc said flatly. "Art will go check for you. You've been busy with the lunch crowd, but now that everyone's gone you might as well sit. He indicated the chair across from him, and Henry pressed her firmly into it. Doc let his jacket fall open so the pistol showed.

"Henry, take care of the door for Miss Clara." Doc's voice was thick with strained southern charm. Henry moved to the door and turned the OPEN sign around. He pulled the shades.

"Anything yet, Arthur?"

"Not yet," Art answered and pushed a stack of plates onto the floor. They shattered on impact sending jagged pieces skimming across the floor. Clara gasped and tried to stand, but Doc lunged across the table and slapped her hard across the face. Instantly, a red welt rose where his hand had connected.

"Who's your supplier?" he shouted.

"I don't know what you're talking about!" She pretended to cry but was not convincing. She seemed used to the routine. Doc settled back in his seat.

"What's your specialty, Clara? Beer, apple wine, white lightning?"

"Well, well," Art called. "Look at the lemonade!" He pulled two bottles from behind a panel in the wall and set them on the counter then took out two more and another two. He continued until there were fourteen bottles set on the counter top.

"Clara!" Doc said with mock surprise.

"Who are you?" she asked.

"Your new supplier, Clara." Doc said, and Art opened a bottle and sniffed its contents. He took a long draw.

"It's a blend," he said and smacked his lips. "I'd say thirty per cent Canadian and the rest shinie." He took another long swallow. Clara watched his adam's apple bob in his short neck. "Not bad though."

Henry leaned against the door with his arms folded across his chest, his face immobile except for a grin that played at the corners of his mouth. He rocked back and forth on his heels and flexed his shoulder muscles, all gestures Doc had taught him. Suddenly, a door to his right flew open, and a tall black man wearing a white apron stepped through, a stack of coffee cups in his hands. Henry pulled a .32 revolver from his jacket, pointed at the man's temple and cocked the hammer.

Everyone watched in silence. The black man set the cups carefully on the counter, walked to a table, sat and lit a cigarette. He acted bored, as if he faced interruptions of this sort regularly. He waved for them to return to their business and blew a smoke ring at the wall.

Doc glanced at the fourteen bottles on the counter then turned to Clara. "Clara, I have a proposition for you..."

"Proposition?" she snapped. "You mean I can refuse?"

"I don't think you will." Doc looked long at her while stretching his fingers against the palms of his hands. "I think I can help you."

Her dark eyes studied him carefully. She wiped her upper lip.

"You are through with the nickel and dime crowd. You're moving up." He called over his shoulder, "Arthur, bring your notebook here."

Clara remained silent.

"Write this down: Starting immediately Miss, or is it Mrs?"

"It's Miss," Clara answered.

He continued, "Miss Clara...I am sorry, I didn't get your last name."

"Bow!"

"If you like. Miss Clara Bow will become the Eastern Nebraska distribution manager for wholesale spirits, details to be discussed at a later time—end of entry."

Art snapped his notebook shut.

"I'm overstocked as it is. I have no reason to buy anything from you."

"Overstocked?" Doc asked. "Henry, did you hear the lady? She says she's overstocked. Maybe you can help her."

Clara leaned heavily on one elbow and watched Henry pick a coffee mug from the stack on the counter. He drew his arm back and let fly at the row of whiskey bottles. Glass and ceramic splinters hit the wall. Two bottles disintegrated. He wound up again with another cup, and two more bottles exploded into slivers and whiskey. He picked off the remaining bottles, one at a time, never missing a shot.

Doc stood. "Expect your first shipment in about a week and a half. You're in business with us now." He picked his hat off the table and moved toward the door stopping in front of the black dishwasher. "Do you know how to drive a truck?" he asked.

"Yes," the man answered.

"Starting next week you're a truck driver. Interested?" Doc handed the man a ten dollar bill.

"My cousin's got a truck. Where you want it?"

"We'll be in touch."

The man nodded then walked behind the counter, took a broom and began to clean up the shards of glass and crockery left from the negotiations. He said nothing. The three men left, and Clara disappeared into a back room.

Outside, Art said, "Less the ten-dollar tip to the colored guy, we now have four thousand, six hundred, twenty-two dollars, and seventy-five cents."

"Yes, but I saved ninety cents on the lunch." Doc smiled. "To the airfield, Henry; we have work to do!" He ushered Art into the front seat and took the back for himself. Art still carried a bottle from the cafe.

Jake Hollow opened one eye and felt the inside of his mouth with his tongue. Teeth are still there, he thought. "I've always been proud of my teeth," he said aloud, or at least what he thought was aloud. He heard something move and turned his head. He was in a bed, sheets tucked to his chin, head bandaged almost covering his eyes. Dull pain ran the length of his body, and he moved his legs. "Sore but functional," he said.

Someone, a girl maybe, was across the room, her back to him. she was folding clothes, his clothes. "Hey," he called, but she ignored him. "Those are mine." He realized his voice was only inside his head. "I don't have much, just the clothes...and an airplane, somewhere..." The room and the girl faded then reappeared. "I'll tell you later," he said, and everything blanked.

Chapter
IV

Art nodded in half sleep, oblivious to the humming throb of the Packard moving down the highway at 60 miles per hour. Henry loved to drive the straight-eight machine, and he especially loved to drive fast. There was only one other car on the road between town and the airfield, and he rolled past giving the other driver a quick wave.

"If they'd pave this damn thing, I bet I could get it over eighty," Henry said.

"Just get us there alive, Henry," Doc called from the back.

The long car pulled alongside the hangar where the three men had been earlier. The sales manager, at his desk, looked up from a folder of papers, saw the car return and dashed from his office. He was outside before anyone had a chance to leave the car.

"What are you doing here?" he asked.

"Need to talk business, Mr. Howard. Your office?" Doc said.

"No, I told you the plane would be ready tomorrow. You shouldn't be around here...this isn't your usual purchase, you know. I've got my neck stuck out a mile—a mile!" Mr. Howard looked nervously about. Doc noticed he sweated like Clara, heavily on the upper lip.

"I hope you're not having second thoughts," Doc said and leaned forward in his seat. "You're making five hundred dollars pure profit on this, and in only one week's time."

"That's what you promise, but how do I know you won't just disappear with the airplane and never return?" He turned to Henry, who had just stepped from the car. "What then? What happens to me?" He turned back to Doc. "I have no collateral, no securities, no references, not even an address— NOTHING! How do I know you'll be back?"

"We have to come back," Doc said. "How else will we pick up our other planes?"

Mr Howard stepped back as if hit by a slow bullet. He wiped the sweat from his upper lip, looked at Henry, then at Art asleep on the front seat. He threw his arms over his head and turned for the hangar, then changed his mind and quickly stepped back to the Packard. With his foot on the running board, he poked his head through the rear window putting his nose within inches of Doc's.

"What?" he whispered.

Doc briefly shut his eyes. "Mr. Howard, I haven't time to go into details, I leave that to my accountant." He indicated Art who stirred in his sleep long enough to utter the word, "fart," and wipe spittle from his mouth. "All I expect from you is to have a second airplane, just like the first, waiting for us upon our return on Monday next. You will not only receive the full purchase price, plus your bonus, as agreed, but you will also receive an additional bonus for the second. Are you following me, Mr. Howard? You seem to be staring at Arthur."

"I'm listening," he answered sharply. "You want another plane, like the first—like the first that you don't own. And you'll give me a second bonus, like the first—like the first that I haven't seen yet!"

"Exactly," Doc said and waved Henry into the car. "A week from Monday, Mr. Howard, a week from Monday." The Packard's motor started.

"A week from Monday," Mr. Howard mumbled and walked slowly into the hangar. A large weight pressed on his shoulders.

"To town, Henry. Find us a hotel—nothing second rate, either." Doc said.

"This is Lincoln, Nebraska—I think everything here's second rate," Henry answered.

Art woke briefly. "It'll cost us."

"You worry too much," Doc replied.

Art slipped back into sleep, and Doc sat back in the Packard and watched the Nebraska landscape slide past outside the window. Henry drove in silence leaving Doc time to think. As in other moments alone he conjured up his own past, a past he shared with few. He found it a soothing routine to review his life with its triumphs and pratfalls from the safety of years removed.

His mind returned to Honduras before the war, but he screened out the filthy heat in the city, and the handful of friends who had passed through his life like meteors, brilliant for a moment, then vanished. In his present situation, with one car, two associates, and less than $5,000 he thought only about the money he had left behind. There had been piles of it, all virtually unaccounted for, and all gone.

As a public relations representative for the Canadian Fruit and Sugar Company, he had once controlled vast sums of cash listed on the ledgers in Montreal under the heading, MISC. It had been his job to smooth over the political obstacles encountered at the various administrative levels from dock-side customs official to plantation tallyman. Competition from American and European firms was stiff, but Doc knew his trade well and flourished until 1914 began to restructure the world's priorities, and his Canadian employer became occupied in Europe.

Unable to adjust to the post-colonial world of 1918, Doc found himself adrift in the American Northeast living off a small cash reserve he had hidden away in New York banks. This, however, soon dwindled, and Doc viewed his own future without hope as he celebrated his forty-sixth birthday, the year the Armistice was signed. He was in Manhattan to watch the Johnnies come marching home. He watched, in despair, a young nation return from a smoldering decayed Europe to take the lead in the economic future, a future with no place for an aging public relations man who had squandered his best years in the jungles of a place nobody cared about.

Doc's fortunes improved when he met Rothstein, "the man uptown". Arnold Rothstein ran everything in the nether world of gambling, prostitution, numbers and protection throughout New York City and most of the East Coast. Starting in 1920, he expanded into the newly opened field of illegal alcohol. The opportunities Prohibition offered the entrepreneur were limitless. Doc soon found himself in upstate New York, Lake Placid, managing a growing bootleg operation from a sixty acre estate owned by Rothstein. He rarely saw the man himself but knew the business was closely watched by the middle management types who shuttled between the Lake and the city, reporting to Rothstein.

Rothstein was pleased with his junior partner, who had organized a loose collection of independent smugglers, pioneers in the new trade, into an efficiently run shipping concern. The Village of Lake Placid was proud of its newest resident and the business he brought. Bootlegging on a grand scale meant jobs. Trucks were purchased new, with cash. Drivers were recruited and paid well, in cash. Accounts were kept of every nickel spent, and a wily accountant, named Arthur, kept guard over the coffers, while shifting the huge profits between local banks and those in New York City, Albany and Boston in an effort to dazzle federal investigators. It seemed to work.

Doc found the Lake Placid authorities easier to handle than those in Honduras. Their greed for the easy money propelled him into the resort town's more genteel society. Their inexperience with big time graft left them vulnerable to a pro like himself.

From his mountain redoubt, Doc monopolized New York's trade in illicit booze. Competition was discouraged, with upstarts, regardless of their scale, coerced into consolidating with Rothstein or simply put out of business with a permanence everyone understood. Rothstein had put the right man in charge.

All this changed overnight when, on October 6, 1928, Rothstein was seated at a card game in New York's Park Central Hotel. Suddenly, a man stepped into the room, removed a pistol from under his coat and fired three times. One bullet ripped through Rothstein's upper arm. The second passed harmlessly by his shoulder and splintered the chair back. The third lodged in his chest causing his lungs, over the next thirty-six hours, to fill with blood. He died without identifying his killer, and his empire fell into the hands of two apprentices, Meyer Lansky and Charles "Lucky" Luciano. A third underling, Jack "Legs" Diamond, quickly moved into the upper New York state liquor trade administered by Doc.

Diamond proved to be a ruthless businessman bent on monopolizing the trade by preventing Doc from striking out on his own with the Rothstein legacy. He began by hijacking Doc's trucks as they rolled down from Canada loaded with whiskey. He terrorized the drivers, and left a dozen mutilated bodies along roadsides where they were certain to be found. He, too, understood the business.

The war reached Lake Placid Village itself when Arthur, the accountant, was snatched out of a lakefront restaurant, in full view of other diners, and pushed into an awaiting car. Two days later, he crawled to the front gates of Doc's headquarters with both kneecaps broken and a message: "All the king's horses and all the king's men won't put the Doctor together again!" Doc decided to fight. He still had six good men and two trucks at the estate.

The gates were locked, windows boarded. In short, the lodge was readied for a siege. That night, he awoke to the acrid odor of smoke filling the room. He leaped from bed and headed for the door, stopping long enough to reach for a small automatic pistol he always kept in the nightstand beside his bed. The drawer was open, and the pistol gone. Smoke filled the room and forced him to crawl to the door in the clear air near the floor.

The building was quiet except for the crackle of flame consuming wood and furniture. He burst into the hallway and called the alarm, but no one answered. He continued along the hallway finding all the doors open and rooms empty, except Arthur's. The injured man lay in feverish torment on his bed gasping for air. Doc pulled him to the floor and dragged him through the lodge, down a flight of stone stairs past burning walls near the entrance and out the double oak doors to the driveway.

The estate was abandoned. Smoke rolled out the doors and windows and billowed into the clear night sky. Doc coughed painfully feeling the sharp cold air sting his smoke-filled

lungs. Arthur was unconscious by his side, breathing errati-
cally.

Where were the guards, Doc wondered. Dead? Burning?

Timbers crashed from the ceiling inside. The flames intensified and engulfed the entire lodge. The sky glowed orange
from the holocaust and allowed Doc to see his car parked
below the front steps in the driveway. He grabbed Arthur and
pulled him toward it. Moving closer, he heard the Packard's
motor running and saw the driver's door standing open, a note
taped to the glass: "Doctor, we make house calls."

A chill shook his body, and he stuffed Art into the back
seat. There was no longer any doubt, the Rothstein empire
was gone. Without the man himself, subalterns like Doc were
powerless. Whoever removed his guards and the pistol at his
bedside while he slept, could just have easily fired a round
into his brain.

His cold feet groped for the clutch, and he mashed the gears
searching for first. The front wheel shot over the curb and
struck an iron railing before it dropped heavily back onto the
gravel. Down the long winding driveway they roared, past
the swimming pool and over the wooden bridge in the trees.
Two great oaks stood as sentries at the front gate, which was
already open. Doc sped through and turned south.

Heading for New York City, he stopped to phone the
various banks Rothstein had used, but discovered all the
accounts closed, the money gone. The only money he could
find was in a personal account he kept in a small bank in New
Paltz, New York. It yielded only about ten thousand dollars
and caused a commotion when he announced to a distraught
bank manager that he wished it all as soon as possible. Doc
had reinvested the bulk of his personal earnings in with
Rothstein on a new bootlegging venture in the New England
area. A young Boston Irishman named Joe Kennedy
managed to grab that portion of the Rothstein largess after the
murder. Doc did not bother to go after it.

He had to shake his head now, thinking about his connection to a maverick airplane pilot. This is truly the damnedest
scheme I've been into yet, he thought. Getting too old for this

30

nonsense. "Airplanes," he said aloud through a sigh. "How's that?" Henry asked.

"Oh, nothing." His thoughts returned to the present. "Henry, let's pull into that filling station."

Henry turned the Packard off the road and stopped alongside a pair of tall pumps in front of a white garage. An enamel SHELL OIL sign hung from a pole overhead. A skinny kid with freckles and acne stepped out the screen door, a Coca Cola bottle in his hand. He whistled at the expensive touring car and ran to fuel it.

"Fill 'er up," Henry growled.

"Yes, sir!" He removed the gas cap and whistled once more before inserting the nozzle.

Doc leaned slightly out the window and called, "Young man, come here a moment."

"Yes, sir!" He ran toward the rear door.

"Are you familiar with this area?" Doc asked.

"Oh yes, sir, lived here all my life."

"Well then, maybe you can help us. You see, we're from out of town..."

"Yea, I saw your New York plates." The redhead beamed. "Sure are a long ways from home."

"Ah yes, how very observant. What's your name?"

"Sam, Sam Nyberg."

"How old are you, Sam?"

"Almost nineteen."

"Well, Sam." Doc leaned forward in his seat and spoke softly, in confidence. "My associates and I are here on business, and being strangers, we are less familiar with some of the local accommodations. Do you understand what I'm referring to?"

Sam grinned. "Oh, you folks mean you're looking for a whore house!" He dropped a foot heavily on the running board. "That's no problem. We've got a couple of good ones just the other side of town..."

"Well, that isn't exactly what we had in mind," Doc said. Sam backed from the car and eyed him suspiciously. "We

31

were more interested in drinking establishments." Sam glanced up and down the Packard once more before he spoke. "You cops?"

"No! Heavens no," Doc said, and Henry laughed.

"No, I didn't think you were," Sam said. "You dress too swell for cops. We had some federal agents through here last March. Now, you could tell from a mile off that they were cops—wore brown suits all the time and drove a Chrysler. Easy to spot."

"You're very observant," Doc said.

Sam pulled a rolled up magazine from his pocket and waved it in front of Doc. "It's in here," he said. "'True Detective Stories', all the facts about prohibition agents and bootleggers." He flipped the magazine open to a dog-eared page. "Says here that Machine gun Jack McGurn always presses a nickel into the palm of his victim's hand." He thumped the passage with his finger. "Anyhow, feds always drive Chryslers.

"We don't get many Chryslers around here. All day long I see folks come and go. I see everyone in town sooner or later, and I watch them, I study them." He leaned close to Doc and whispered, "and I listen. Not many folks know how to listen. But me? I listen! And that's how come I know what's what in this town." He stared at Doc as if he had instantly discovered who they were and why they were there. The gas tank suddenly overflowed, and he ran to disconnect the nozzle.

"Well, where the hell do we go for a drink?" Doc hollered.

"Clara's," Sam answered.

"Is that the only place?"

"Oh, no. You can go to Jan and Dean's Restaurant over on South 'G', or Sullivan's Barber Shop on Euclid. And there's a handful of roadhouses about can give you a snort. The coloreds got their own places out west o' here. Course the Hotel Lindell has the best selection, but they're awfully picky about who they let in. I can give you the name of a bellhop over there can help you out."

Doc looked at Henry with a grin. "Oh?"

32

Sam continued while he mopped gasoline from the car. "Then, of course, there's right here. I can get you a fifth of whiskey if you like."

"Canadian?"

"Mostly."

"Arthur? Henry? Shall we see what the gentleman has in stock?" The three men followed Sam through the garage and into a room at the rear of the building. The door was bolted and locked. The room was windowless. Sam switched on a light.

"Liquor's two dollars a bottle and tastes real smooth. Randall don't let me sample the product, but occasionally I take a test."

Doc walked past Sam toward a stack of wooden crates against the wall. Henry stood in the doorway, arms folded. Arthur removed a notepad from his vest pocket.

"You run this all by yourself?" Doc asked.

"Mostly."

"Mostly?" Doc snapped. "Can't you be a bit more specific?"

"Well, Randall's in town," Sam answered and slowly backed himself against the wall where a sawed off shotgun hung.

"Who's Randall?"

"He's the owner—Randall Ohlsen."

"What's he pay you?" Doc asked.

"Three dollars a day, it's good money. I don't have to do much, just watch the pumps and peddle this stuff."

"Did he teach you how to use that?" Doc pointed at the shotgun. Sam glanced over his shoulder at it, a reflex action, then turned back.

"I know how," he said and noticed Henry was holding a pistol in his left hand. To reach for the shotgun now would be out of the question. He knew this day would come, eventually. Randall had warned him. They had been too successful, bound to attract attention.

"We won't take much of your time; you have a business to run," Doc said and stepped closer to remove the shotgun from the wall. He handed it to Henry.

"Where does Randall buy?" There was a long pause, the only sound came from Art who uncorked a bottle and sniffed the contents before he drank. He spat violently on the floor and tossed the bottle in the corner where it shattered.

"Crap tastes like hair tonic." He spát again.

"Randall cuts it!" Sam screeched.

"With what?"

"Whatever he can find—corn liquor, wine, sugar water. Whatever he can find. He makes about two hundred percent profit on each bottle doing that. Folks around here don't seem to notice. Except over at the Hotel, they buy the straight stuff—off the truck." He spoke quickly.

"Where's he get it?" Doc asked. Sam paused, and Henry pumped a round into the shotgun's chamber. The dry click of the action echoed in the tiny room.

"Nigger from Omaha drives it down, once a week, leaves it here, and everybody in town buys from us...mostly." Sam's eyes darted from Doc's face to the short barrel of the shotgun, it looked small in Henry's thick hands.

"And who does he drive for?" Doc persisted.

"The nigger? Sam asked.

"I wish you wouldn't use that term, it's so...vulgar. Yes, the Negro. Who's his boss?"

"Don't know, never seen anyone but him. He doesn't say much, just takes the money and drives back. There's usually two of them."

"Two trucks?"

"No, two nig...ah, negroes." Sam stepped closer to Doc. "Say, Randall'd shoot me, if he finds out I'm telling you this."

"Henry here will shoot you if you don't!" Doc smiled and turned to leave. "Oh, tell your delivery man you'll not be needing him anymore. Starting Monday you work for us. Straight Canadian, we cut it—you don't. We set the prices, quotas and customers. We provide protection, so you don't have to worry. I'll be back to discuss details."

34

"What about Randall?" Sam asked.

"Tell him I look forward to a profitable relationship."

"He isn't going to like this. Folks are used to him, you know. He isn't going to just hand this over to you."

Doc removed a twenty dollar bill from his billfold and stuffed it into Sam's shirt pocket.

"You'll help me convince him, won't you?" He left the room with Art close behind. Henry began to leave, then turned, pointed the shotgun at the stack of crates and fired four times.

Glass and wood splinters flew through the air, and brown liquid splattered the walls and oozed to the floor. He tossed the gun to Sam who sat huddled near the floor covering his head. Blue smoke hung near the ceiling as quiet returned. Sam's ears were ringing too badly to hear the Packard start and leave. He stayed immobile on the floor for a few minutes waiting for his hearing to return. He took the twenty dollar bill from his pocket, stared at it, grinned and replaced it.

Randall returned about an hour later and found Sam sitting on a chair outside the station, feet up, drinking a Coke. He gave his usual cursory wave as he passed him and said, "Anything happen while I was out?"

"Not much," Sam answered. Randall disappeared inside.

"DAMN IT!" Randall screamed. The screen door flew open and out he came, eyes wide and panting. "They've been here! I told you this would happen! I said you were selling to too many people! I told you!" He wagged a bony finger at Sam and paced the porch in tight circles running his hand across his bald head.

"Now they're after me. I'm dead! Dead meat! I know it. And it's all your fault!" He thrust the finger at Sam's face.

"It's...it's Capone, I know it...from Chicago..." He turned back again to Sam who blew across the top of the empty Coke bottle to make it whistle.

"Did you see their car?" Randall pleaded.

Sam nodded.

"Was it from Chicago?"

"I don't think Chicago has license plates—that would be Illinois."

"Were they Illinois?" Randall appeared ready to cry.

Sam shook his head, no. He continued to blow across the bottle.

"Christ! Big Damn Christ! It's Capone!" He slumped to the porch steps with his face in his hands, then just as quickly stood turning on Sam.

"The only reason they didn't kill you is 'cause you're a kid; a lazy, stupid kid!" He ran toward his car and waved his finger once more at Sam on the porch. "A GOD DAMN KID!"

Randall left. Sam sat for a minute blowing on the empty bottle. When Randall's car had disappeared he put the bottle in a box near the door, looked once more at the twenty dollar bill and walked home. It was suppertime.

Chapter
V

Jake Hollow leaned against the low stone wall and stared at his airplane. It was on a farm wagon, its landing gear crushed and folded beneath the fuselage. The wings were intact, but the lower one had several large tears in the fabric. Two ribs on one side were smashed. The engine, however, was unscathed, its long wooden propeller turned parallel to the ground. He reviewed the mess and wondered how he would get it out. He was broke.

Jake stepped slowly around the wagon, his body sore all over from the crash. He slid his butt onto the wagon and pulled his legs behind him.

"I left everything in the airplane the way I found it," a girl with short blonde hair said. She stood behind Jake and helped him onto the wagon. It was a painfully slow process. He had spent one whole day in bed and, upon waking, insisted to see the plane. She had helped him out of bed, but he fainted almost immediately, leaving her to lift his limp form back and let him sleep another six hours. When he awoke again she led him to the Standard.

"How did you get it up here?" Jake asked.

"My father and Mr. Grimes—he lives up the road—they put a couple of ropes around it, and a team of horses pulled it up. I think your wheels were already broken."

"Yes," Jake said and grinned. He sat heavily on the wagon, his head buzzing with hissing sounds as red lights flashed before his eyes.

"Are you all right?" she asked. "You're so pale. I don't think you should be up yet. I've never seen anyone crash an airplane before. I didn't think you could live through that."

"Some live; some don't. This time I got lucky, I guess." He leaned his head over.

"Are you all right?"

"Not completely; let me sit a minute." She jumped onto the wagon and took his arm. She could feel through his sleeve that his arms were thin but strong. He had the lean build of an athlete and was handsome, she decided. He was so different from the farm boy types she had known all her life. She had tried to guess from his accent where he might be from, but he sounded like he was from everywhere.

"You shouldn't be out yet," she said. "You're not well."

"I had to see the machine—it's the only thing I own. It's how I make a living." He looked at her. "Do you think your father would stay in bed if all his cows were out wandering around loose?"

"We only have one cow, she never goes anywhere."

"Pigs then," Jake said. He smiled at the girl and wondered how old she was. He guessed sixteen.

"What kind of pilot are you?" she asked.

"Not a very good one." His answer was wistful, he looked from her to the broken biplane. It was a mountain of wires and wooden braces sitting atop a dirty wagon, no longer a flying machine.

"Oh, I didn't mean that! What kind of flying do you do? I didn't see any mailbags; you're not flying the mails are you?" She cocked her head slightly to the right. Her pale blue eyes were innocent; her smile tight.

Jake squinted at the sun and gently touched the bandage on his head. "Let me see, what kind of pilot am I?" He shifted to face her, slightly closer than the situation demanded. I guess I'm a barnstormer, although I hate the term. It's corny, something made up by the newspapers. I prefer to think of myself as a freelance aviator."

"Freelance?" Her head tipped to the other side.

"Yes. I am an aviator operating at will, open to opportunity, ready to grasp the leading edge of the ever-changing technological world of modern aviation." His chin jutted out in a pose he often practiced to resemble John Barrymore. His head was swathed in bandage, and his chin still swollen with purple bruises, so the effect was more Lon Chaney than Barrymore.

"Are you ill?" she asked.

"No," he said deflated. "You said earlier your name was Susan. Do people call you Susan or Sue?"

"My father calls me Susan, or Susie; my friends at school used to call me Sue. The teachers just called me Miss Bowers."

"Miss Bowers? Sounds formal." Jake teased her.

"You can call me Sue."

"How old are you?"

"Nineteen."

"Really?"

"Well, almost...I will be in August. Your bandage is coming loose." She reached a strong hand toward his face and tucked a stray corner under a fold. Jake winced with pain.

"Hurt?"

"No, I just stuck my hand on a splinter on the wagon." He held his hand before her. She took it and looked for the sliver.

"I can see you've never worked on a farm," she said. "Your hands are so soft, no callouses, no broken bones. Except for the grease stains, I'd say you'd never worked at all."

Jake was selfconscious with Susan stroking his pale hand looking for the splinter. Her hands, by contrast, were strong, still feminine, but the hands of a worker.

"I see it," he said and pulled away. They looked at each other in silence while he picked at the splinter with his teeth.

"See, I'll live!" he said and held his hand up.

"What's it like?" she asked, her eyes sweeping over the mass of wrecked biplane behind him.

"Flying?"

"Flying—being an aviator; floating from town to town, anywhere you want to go. What's it like?"

Jake thought of the lost fortune in the wrecked biplane. He thought of the many nights alone, sleeping on cold hard earth, hungry and often wet, hoping the next day, and the next tank of gas, would find a town with a population ready to part with its money, to be introduced to the world of flight. The gypsy life had claimed yet another year, leaving him little to show.

"It's the only way to live."

39

She stood and ran her fingertips along the plane's skin. The wagon teetered on the uneven load.

"Have you ever been flying?" Jake asked.

"Oh, no," she answered. "This is only the second airplane I've seen up close. I saw one in Des Moines once, at the fair. He did loop de loops, and tailspins, and who knows what else. It was all very impressive. I don't think I'd be scared in one." She leaned against the fuselage, eyes half-closed, her arms described the loops and spins. "Will you be able to fix your machine. It looks awful bad."

"I can't stay," Jake said. "I'm due in Lincoln. I have something lined up. I can't stay."

"What about your plane?"

"Is there a train station nearby?"

"Yes," she said. "In Luther; the westbound comes through every morning about four." She moved closer to him, eyes vaguely sad. He was a bandaged up mess without prospects, but he was the first real hero to step into her life.

Jake caught her stare, the sad eyes. He hesitated, then touched her yellow hair. She was young, the hair still soft. Her figure showed through the plain green dress. She was a healthy pleasantly-shaped girl—a farm girl.

Her hand shot up to touch his. Already her hands showed the wear of a woman destined to spend her life near the earth, wiping noses and scrubbing. Jake found it all pointless, wasteful.

"Have you ever been anywhere other than Des Moines and the fair?" He still touched her hair, but his voice was tinged with annoyance for the situation that would keep this pleasant, healthy girl stuck amid the corn.

"Not really," she said. "We go to Luther regularly."

Oh, she is cute, Jake thought. His first experience with a woman had been in the Army while stationed at Fort Sam Houston, Texas.

The prostitutes in San Antonio were dark, at least the ones a soldier could afford.

Their faces were grotesque masks of red and white paint, intended to make them look seductive or beautiful, Jake supposed. The effect was repulsive at first, but after three months at the desert post they began to look better. The make-up hid the pock marks and leathery skin, abused by a thousand forgotten hands. They were whores dressed like ladies, and Jake would forever associate over-groomed women with the trollops of San Antonio. Susan wore simple clothes and no make-up.

They stared at each other, and he blurted, "would you like to go to Lincoln with me?" He immediately corrected himself.

"Oh, I am ridiculous!" He dropped his hand from her hair. "I am sorry, my head's still spinning from the crash..."

"No," she said. "It's all right."

"I'll be gone tomorrow...I have no money. I don't know you, really...excuse me," he stammered and slid off the wagon.

"Don't worry," Susan uttered. She reached for him, but he dropped to the ground and limped toward the truck.

"We should get back; I have to catch a train," Jake called over his shoulder. His progress across the field was painfully slow, no bones had broken in the accident, but his neck and back were stiff and sore. She remained on the wagon, admiring him. The man had fallen to earth and was about to leave her life before she could experience anything.

"I wish I could fly," she said.

Jake stopped. Susan jumped from the wagon and ran toward him. A gust of warm air swept through the field rocking the airplane and pressing her skirt tight against her thighs. She stopped in front of him taking his outstretched hand.

"You are cute," he said.

She released his hand and slipped her arm around his waist. His arm encircled her, and they walked to the truck. To the west cumulus clouds grew in great towers, she knew it would rain before nightfall.

Inside the truck, he said, "I don't think I'll be gone for long. I can't leave my plane forever." She drove the truck skillfully through the ruts toward the house. "I'll need help rebuilding it," he said.

"I can help; will you take me up?" she asked.

"I'll do better, I'll teach you to fly!"

She smiled in her tight-lip way and took his hand and squeezed it. A wave of regret flooded over Jake as he realized the entire situation was impossible. I'm broke, he thought. He squeezed her hand hard. His eyes burned.

"Where have you two been?" Susan's father, John Bowers, called as they stepped onto the back porch. His strong arm held the screen door open. Jake passed him with some concern. Although the father stood only an inch taller than Jake, his thick arms and shoulders presented a severe countenance. His face, however, was gentle and warm.

"We've been out to the airplane, Daddy," Susan chirped and kissed him on the cheek standing on her toes to reach.

"Can you fix the machine?" John Bowers had a high pitched flat voice, out of keeping with his appearance. "I set it where we're building the new barn. Those walls might help block the wind some."

"Is that what those stone walls were for?" Jake asked.

"That's just the foundation. We'll be raising the walls soon." John said. The wind sure likes to pick up that kite of yours. I put some ropes around it, should hold."

"I hope it's not in the way," Jake said.

"No. Truth is, I've been slow about finishing that barn. Real pain in the ass job. Might give me motivation to finish up now. I've only got enough lumber for three walls, no roof yet. Maybe next year. In the mean time you'll have a wind break."

"Thank-you," Jake said. "I have a job waiting for me in Lincoln." He saw John's face wrinkle into a frown at the thought of the stranger leaving the broken useless airplane behind.

"I'll be back, though."

"You certainly can't travel, yet. You're all banged up— look like hell!" John pronounced. "I'll have to see what the doctor says. He walked into the kitchen changing subjects.

42

"Now, supper's all ready. Have you got a good appetite? I consider myself a fair cook."

He plucked a ladle off a hook and said, "Your head took quite a knock. I'm supposed to report to the doctor if you have any dizzy spells. You dizzy?"

"I'd have to be to want to fly for a living!" Jake grinned, but the remark fell flat. Susan and her father looked at him puzzled. He felt his face flush under the bruises and gauze. "Say, that smells good! What is it?"

"Stew," John answered and spooned out a thick mixture of potatoes, meat and something green onto a chipped plate. The pile sat unmoving, and Jake stared. An equal portion hit Susan's plate, and she immediately fell to it. "We don't stand on ceremony here—eat!"

"My father doesn't allow praying at the table, if that's what you're waiting for." Susan tore a ragged piece of bread from a black loaf.

"I wasn't planning to," Jake answered.

"Good!" John bellowed. "Can't stand people praying over dead meat."

Jake stared at the large man with strong arms and high voice and decided he liked him. He glanced at Sue to see if her father's impious speech bothered her, but she ate untouched by the tirade.

He burned his tongue on the first forkful and quietly sucked air through the corners of his mouth deciding not to insult his host by spitting stew across the table after only one bite. A glass of beer appeared in front of him, and he drank it off.

"Susan can eat things straight off the stove," John said. "Me, I have to wait for them to cool." He grinned at Jake and waved over his plate at the steam. He ate slowly, talking between bites.

"What kind of job do you have lined up in Lincoln?" He poked stew into his mouth and chewed slowly.

"I have an investor lined up. He plans to convert my Standard into a cargo hauling machine." Jake ventured another try at the stew.

"How are they going to do that?"

43

"Well, right now the plane's a two-seater with a ninety horsepower motor. The Army used them during the war for trainers."

"Were you in the Army?" Susan asked.

"Yes, but after the war," Jake answered. "I spent two very boring years in Texas watching officers learn to fly in planes like the Standard."

"How come you didn't fly?" she asked.

"There wasn't much call for pilots after the war. I never finished high school, so I had no hope of becoming a pilot."

"How did you learn to fly?" she persisted.

"After I got out, I had a couple of hundred dollars in muster-out pay, and I heard the Army was selling off Standard J-1's for 200 dollars, so I bought one." The stew had cooled enough to eat without fear.

"Well, who taught you to fly?" John asked.

"Oh, I had seen enough of other people flying, plus I worked on Standards as a rigger at Fort Sam. We used to taxi them around for maintenance, and every now and then I would accidentally take off and fly a few feet."

"Accidently?" Sue asked.

"Yes. So when I saw them selling them off, I knew I could fly one, so I bought it. They threw in a tank of gas, and I took off for real."

"But how did you get a pilot's license?" John asked.

"Oh, never did!"

"And they let you take off?" Sue asked.

"After I gave them the money it was my airplane. No one asked if I was a pilot, so I didn't offer to tell!"

"Ha!" John Bowers laughed. "I like that! Ha Har!"

Jake smiled, he knew Susan's father liked him.

"How can you fly around without a pilot's license?" Susan asked.

"Airplane doesn't know I don't have one."

"God Damn! I like your attitude." Her father bellowed and slapped Jake on the back. "Have another beer." He opened a bottle and slid it in front of Jake. "You see, Susan, the government doesn't have to go sticking its nose in everybody's affairs."

Jake took a swallow. "Thanks, this is good beer. Where do you get it?"

"I make it. It's not hard. Beer can be made out of just about anything, corn, rice, barley, and we've got plenty of corn around here, God knows!" John leaned back in his chair and added, "and I don't tell the government about it either. None of their goddamn business!"

"Have you ever thought of selling this?" Jake asked.

"Already do," John answered. "I make about five, ten gallons at a time, sell it to the neighbors and a few folks in town. It helps to pay the bills, there's no money in farming these days. Rest of the world's making money ass over tea kettle in the stock market, but farmers can't make ends meet."

John popped another beer open, it had received too much sugar in the priming stage, and the cap flew off and struck the plaster ceiling. Foam ran down the neck.

"Some are a bit foamy," he said. Jake looked up and saw over a dozen nicks in the ceiling from bottle caps.

Jake finished the second beer and another was placed in front of him. Susan did not drink, but ran to the cellar to fetch more beer for the other two. Jake made an effort to keep up with her father, beer for beer, and found himself drunk. Her father, after the fourth beer, sat completely still in his chair while he kept up a steady conversation with Jake. Jake, however, found himself sliding beneath the table.

"You were telling us how to convert your two seat biplane into a cargo hauler," John said.

"I was?" Jake slurred.

"Yes, but Susan here interrupted you. She's a very rude woman at times, pretty though, or don't you think so?"

"Oh, I do!" Jake answered without really knowing what the question was. "She's very pretty; cute, too!" He closed one eye to focus on her. She blushed.

"Tell us about the airplane," she said to redirect attention. "How do you convert yours?"

"To what?"

45

"A cargo plane!" she said.

"Ha, Har, Ha!" John laughed and took a deep swallow from his beer.

"Well, first I take my plane to Lincoln, Nebraska to the Standard factory."

"Is that where it was made?" she asked.

"Susan! You're interrupting." John's high voice snapped. "Rude woman—pretty, though."

"No," Jake said. "I mean, yes, you are pretty—she is pretty—but no, the Standard was not made in Lincoln...They were actually made some place in New Jersey, but the whole factory packed up one night and moved to Nebraska. Can't say I blame them."

"So, what do they do once you get the plane to the factory?" she asked.

Jake toyed with a loose end of the head bandage. "Oh, they take off that little ninety horse OX-5 engine I got on there now, and they slap on a bigger Hispano-Suiza—hundred and fifty or hundred and eighty horse."

"Sounds like a big bastard," John mumbled, his eyes half-closed.

"It is," Jake said. "Then they cut up the front seat, take out the one seat and put in four, add a couple of round windows, too. I guess that's so the passengers have something to look at when they're not too busy throwing up."

"I thought you were going to haul cargo?" Sue said.

"We are, so instead..."

"Always interrupting, that woman," John quipped.

"So instead of seats up front, there'll just be this big hole—the cargo hold." Jake leaned heavily on the table, his voice thick, eyes unfocused. "Then they do something like change the landing gear from wood to steel, add some instruments, bigger gas tank...and probably a bunch of other crap."

"Sounds like a big bastard," John announced again.

"It is!"

"Does this plane go faster than yours?" Sue asked.

Jake struggled to remember the data on the two aircraft. "Well, it's not so much speed I'm after; it's load, how much more can I carry. That's where the money is. I want to be able to pack that front cargo area full of...of cargo and fly a couple of hundred miles without refueling. Whether I go eighty miles per hour or ninety really isn't much of a difference. The secret to hauling cargo is load."

He held an eye closed with his finger to keep the number of Susans at one. "Remember that! It's load that counts!" He thrust a finger in the air to emphasize his point, and upset the bottle in front of him. Susan was mopping the foamy mess before he realized what he had done.

"Biggest bastard," the father muttered, though his eyes were closed, his chin planted on his chest.

"I think it's time for bed," Sue said pulling Jake to his feet.

"Best offer I've had in months," he said and slid his arms around her waist trying to kiss her. She withdrew her face at the advance of his yeasty breath. He lost his balance as her face receded, and the room began to spin. First one wall rose, then the other. He dropped his arms to steady himself on a chair.

"Think I could use some air," he said. "Real soon."

She yanked him toward the door. "Don't you dare! Not in here!" she yelled and shoved him outside.

He bounded down the steps missing the last one. He hit the dirt knees first and crawled to a tree near the corner of the porch. From the shadows, Susan heard a series of animal retching noises interspersed with gagged apologies. Her father was the same way when he had the neighbors over. She never complained, but she never approved.

Jake stumbled to the porch and dropped on the top step at her feet. "Sorry," he said, and wiped his mouth.

She went inside and returned with a coarse blanket throwing it over his prone body. He snored quietly, the sour odor of vomit and beer rose to her. She considered dragging him inside but, instead, left him to sleep on the porch. She did tuck the loose strand of bandage in place across his face, hesitating long enough to caress the thin line of his jaw. Lightning flashed in diffused sheets to the west, the deep rumble of thunder rolled toward the farm.

A loud crack of thunder, like an artillery round exploding over the house, awoke Jake on the porch. Another report of thunder was quickly followed by a spidery flash of lightning, and the sky opened up with fat raindrops. Jake pulled himself off the hard floor, felt his back twitch with pain and dragged himself and the blanket into the house.

It was dark inside, except for a light from an upstairs room, where someone stirred. He guessed it was Susan. He walked slowly back into the kitchen and found it clean, dishes washed and stacked, Susan's father no longer asleep at the table. A regulator clock on the wall bonged three a.m.. A burst of lightning froze the swing of the pendulum in the half second it took to read the time. His head pulsated with loud torment, and his empty stomach alternately called for food and convulsed with nausea. He remembered the oak tree.

"Good morning," Susan called softly from behind him. "Sleep well?"

"Very, thank-you!" He forced a smile, embarrassed at his condition.

"I'll take that," she said and took the blanket from him. Another flash of light from the storm showed her smiling, her round face fresh, eyes bright. She moved quickly through the dark to the far side of the kitchen and lit a lamp. After rummaging through a wood box near the stove, she soon had a fire crackling.

"Step out on the porch for me and bring in some wood, small pieces," she ordered without looking up. He heard her rattling pans on the iron stove as he filled his arms with split hickory. He dropped the wood into the box near the stove and stood close to her. She broke eggs into a bowl.

"I'm sorry to get you up this early just to take me to the station," he said.

"Don't flatter yourself; I'm usually up this time every morning." He stared at her soft round face made rosy by the glow from the stove.

"I take the truck and pick up milk cans along the road from here to Luther every morning" she said. "That's how I know there's a four o'clock train to Lincoln."

48

She turned away. "Now go wash yourself—you smell bad!"

She beat eggs vigorously with a fork then poured them onto a hot skillet covered in grease. The eggs hissed, and the coffee pot rumbled. Strips of thick bacon sizzled in bubbling grease on a second pan. Jake became painfully hungry with the breakfast smells.

Susan's father came down, ate silently, then disappeared through a back door without a word. Jake saw him carry four wooden crates from the cellar and place them on the rear of the truck.

"Tell Andersen his bill's getting up there. Owes me fourteen dollars." It was all he said and walked off toward the barn. Jake had forgotten to say good-bye.

"Finish your breakfast, we have work to do before you get that train," she said.

Jake unwound the sticky bandage from his head and stared at the dried blood.

"Give me that," she said, and stuffed the bandage into a garbage pail.

One storm passed, and another moved slowly in from the west. For the moment, the rain had stopped. Jake sat beside Sue as she drove along the dark road. They stopped every mile or so at farmsteads to pick up milk cans left near the road. The cans were full and weighed close to eighty pounds, Jake guessed. Susan had little trouble lifting them onto the truck. Jake, feeling useless, lent a hand but could only hoist one can for every three of hers. She smiled each time they passed.

The rain had begun again, blurring the windshield. At four houses along the route she traded milk cans for the crates her father had placed on the truckbed. Jake guessed correctly that the boxes contained his home brew, but said nothing.

By the time they reached the station in Luther, Jake was thoroughly soaked from the rain. Sue backed the truck to a loading dock, and the two unloaded the cans onto a rail car. She tactfully suggested Jake load the empty cans waiting on the platform onto the truck while she carried the full ones.

They worked together like this for thirty minutes, when someone called from alongside the train.

"Morning, Susan. I see you got help this morning." It was the brakeman, a short fat man in blue overalls and cap. He walked along the train, striking the wheels with a long hammer, testing for cracks. Any change in ring would indicate trouble.

"New hand?" he asked.

"Good morning, Elliot," she answered. "No, this is Jake. He's a pilot, flies airplanes for a company in Lincoln, needs a ride this morning. Think you can help him out?" She squatted near the edge of the platform overlooking Elliot.

"Little short of funds, are ya?" he asked Jake.

"Yes, somewhat. I have a job waiting there," Jake said. "What'll it cost me?"

Susan scowled at him as if he had been rude. Elliot answered with a slight chuckle. "Oh, you're a friend of Susan's. She's a good kid. You can ride with me in the back." He turned to leave, and Susan jumped from the platform and ran toward the truck.

"Just a minute, Elliot. I've got something for you." She reached into the cab and pulled out two bottles of beer and handed them to him. "Thanks for your help."

"My pleasure, now get on board. We leave in five minutes." He continued along the tracks, the clink of his hammer faded.

Sue slid the last of the full cans onto the train while Jake gathered his bag from the truck.

He spoke first. "I don't think I can properly thank you, you've been such a help. I'll be back real soon...to get the airplane. I promise." The words gushed from him.

She smiled. "Remember, you promised to teach me to fly." The station was poorly lit, and dawn still to come, but he could see her eyes blink with tears. A choking lump filled his own throat, and they both said nothing. He reached out for both of her hands, and she pressed against his body. He could feel her soft breast against his jacket. They kissed.

"Remember, you promised to teach me how to fly!"

50

She drove away quickly, empty milk cans rattled and clanked in the rear of the truck. He saw her disappear behind a grove of trees as the train pulled out. What he could not see was her parked behind those trees watching the train through moist eyes.

Chapter
VI

"What's seven hundred and twenty divided by eight, Arthur?" Doc asked.

"Ninety," Art answered.

"Ninety gallons of whiskey..." Doc said slowly. "That should be a fair start for a first load."

"You're something the way you can figure those numbers, Art," Henry said.

They stood beside a Standard LS-5, outside a hangar at the factory. The day was hot and sticky. It had rained the night before, and clouds were already forming from the afternoon heat.

Jake Hollow walked around the biplane, carefully inspecting the entire structure. The LS-5 was a stout airplane, its steel landing gear looked tough enough to take the anticipated punishment it would receive from operating off rough strips between Lincoln and Canada. There was no front seat, the cockpit having been left empty for cargo. The machine was painted the factory's usual maroon and white with ALMINCO stenciled on the side.

"What's this, ALMINCO?" Jake asked.

"Alaska Mining Company," Doc answered. "Don't worry about it, just a financial arrangement."

Jake shook his head. It still hurt from the accident, and he reached a hand to pick at the fresh scab above his eye. He climbed into the rear cockpit. Doc stepped closer and put his foot on the step that hung below the fuselage. Jake reviewed the walnut instrument panel, shiny with new gauges. The plane smelled of fresh dope and leather, a smell he enjoyed.

"Does it meet with your approval, Jake?" Doc asked.

"What size motor did they install?" Jake asked. He spoke without looking at Doc, who found this vaguely annoying.

"I'll defer that question to Mr. Howard, the salesman." He turned. "Oh, Mr. Howard, could you come over here a moment, please?"

The salesman stood talking with a photographer near the propeller. He motioned for the cameraman to remain and trotted toward his clients.

"What motor is installed here?" Doc asked.

"The One-fifty Hisso," Howard answered.

Doc smiled and looked at Jake. "One-fifty..."

"Is it new?" Jake asked in that same rude voice that irritated Doc. His head was buried beneath the instrument panel.

Mr. Howard moved closer. "Oh yes, it's new. You are seated in the finest airliner in the business right now. Why these machines are in use from Alaska to the jungles of Central America, and..."

"Sounds like one of your brochures," Jake muttered from deep within the cockpit. His head popped up, and he looked straight at Howard, who was still worried about the loose financial arrangement he had with the buyers.

"Let's see how it flies," Jake said and pulled a leather helmet over his head. Mr. Howard ran around to the propeller, and soon the motor was running, the water temperature crept into operating range.

"I'd like to see how it does with some weight on board!" Jake hollered over the noise of the engine. "Got anything we can put up front?"

"How 'bout me?" Henry called. He looked anxiously from Jake to Doc for permission. Mr. Howard scowled and turned on Henry.

"No, you can't go up; there's no seat up front."

"You'd be fine," Jake called and motioned him into the front cockpit. Henry gave his hat to Doc and scrambled into the plane, a wide grin spread across his usually stern face.

"Ever been up before?" Jake yelled through cupped hands.

"No, but don't worry about me!"

"Well, there's no seat so just squat down and find something to hang on to."

Henry nodded, the propeller blast whipped at his trousers and tugged his hair and necktie. He dropped into the hole until just the top of his head showed. He waved proudly at Arthur seated on the Packard's runningboard.

Jake nodded to Howard, and two lineboys pulled chocks from under the wheels. He advanced the throttle, and the biplane rolled over the grass into the wind, bounced twice and lifted gently into the sky. The racket of the engine changed from clacking roar to distant hum to those on the ground as the Standard climbed. Sunlight glinted off the waxed skin of the only machine in the sky.

Doc and Mr. Howard squinted with necks craned watching the airplane climb in lazy circles.

"How much does Henry weigh, do you think, Arthur?" Doc asked.

"I'd say two fifty, two fifty-five," Arthur answered from the Packard. He spent as little time as necessary on his feet, his knees had mended so poorly after Lake Placid. He still took medication for the pain, this, with the alcohol, limited his waking hours.

"Two fifty-five, huh?" Doc uttered to himself. "Seems to perform well with two hundred and fifty pounds. Will it carry the full seven hundred?" He was thinking out loud, but Howard nervously listened.

"Seven hundred and twenty," he added.

"Yes, and twenty," Doc said.

"And, if you ask me, it'll carry much more—much more!"

Doc looked briefly at the salesman trying to sort the facts from expected sales pitch. "More?"

"Much more."

"Hmmm!"

The Standard flew around the field for almost twenty minutes, Jake at the controls and Henry squatting in the cargo hold. The airplane banked, and turned both steep and shallow. The motor howled with power one instant, then suddenly was throttled to idle, and the nose would pitch up, and the whole machine would hover shuddering on the verge of a stall.

Jake pointed the nose down and raced toward the ground, then pulled up sharply over the hangars before banking into a tight turn with the power back and glided to the field to bounce several times and stop before the Packard.

Henry was out of the cockpit gesturing to Doc before he and Mr. Howard reached the airplane.

"I know I could fly one of these things, Doc!" Henry shouted. Doc only smiled an indulgent response and continued toward Jake, still strapped in his seat.

"I could see everybody on the ground from up there!" Henry said. "'Course, it was hard looking over the edge of the side there, but I stood up, sort of, and could see the Packard. It looked so small—like a toy!"

"Glad you enjoyed it, Henry," Doc said as he would to a child. "Now, see if Arthur needs any help getting in the car, we'll be leaving soon...comb your hair." He turned to Jake.

"Yes, sir," Henry said, a note of disappointment in his voice. He ran toward Arthur. The others heard him retell his story. "You should have been there, Art! It was like nothing you'll ever do."

"Jake?" Doc said. "Do we have an airplane?"

Jake nodded and looked at Mr. Howard, who was anxious to move the customers out, lest someone from the front office interfere.

He was desperately afraid a representative of the real Alaska Mining Company might show up or call, and the scheme would disintegrate with him out of a job, or worse, in jail.

"When do I leave?" Jake asked.

"As soon as you're ready," Doc answered. "Mr. Howard, is the machine completely airworthy?"

"Certainly! We need only top off the tank, and you can be off—say, half an hour."

"Jake, let's have lunch. There are a few details we need to discuss," Doc said. "Mr. Howard, one-half hour. Have the machine ready!" He turned.

"It will be! Please don't be late," Howard called after them, but they seemed to ignore him. He waved to the lineboys to bring the fuel truck. They moved slowly adding to his anxiety.

56

The Packard stopped in front of Clara's restaurant. The four men walked into the room filled with lunchtime eaters. Clara was busy with a customer at a corner table when they took their seats near the door. It was several minutes before she noticed them, but when she did, her normal easy manner turned stiff. She was soon at their side, their table isolated from the other diners.

"What are you here for?" she hissed.

"Oh, Clara, we're just here for lunch," Doc said. "What's the special today?"

"I'm out of whiskey; you can take a look if you like, but I think people will get a little suspicious seeing you strangers rummaging around behind the counters. You know, I don't know who you people think you are!" Her voice grew agitated with courage as she spoke. "I run a respectable establishment, people know what I sell, and they want it! Now, you come along, Mister Big Shots, and upset a nice thing.

"I don't see you getting tough now! Not with a room full of people, I don't think so!" She stood over them, arms akimbo, her weak chin thrust forward, her face greasy with sweat.

"Clara, tell me something." Doc motioned for her to take a seat beside him. She remained standing. "How much do you make per bottle?"

No answer.

"Arthur?" he said and held out his hand.

The accountant took a small sheet of paper from his brief-case and handed it to her. She accepted it cautiously never taking her eyes from Doc. Several other diners began to notice the strangers.

"Now tell me, Clara. Can you honestly tell me you're making profits like that?" Doc's face was soft, a trusting frown above his eyes. "I know you're not. I've spoken to Sam over at Randall's Garage." He sat back in the chair and watched her face register surprise, a response she was unaware of.

"Sam's my new partner; Randall's out. Sam's been a good kid helping me set up accounts throughout town. I'm the source, Clara. I'm it!"

Cautiously, Clara looked from Doc to the figures on the sheet. They were remarkably accurate detailing her liquor sales, cost of goods sold, mark-up, overhead and ultimate profit. The paper was divided neatly in half down the center. On the left side was Clara's present consumption, and on the right Art had figured her projected profits based on a new supplier, namely Doc. The right side showed a clear increase in profit.

"Sam gave you these figures?" she asked.

"He's a smart boy," Doc said. "He should be a well-to-do young man, soon."

"What about Randall?"

"Don't worry about Randall; worry about me." Doc's face was cool but smiling, his friendly bully face.

Clara laughed in a quick heave. "Sam! Sam Nyberg—I always suspected that kid had something under the pimples and red hair." She laughed again. She knew how to interpret a balance sheet. She knew business.

"Hey, Clara!" a customer called from behind. "We've been waitin' a while; those folks promisin' you a big tip?" The room broke into rumbled laughter.

"Oh, hush, Thomas! You got nowhere to go that's important!" The room laughed even more sharing in some privileged gossip.

"You can deliver by next week?" she asked Doc.

"Next week."

She smiled, folded the paper and slid it into her apron. "What'll you have?"

They ordered the special, and she served double portions. Henry ate all of his and most of Art's. Doc picked carefully at his and spoke with Jake.

"Arthur has a map for you."

Art unfolded a dimestore map of the northern plains states and southern Canada, showing lower Manitoba and Ontario. Doc's finger ran over a penciled route from Lincoln, Nebraska, across western Iowa, and straight across Minnesota and into Canada. Several towns were circled along the route, and a star marked the end of the route near the town of Sprague, west of the Lake of the Woods.

"The route, of course, is not mandatory," Doc said. "You are the pilot and will have to make your own decisions on weather, terrain, etc."

"Goes without saying," Jake said through his meatloaf. "The destination is, of course, fixed." Doc tapped the star in Canada several times.

"Pass the pepper, please," Henry interrupted. Doc handed him the shaker without comment.

"You must..."

"Thank-you," Henry said.

Doc glowered. "You must land here. You can see this railway runs east to west through here." His finger traced a black line. "There is a watering stop here." He tapped the star.

"Watering stop?" Jake asked. "A town?"

"I don't think so...Art?"

"Water tank and a couple of wood and tarpaper shacks," Art said. Henry leaned over the map still chewing and grunted once.

"A man from the Canadian National Railroad will meet you there tomorrow afternoon. He said he will hold the whiskey until the afternoon train comes through. Now, this is important: If you don't show—he puts the whole load back on the afternoon train, and we are out of luck. He won't hold it—you must show with the money and make the trade."

Jake looked at the map. "What's his name?"

"Moyer, Phillip Moyer," Doc answered. "He's Canadian; I knew him in upstate New York. He worked for me up there. He's a good man, the honest type, but don't ever think of crossing him!"

"I gather you haven't paid him anything, yet."

"Half," Doc said. "Nothing happens for nothing in the business world. You will deliver the other half. Arthur?"

Art handed an envelope to Doc, who passed it to Jake. "Needless to say, you had better not misplace this."

Jake thumbed through the contents and tried to whistle, but only spit bits of meatloaf across the table.

"There's almost one thousand dollars there; I suspect you've never touched that much money before." Doc looked straight at Jake, their eyes met, an understanding began to form between them. Henry burped.

Doc spoke. "You'll pay Moyer, and pay for your gas and expenses out of the change."

"My fee?"

"Half your two hundred is in that thousand, provided you budget yourself on...expenses. The other half will be awaiting your return."

Jake pushed the envelope into his jacket and studied the map sliding it closer to himself. He knew the map would be woefully inaccurate in the remote areas, especially in northern Minnesota and Canada, where he suspected the cartographers had merely guessed at terrain features knowing there was no one to question their work. It was a cheap road map, meant for automobiles, but it would have to do.

Despite the lack of detail on the map, he quickly estimated headings off major landmarks, such as towns and rivers, and figured the trip would be little trouble—an easy two hundred dollars.

"What's the airfield in Canada like?" he asked.

"Moyer sent this telegram," Art said and handed Jake the note.

> GOOD FIELD SOUTH OF TRACKS, STOP
> LONG, STOP
> WATCH FOR TREES, STOP
> AND WIRES, STOP

Jake moaned.

"Now, the return route," Doc said, and Art took a second map from his briefcase.

"Can't I just use this?" asked Jake.

"No, you're not coming back here right away. The first load gets delivered in Minnesota," Doc said and spread the second map across the table and dishes.

"Here," he said.   A pencil line ran from Moyer's in Canada, now indicated with a circle, to a star west of Bemidji. The distance was about half that from Lincoln to Moyer's.

"I drop the liquor here?" asked Jake.

"Yes," Doc said. "At a farm west of town...here." He tapped the star. "A woman named Adair, no last name, will meet you. The farm has a silo..."

"Don't most farms have silos?"

"This silo will be marked with a white cross painted on top. She says it can't be seen from the ground."

"The silo?"

"The cross! Don't push me, Hollow!"

They both smiled with eyes locked.

"Find the cross and deliver," Doc said.

"I take it I don't stay too long with Adair." Jake leaned back with a sigh and slipped a cigarette from his pocket. He tapped his jacket looking for a match.

"Correct," Doc said. "Maybe the night, but don't get any ideas, she's ugly as river mud, and her husband's...untrustworthy." He grinned.

"Charming," Jake said.

"You don't have to like them, just deliver the whiskey, and get the money. Make sure you have that first before you hand over the goods."

"How do I do that? Are these people to be trusted?"

"Is anybody?" Doc asked. "Are you? Am I?" He laughed. "Once you have the money, you fly back to Canada..."

"Do I get to use the same map?"

"Same map," Doc snapped. "You return to Canada, you meet Moyer, again, buy a second load..."

"With the money from Adair?"

Doc nodded, annoyed with the interruptions. "It's the same deal as the first time through, only this time you give him the full two grand."

"How much does Adair give me?" Jake asked.

"Six."

"Six thousand dollars?" Jake asked. Two customers at a nearby table turned.

"Something the matter gentlemen?" Doc asked casting a venomous glare at them. They quickly looked away.

He turned back to Jake. "Yes, six thousand dollars; do be careful."

Jake still held his cigarette, unlit. "That's a lot of moola to be flying around with. Suppose I got a hole in my pocket and the wind...PHITT!" He flashed his fingers to demonstrate the money vanishing in a wind.

Doc struck a match and held it halfway across the table. Jake leaned forward, the cigarette between his lips pointed toward the flame. Henry suddenly grabbed Jake's shirt front and pulled his face closer to the flame until it curled a stray hair off his forehead.

Doc twirled the wood match as it burned toward his fingertips, then died, the smoke rising past Jake's face.

"You may ask yourself how I know you will come back at all. After all, you might ask, you will have six thousand dollars of my money. A gypsy pilot, used to living on twenty dollars a month, could do nicely on six grand."

Jake tried to straighten himself but felt Henry's strong grip pull him closer to the table top. The men at the next table stood quickly and left.

Doc crumbled the burnt match slowly between his fingers and dropped the ashes beneath Jake's nose.

"Trust, Mr. Hollow. Trust!" Doc spoke slowly, eloquently. "Trust is the key to this whole enterprise. I trust Henry with my very life. I trust Arthur with every penny I have. So, what keeps him from running out with it all?"

"Henry?" Jake offered.

"Exactly, Mr. Hollow!" Doc said. "Trust—enforced!"

Jake pawed at Henry's wrist, but the grip was firm without being painful. "But I'll be gone with your airplane, your money," he said. "Even your maps! You won't have any idea where I am." He dug his thumb deep into the soft underside of Henry's wrist pinching the sensitive nerves there relaxing the grip until he could wiggle free.

Henry massaged his wrist confused at how he could have lost his grip. Doc waved him to remain silent. Jake leaned back and poked another cigarette into his mouth.

Jake spoke. "Mind you, I mention this only out of curiosity, I'll be back."

"I know you will, Mr. Hollow. When I first met you I judged you to be a reliable sort. I trusted you then with 50 dollars to fly your plane up here for this venture. And here you are."

"Without airplane, however," Jake said.

"Through no fault of your own," Doc offered. "No, Mr. Hollow, you could skip with the money, but you won't.

"You are a flyer, and from what I know of the trade, your type is never big on business, harboring only enough acumen to keep yourself in the air from sponsor to sponsor."

Jake seemed embarrassed but said nothing.

"You'll do anything to keep flying. You don't want money—only flying—airplanes and none of the obligations of business. I represent that freedom from responsibility, that freedom to fly and fly. With me you can fly forever, and in new machines, not the junk you had before. Probably a good thing you crashed it!"

"It was a good airplane." Jake was offended.

"You'll fly better, and in spite of yourself, will make real money doing it!" Doc shifted in his seat. "Oh, sure, you can skip, but you'd have everybody and their cousin out looking for you; think about it."

He held his hand up listing names on his fingers. "First there's ALMINCO, they don't know it, but you vanish, and they get a bill for the airplane. They won't be pleased—big company, too, lots of subsidiaries in the states."

He ticked another finger. "Then there's the Standard factory; they have yet to be paid. You leave; they'll look for you. Word will spread through the entire aviation industry to watch for the thief, Hollow."

He tapped another finger. "Then there's Moyer, nice guy but very well connected in Canada, so hiding up north isn't practical. You don't want him after you."

Doc leaned until his face was inches from Jake's. "Then there's me," he hissed. "You screw me, and I guarantee you I've still got enough friends to spread the word. You'd end up in a meat packing plant somewhere, ground up into Tuesday's hamburger!"

He leaned back. "It's a very small world, Mr. Hollow, and all operated on a foundation of trust."

"Just making conversation," Jake said flatly. "You can depend on me."

"Good."

"One question," Jake asked.

"Yes?"

"Why use an airplane? Why not trucks?"

"Simple," Doc said and scanned the room for Clara. She stood near the counter giving orders to the black dishwasher. Doc waved at her, and she held up a finger to indicate she would be over in a minute.

"Trucks get hijacked," he said. "Too much competition out there on the roads. We just don't have the personnel to operate a trucking concern. True, trucks can haul a great deal more, but they're subjected to running the borders—that means the Feds.

"Then there's the gentlemen who already operate in this region—the rural midwest—they would not take kindly to our transiting their territory. There would be tribute to pay, to say the least. With airplanes we merely override the competition."

He waved at Clara, again. "No, using airplanes is an experiment, a venture into the unknown! Why, you're an aviation pioneer, Mr. Hollow. Granted, we shan't make it public just yet. I'm surprised you didn't ask these questions sooner."

"I'm slow," Jake said.

"You boys finished?" Clara stood beside them, all smiles. She had, again, reviewed the figures from Art. "Care for some

64

pie? Apple or peach; peach is best." Henry began to speak in favor of peach when Doc interrupted.

"No, my dear, we must be off. What do we owe you?"

"Oh, I should say, nothing," she whispered. "We're business associates, now; you come back whenever." She smiled brightly showing a row of crooked yellow teeth.

"Thank-you, and until next time..." Doc took her hand and kissed it with much display. The few remaining diners exchanged sneers. The four men left, and the Packard drove off.

"Just move in a little closer! Now, you, big fellow, you're blocking the man next to you. That's it, turn yourself a little sideways, that's it!" The photographer called his instructions from under the black hood behind the camera. The bellows moved in and out as he focused.

"Now, the big fellow, you might want to lift those goggles off your eyes, can't see your face."

Henry pushed the goggles onto his forehead. He had taken Jake's leather helmet and put it on when he heard a photograph was to be taken.

"Is everybody ready?" The photographer's voice sang high with strained enthusiasm.

"Just take the damn picture," Jake said curtly. "I'm not sure getting our picture taken is such a hot idea."

"Oh, nonsense, Jake," Doc said through a fixed smile, one hand resting in a stiff pose against the wing of the Standard. "We've nothing to hide, all new customers get their picture taken with a new airplane—would be suspicious if we didn't. Now smile, damn it!"

CLICK

It took the film one sixteenth of a second to record forever Henry with the leather helmet supporting Arthur, who smiled, beside Doc dressed in white linen, head held high, beside a figure whose face was blurred, because he moved during the exposure. Jake swore it was accidental, but it was only noticed after the film plate was developed.

"Well, Mr. Hollow, have a safe journey..." Doc started but was interrupted when Jake ran over to the photographer as he folded his tripod. A large brown box sat open at his feet, several lenses nestled in the red felt compartments.

"Where-the-hell is he off to?" Doc asked. He watched Jake speak with the photographer, point to the biplane, write something on a scrap of paper and hand it with some money to him.

"What was that all about?" Doc asked, annoyed as Jake returned and hoisted himself into the plane. His movements were slow, still stiff from the accident days before.

"Personal!"

Doc clasped the lip of the cockpit. "Listen, you rude son of a bitch, don't ever think about crossing me, because I'll come down on you like a ton of crap!" The venom spit with his words.

"Trust, Doctor! Ours is a relationship built on trust," Jake said and waved to the lineboy to start the engine.

Doc and Art sat behind closed doors in the Packard while Henry watched the Standard fly off to the north. Other than a single towering cumulus cloud to the southwest, the sky was clear blue.

One week later, Susan Bowers was in the town of Luther to pick up the mail. The clerk handed her a large flat yellow envelope with SHAFFER PHOTOGRAPHIC STUDIOS, LINCOLN, NE., as a return address. Inside was the photograph of Art, Henry, Doc and a blurred face she assumed was Jake's posed in front of a large biplane. A short note was attached:

> *Susan,*
> *Do you still want to learn to fly?*
> *Jake*

"Oh, yes," she whispered. "Oh yes!"

Chapter
VII

With 1000 dollars, mostly in small bills, Jake and the Standard biplane flew north to Canada. The air was still warm high above the terrain, and the urge to unbutton his collar was suppressed by the horrifying vision of the money being sucked into the slipstream and seeded across Iowa.

A wind from the southwest added ten miles per hour to his ground speed, and he arrived over his fueling stop sooner than estimated. There was no airport, only a small town on the map with an Indian-sounding name. Jake forgot the name as soon as he read it and circled the town once to attract attention. His years of barnstorming told him there was never a need to walk into a town after making a pass or two over the square. Aviation was still a novelty to rural America, and the sound of an airplane motor overhead was certain to draw a crowd.

Jake saw the faces turn in the street, hands shaded eyes, and doors opened. Everyone pointed.

Good crowd, Jake thought. Thirty-five, fifty bucks easy before sunset! Instantly, he realized how silly his thoughts were. With a thousand dollars stashed under his jacket and six thousand waiting further along the route, he was foolish to consider hopping rides for a buck apiece.

"This is a fuel stop!" he reminded himself and buzzed the town square just one more time, straight over the flagpole next to the civil war cannon and over the houses and upturned faces.

A dirt road led off the square, up a hill and curved past a long green field bordered by trees and fences. Jake sized the field up with one glance, planning his approach from the east over the road, landing uphill.

The Standard planted itself firmly on the tall grass and rolled to a stop at the crest of the hill. Applying throttle he swiveled the tail around so the nose pointed back down the slope.

Regardless of how empty his bladder may have been on take-off, a vibrating ride over two hours left him anxious to get unseated.

He undid his seat belt and scrambled from the cockpit. With one eye watching the road he unbuttoned his fly and, "Phwew!" He stood beside the fuselage, one hand on his hip. Suddenly, he heard the rattle and squeak of a tall fliver pull to a stop behind him.

"Hello!" a man's voice called. Jake looked over his shoulder, unable to greet them full on. He saw the driver step out followed by what he guessed to be the wife and a collection of daughters. They moved cautiously toward him, shy, but anxious to be the first to meet the aviator. They rounded the airplane's nose, and Jake waddled away past the tail, his back to them.

He slid along the fuselage calling over his shoulder, "Ahh, hello! Be right with you...I have to check...the wing!"

"Can we help?"

"Oh, no!" Jake yelled.

"We heard you go over; lots of folks did, I suspect. We live outside town just over the hill. Name's Wayne; this is my wife, Ruth; girls—Janice and Wilma. Say, hello, girls."

"Hello," they piped.

"Pleased to meet..." Jake looked down the road where a column of trucks, wagons and cars made their way up the grade and across the field straight toward him. He waved with his one free hand at the crowd now being intercepted by Wayne.

"Now just a minute, folks, let the man finish his business!" Wayne called, and the crowd stopped a respectful distance from Jake waiting in silence for him to finish.

Jake buttoned his trousers and held up his hands. "Thank-you." They surged toward him surrounding the plane.

"Where you from? Gee, it's a big airplane. You an ace? Yep, that's what I heard—this fella's an ace, shot down five Germans! You did say five didn't you? Who's a German?"

The crowd was friendly, and only the smallest children ventured to touch the plane. One boy repeatedly struck the lower wing with a stick. A thin woman with gray skin made ineffective attempts to correct him.

WHACK! WHACK! WHACK!

"Now stop that!"

WHACK! WHACK!

Most stood in polite awe staring at the man with the goggles and leather jacket. From the crowd a camera appeared, and Jake was soon posing with several families or holding babies. Someone offered him a cigarette, he took it and six matches flamed in his face.

"Thank-you. I need gas..."

"Ed, you and Budd run into town and get Bob to bring out the gas truck!" Jake had no idea who Ed or Budd were, nor could he tell who ordered them into town, but two cars started up and raced each other across the field toward town.

A square-looking man dressed in a threadbare white suit pressed against Jake clasping his hand. His breath smelled heavily of liquor. "On behalf of the town council..."

"Oh beat it, Warren! Get that drunk outa here!" The crowd ushered Warren off before he could finish his presentation. "Don't mind him!"

Jake peered through the faces toward the road where several minutes later a yellow fuel truck crawled up the hill from town. The gears crunched, and the fuel tank heaved from side to side as it crossed the field toward the biplane. The crowd opened a gap for it with all the men waving directions at once, all wanting to somehow associate themselves with the airplane.

"You the fella wanted the gas?" the truck driver asked.

"'Course he wants the gas, Bob! Who ya think might want it up here? What're ya crazy or something!" The comments were shouted from the group while three men fought over who would carry the ladder to the plane.

"What's your hi-test?" Jake asked.

"Sinclair HC."

"Great, get the hose." Jake slipped past the three men with the ladder and climbed up the airframe to the fuel tank in the upper wing. "Let's have it here," he called, and thirty hands passed the nozzle across their heads.

"Are you from the airlines?" someone asked.

"No," Jake answered from above.

"Mail?"

"Not really; I'm just flying cargo."

"Is this going to be a regular stop for you folks?" a short man with round glasses asked.

"Possibly, right now I'm just scouting routes," Jake said. The crowd was immediately pleased with this, and they fell into speculating about the future of their town in the air transport system.

"How much would it cost to go for a ride with you?" a fat teenager on a bicycle called.

The crowd's playing into my hands, Jake thought. This is a hundred dollar crowd if it's worth a dime. He stared into the gas tank watching the level rise. The barnstormer lived for the crowd pleading to hand over its money, but he fought the temptation and said, "Sorry, company policy—no riders. You'll understand."

The crowd agreed in unison, its voice laced with disappointment.

The sound of the gas running into the tank changed pitch, and Jake eased the nozzle out squeezing in that little extra that could prove welcome in the sky over the Canadian woods. The gas spilled over, and he twisted the cap back on wiping at the spill with a rag. Thirty hands groped for the hose, and Jake slid off the plane.

"What do I owe you?" he asked.

"Well, let's see," the driver pulled a stub of a pencil from his overalls and slowly scratched some numbers on the scrap of paper he had spread on the truck's hood. "At twenty-six cents per gallon...you need any oil?"

"No. How much for the gas?" Jake was becoming impatient, anxious to be away from the questions.

The driver finally figured the tab, and Jake counted the money from the envelope that held the thousand dollars. More than one person saw the wad of money and a ripple of "ooo's and ahh's" ran through the crowd making Jake more uneasy about his stay.

The driver slowly counted the change and just as slowly tore off a neatly printed receipt for him.

"Thanks," Jake said. "Now if you will all stand back, I have to be off again. You've all been wonderful, really, and I'll certainly put in a good word back at the home office..."

"Where is that?"

"Ahh, St. Louis," Jake said.

"You know Lindbergh?"

"No, not personally." This seemed to disappoint them, so Jake added, "But I've met him—nice guy." They liked this answer.

He motioned for them to step back, and some of the more helpful in the crowd began urging the crowd away from the plane, while Jake pulled his leather helmet on and climbed into the rear cockpit with quick easy movements. He glimpsed the simple faces around him watching his every move, memorizing everything he did, so they could retell their version of his visit for years to come.

"Thank-you!" he called and waved like a politician.

"Where you off to, now?"

"Ahh, Bismarck. Bismarck, South Dakota!"

"Don't you mean, North Dakota?"

"Yes! But I have to go...via South Dakota!" He waved again, and they seemed pleased with his answer stepping aside as the engine fired. The propeller blast laid the grass flat behind the plane's tail.

Jake locked the brakes and ran the engine up in a clacking roar that caused children to grab parent's legs and parents to cover their ears and stare. He waved, released the brakes, and the double winged machine hopped down the slope and lifted into the air. He turned back toward the field and passed low over the crowd, about half of whom ducked with his passing. He then disappeared to the north.

"Damn fool's never going to find any of the Dakotas going that way!" Wayne announced from the running board of his fliver. The crowd agreed and returned to town. It was almost suppertime.

There was no dotted line like on the map to show when Jake had crossed from the United States into Canada. Somewhere in the trees and vast swamps below was a boundary recognized by the entire world on the one side of which whiskey was legal and on the other illegal and in demand.

With the rudimentary navigation equipment, namely a compass and his watch, Jake estimated headings and times between checkpoints. Sometimes the checkpoints would appear, and other times he picked alternates.

He passed over Roseau, Minnesota and noted the time: 9:22 p.m. He pressed two fingers to the map—each finger width measured ten miles—the distance to Sprague, Canada was about eighteen miles. He knew he was getting about ninety miles per hour ground speed, and that worked out to a mile and a half every minute. Sprague by 9:36 p.m., he figured— roughly.

The distance on the map from Roseau to Sprague looked insignificant, barely an inch, but with less than a quarter tank of gas, and the lower half of his body numb from the many hours strapped in the seat that now felt like a slab of granite, he was extremely anxious to find the strip.

The sun was a half hour above the horizon, slow to set with the approach of summer. Its light sparkled off the dozens of small creeks and ponds stretching across the landscape. Here and there, the marshland was interrupted by thick patches of conifer forest.

The map showed a thin blue line leading to Sprague labeled: ROSEAU RIVER. The landscape, however, showed a number of greenish fingers leading in all directions, any one of which could be the Roseau. Jake held his heading and watched for the railroad running through Sprague.

Time crawled, and he checked his watch at 9:25, 9:27 and 9:34.

There! To his right, he saw the thin steel rails leading across the swamp. His eye followed it to the left where the

72

terrain rose gently giving way to dryer ground and prairie before it disappeared into a small woods.

The left wing dropped, and he intercepted the tracks and saw Sprague.

"Now, to find the damn field," he said, but the words were sucked into the wind.

Gradually the Standard descended, crossing the small town less than 500 feet above the tallest building. Jake was too busy staring into the woods looking for the airfield to notice the handful of faces watching him go over.

His map displayed the star roughly five miles west of town. However, the star itself covered about three miles. Jake flew on, straight into the sun, its glow turning the visibility ahead into an orange smear.

He was low over the treetops, his shadow racing along behind him in an elongated blur holding the tracks. Outside Sprague there was an instant lack of civilization, only trees, and prairie with the one set of rails leading west.

He pushed his jacket sleeve back to see the watch: 9:38 p.m. Sprague was well behind him and no sign of the watering stop, the airfield or the man named Moyer. Jake pressed on wondering how much gas he really had. The gauge, at best, gave a rough estimate, and he was still unfamiliar with the plane's real performance figures.

Something darted across the tracks and was hit by his shadow—a deer. Jake turned, and with his back to the sunlight he saw the glint of a metal roof hidden by the trees.

He banked. The right wingtip pointed straight at the roof beside the tracks. Suddenly, the water tank appeared only a short distance beyond that, and Jake knew he had found Moyer. He had to have found him, the gas gauge was almost on empty. He thanked the deer for appearing when it had or he might have wandered another dozen miles into the woods and marshes of southern Manitoba.

The shack and the water tank were close to the tracks, but hidden below a small ridge in the otherwise flat landscape. Had he been higher he might have spotted it sooner, but none of this mattered. He picked out the landing strip, a short rough field carved from the forest and studded with tree stumps. It ran parallel to the tracks on the south side. The shack was across the tracks.

Jake stared in disbelief at what was possibly the worst place he had ever had to land an airplane. Two figures stepped from the shack, their round faces turned up.

"Ridiculous!" Jake said and passed over the field. Those on the ground, of course, heard nothing except the roar of the Hisso engine as the Standard flew back and forth, the pilot waving and pointing wildly.

"You expect me to land on that?" Jake shouted into the wind. "Stupid Eskimos! That's no place to land an airplane! Those stumps'll rip the wings off! Idiots! Morons! Morons and idiots!"

As his anger peaked, he saw the clear strip of dirt running between the stumps that would be the runway. About a thousand feet long, it offered the only choice, and he knew he could make it with careful attention to his approach.

The figures below watched the biplane make one more pass, turn parallel to the field, downwind, then bank into the wind over the trees and settle toward the strip.

Jake stuck his head into the slipstream to see around the nose. His approach was good, interrupted briefly by the right wing striking the telegraph wire along the tracks.

"Christ!" He looked quickly to the right and corrected, a telegraph pole slid by. He remembered something in Moyer's telegram about "wires". The left wing brushed a pine tree. Jake held the aircraft on its descent to the strip. The water tank whizzed past his peripheral vision, and the Standard's left wheel rumbled across the uneven ground nicking a stump.

The two on the ground saw the biplane drop softly to the ground, roll less than two hundred feet, turn, and, with a spurt of power, taxi back picking its way between the stumps. The propeller blast blew chips of wood and dust into the woods. A third man stepped from the shack.

He was a large white man with a thick red beard. The other two, Jake guessed, were locals—Indians. They followed the white man but without his enthusiasm in stride.

"Excellent!" the white man called. "Excellent! Always wanted to fly one of these things; never could muster the courage, though, eh. Don't worry about the telegraph wire, eh! One of my boys here'll patch it. I'm Moyer; you're Hollow!" He thrust a hand, more like a paw, at Jake as he stepped from the plane. Jake's thin hand disappeared into Moyer's. The grip proved surprisingly gentle, friendly.

"I didn't think I'd see you until tomorrow morning," Moyer said. His voice wheezed, and he panted heavily, his red cheeks puffing for air. "Have you a cigarette?" I left mine back there," he pointed, "at the house."

Jake removed a pack of Camels from his jacket and offered. Moyer took one and said, "How 'bout my friends here?" He indicated the two Indians who remained silent behind him. One stared at Jake, the other at the plane. Both seemed bored.

"Certainly," Jake said, and Moyer took the whole pack calling something to the Indians that sounded to Jake like, "Mud wallow and bug dump." They muttered a similar reply emphasizing, "bug dump", took the pack and left.

"Anytime, fellas!" Jake called after them.

"Don't mind them," Moyer said. He combed through his pockets for a match. Jake fished one from his jacket and struck it on a strut and held it for Moyer.

"Thanks," Moyer said. "They're good folks." He waved vaguely toward the Indians. Blue smoke rolled from his nostrils. "This is their country; white man's just a visitor to them, an unwelcome visitor." He seemed to calm as the smoke worked into his system. "Ever read any Hemingway?"

"No."

"Too bad. Good man, that Hemingway. Were you in the war? No, you look too young. Were you?"

"Was I what?"

"In the war?" Moyer repeated.

Jake massaged his numbed bottom pressing life back into it. "No."

"Too bad; I was." He strode around the aircraft, occasionally tugging on a flying wire or poking the fabric. He flicked ashes carelessly on the tail. "I killed people."

75

"In the war," Jake said more as a statement than a question.

"Mostly!" Moyer roared a great wheezing laugh, smoke poured through several openings in his wide red face. "So the Doctor thinks he can make money with this thing, eh?"

He thumped the fabric with his palm so it rumbled like a kettle drum. "Smart man that Doctor." He shook his head slowly as if he thought the whole scheme ludicrous.

"Why do they call him Doc?"

"Who? The Doctor?"

"Yes."

Moyer scratched his head in thought. "Well, he must be a doctor of some sort I would guess," he said. "That's the only name I've ever known him by." The question seemed to puzzle him, as if no one had ever thought to ask.

Jake took three ropes from the plane, looped two around a strut on each wing and tied the ends to nearby stumps. He did the same with the tail. He then took a small bag from the forward cockpit and asked, "Do you have a place for me to stay?"

"What's that?" Moyer asked, still considering Doc's name. "Oh, yes. Right this way, eh." They crossed the tracks and headed for the shack.

"He must have been a doctor, or no one would have called him one, don't you think?"

"Sounds reasonable," Jake said.

"Doesn't mean he had to be a medical doctor, of course. He could have been one of those philosopher doctors, eh." He turned suddenly. "Ever read any Plato?" They arrived at the shack.

Jake stood in the doorway and peered into the dark room debating whether it would be more comfortable under the biplane's wing for the night. The room was damp and smelled of wood smoke and body sweat long since absorbed into the many blankets and clothes that lay scattered around the room.

There was one bed that looked far too small to hold someone Moyer's size, also a table with a pistol and whiskey bottle on it. The bottle was empty. The pistol, Jake noted, was loaded, the soft lead slugs visible in the chambers. The table had four chairs in good repair, and with closer inspection, Jake noted the chairs were obviously from an expensive dining set. He could only imagine how they ended up there.

Under the table, a rust-colored dog slept, never once raising its head. A photograph of Mae Murray, the movie actress, was tacked to the wall above the table. She stared at the viewer, eyes half closed, mouth puckered, her hand seductively tugging at the low neckline of her gown. What pleasures lurked after the shutter clicked?

Moyer saw Jake stare at the ragged photograph. He had torn it out of a "Photoplay" magazine found on one of the trains. He waved at the picture with an embarrassed gesture.

"I...I thought it might brighten up the room a little," he said. "It's not much of a photo, I know, but I don't see many folks up here...don't go into town much..." His voice trailed off.

"That's Mae Murray, isn't it?" Jake asked.

Moyer brightened. "Yes. Are you a fan of hers?"

"No," Jake said. "I saw her in a movie once; can't remember what it was about—desert or something. Maybe that was Vilma Banky. Oh well, doesn't matter I suppose; it was pretty awful."

"I liked her," Moyer said, deflated. "I met her once."

"You did?" Jake feigned interest. "How?"

"Met her several times, actually," Moyer said and gazed at Mae. "In New York, about two, maybe three years ago. I was working pretty steady in those days—for the Doctor. We made good money then—new cars, clothes; went to all the nightclubs. Everybody liked the Doctor. He treated us good, too."

"So how did you meet Mae Murray?" Jake asked. He stepped closer to the photograph.

Moyer reached to smooth a wrinkled corner and brushed Mae's cheek. "Oh, I met lots of big shots in those days." He turned away from the picture and sat heavily at the table, picked up the revolver and spun the cylinders. The flat clicking of the pistol was the only sound in the room. Outside, the wind began to blow, the pines moaning in whispers.

"Nicky Arnstein," Moyer said. "I met him. That was Fanny Brice's husband, boyfriend...I don't really know what the hell their relationship was except he was baggin' her pretty regular. Funny broad."

He spun the cylinders again. "I think Arnstein owed a lot of dough to Rothstein in those days. Rothstein," he looked up at Jake, "that was Doc's boss." As he reminisced his voice sounded tougher, more the accent of a city dweller than backwoods Canada. He suddenly cocked the hammer, aimed at the back wall and fired.

BLAM!

Jake's ears rang painfully as the explosion ripped through the tiny cabin. Moyer lowered the gun slowly and squinted at the hole in the wall. Jake noted several others throughout the room, although, none in the roof. Moyer set the pistol on the table. Gray smoke hovered in the air.

Jake considered leaving, but Moyer sat between himself and the door. Besides, he was miles from anywhere and to run would be futile. Moyer, at any rate, showed no further interest in the gun and left it on the table.

"Funniest person I ever met though," he said. "Oh, this man was a gem!" He laughed. "John Barrymore."

"Oh," Jake said, his mouth dry.

"I saw him at a number of Doc's parties at the Park Central Hotel. We used to reserve the whole top floor whenever we'd come to town, and I tell you New York would turn out for that man—the Doctor that is."

Moyer toyed with the pistol on the table top. Jake froze.

"Barrymore once rushed over from the theater after a performance, Hamlet, I think. Anyhow, he has on these tight trousers and makeup, rushed all the way over like that just to see Doc. Spent the whole night drinking and playing poker. Left the next morning; still had the makeup on."

Moyer scooped the pistol off the table by the short barrel and held it out to Jake. "Know how to use one of these?"

Jake looked at the black handled revolver, shook his head and took it by the grip, the barrel now pointed at Moyer. It was heavier than he expected and seemed more like a greasy engine part than a gun. From a cabinet above the table, Moyer took a small box and emptied a handful of fat bullets. Two dropped to the floor. Quickly, he scooped them up and dropped them with the others.

"Move the lever on the side." He indicated a release on the pistol, and Jake flicked it. The cylinder with six chambers swung away from the barrel.

"Now shake out the empties, yea, like that. Here, put these in—pointy end facing away from the grip."

Jake looked up, his face blank.

"That's a joke!" Moyer said. "Are you always this serious? Now, click it shut again. There, you've loaded your first pistol, eh."

"Heavy," Jake said and bounced the pistol in his hand and sighted across the barrel.

"Ah, point that somewhere else, friend!" Moyer pushed the barrel away from his face with a fingertip. The dog stirred from under the table and trotted out the door.

"I've never fired one of these," Jake said.

"Follow me," Moyer said and took the empty bottle from the table and slipped through the door into the pale evening. He set the bottle on a log about twenty feet away and walked back and stood behind Jake.

"Now, let's see you do something."

Jake lifted the pistol, heavier now held at arm's length. He pointed at the bottle, the barrel wobbling in a small uneven circle around the target.

"Use two hands," Moyer commanded.

Jake gripped the pistol with both hands and held it steady. Moyer reached from behind and pointed down the sights.

"Just line this front post with this notch in the rear here. Put the front post on the bottle, hold your breath and squeeze the trigger."

Jake took a deep breath, cheeks puffed out, and spread his feet wide apart. The pistol's sight passed back and forth across the bottle, and he kept his left eye clamped shut.

"What are you waiting for, eh, the wind to blow it over for you?"

The target made another pass before the sights, Jake closed both eyes and pulled the trigger. BAM! The pistol barked and kicked his arms up. A chunk of wood splintered off the log eight inches from the bottle.

"Try it again, this time with eyes open!" Moyer chided.

Jake, again, held the gun aimed at the bottle, took a breath, closed his eyes and jerked the trigger. Again, the gun fired, his arms kicked, and a branch from above the log dropped behind the target.

"Again!" Moyer said.

BAM! A third shot fired and a small clod of earth exploded in front of the bottle.

"All right, you're getting your coordinates. Now just average the three and you'll hit it," Moyer said from behind him standing like an umpire behind the catcher. Jake's ears rang from the shots, and he considered the exercise a silly waste of time, but he was the man's guest, and it was his first trip to the country and he hated to appear rude. Maybe, this was a local custom and to refuse would bring great shame upon his host.

BLAM! The pistol snapped back in his hand. His eyes were open this time, and he saw the whiskey bottle disintegrate into countless shards.

"Whoa!" he said. A strange thrill rushed through him the instant of impact—a feeling of power, vicious power combined with the satisfaction of acquiring a new skill. He turned to Moyer and smiled, but the man had disappeared into the cabin returning moments later loaded with empty bottles. He set them in a line on the log.

"All right, again," Moyer said. "Start with the one on your left and work your way down the line. Take a breath, sight, squeeze..."

For the next ten minutes Jake blasted away at Canada, severing tree limbs, knocking down pine cones and occasionally potting a whiskey bottle with a stray shot.

"That should be enough damage," Moyer said and rubbed his face.

"Well, how'd I do?" Jake asked with a proud grin. He twirled the pistol, gunslinger style, but it slipped from his fingers and landed barrel first in the mud.

"Good thing we didn't have you on our side at the Somme!" Moyer picked the gun up.

Jake followed him into the shack.

"You were at the Somme?" he asked.

"Mostly," Moyer answered. "I was gassed twice there, both times in '16, once in March, then again in July. Each time I thought they would discharge me, send me home, but each time I was sent back in the line." He pulled a rag from under the bed and started to wipe the pistol.

"What am I doing?" he asked. "Here, you do this, eh." He handed the gun to Jake. "Get all the dirt out of the barrel, and I'll show you how to oil it."

Jake sat at the table under Mae Murray and wiped the gun clean. Moyer led the conversation through the smoke from a cigarette.

"First time they sent me back to the front they gave me a gas mask; second time they sent me back they showed me how to use it." He drew heavily on the cigarette. "My lungs have hurt ever since then, but these seem to help." He exhaled blue smoke.

"Were you at the front until the Armistice?" Jake asked and set the gun on the table to light a cigarette. Moyer casually picked it up, staring at it absently turning it in the dim light.

"No, in April of 1917 I had my big toe on my left foot shot off. It felt like I had just stubbed it, but the pain grew worse, so I took my boot off, and well, it was a mess. When I woke up, someone had carried me to an aid station. They accused me of shooting my own toe—self-inflicted, they said, to get a ticket out. But, hell they didn't pursue it, and I finally got off the line."

"Did they discharge you?"

"No, I sat around some hospital—nothing but a crowded tent, bunch of guys moaning and farting—then one day they came in, called a bunch of names, told us to get our kit together and line up outside."

81

"To go home?"

"No, back to the line." Moyer sucked on the cigarette and let out a wheezing cough. "I told them my foot wasn't healed. This big sergeant-major, he laughs at me and says, 'Ya don gonna hold your weupun wif yo feet, laddie!' I think he was a Scotsman, or something."

"So how long did you stay this time?"

"Not ten minutes!" He laughed, and inserted a single round into the cylinder and spun it. "As we were riding up in the trucks—they figured they were being generous by not making us walk—I thought about all the different faces I kept seeing each time I'd been back. And worse, I kept thinking about all the faces I didn't see anymore.

"Well, sir, we weren't in the trenches more than ten minutes and wouldn't you know it—a sniper pots me right in the foot, again!"

"Same foot?"

"The very same!" Moyer plunked his left foot on a chair, aimed the pistol at it, cocked the hammer, and squeezed the trigger.

Click. He smiled.

Jake smiled. "A sniper?"

"Amazing, isn't it, and in the same foot that was already numbed from the first wound." He coughed violently and spit through the open door. Reaching into a dark corner he pulled out a whiskey bottle, uncorked it and poured for Jake and himself.

"Well, what happened?" Jake asked.

Moyer snapped open the pistol and out dropped one bullet. He held it before his eyes and twirled it slowly sipping on his drink. Jake felt a chill creep through him and took another swallow.

"I think they figured I was nuts by this time, so they packed me up and sent me back to Canada. Gave me a new pair of boots, though.

"They put me in a training camp, promoted me to sergeant, and I spent the remainder of the war telling recruits to 'squeeze the trigger'! Did that six days a week."

"Sounds boring," Jake said.

"It was, but it beat the hell out of seven days a week in the French mud."

Moyer drank quietly for a moment then stood and stared at the picture of Mae Murray. It was almost dark outside, so he lit a lantern and set it below her. The yellow light fanned up at her like a footlight, and he drank.

Jake sat quietly, drinking and listening. His eyes started to close from whiskey and a full day of flying. Moyer spoke.

"She was married to a movie director at the time, a fellow named...Leonard, nice guy, too. But she dumped him for some foreign guy, a prince—Prince David. Can't remember his last name, started with an M, unpronouncable, whatever it was.

"I only met Leonard once in New York. Mae was always out there without him. I guess he would earn the bacon back in Hollywood while she screwed around in New York."

"Sounds like you liked her."

"I did—everybody did!" Moyer sat on the bed, swallowed the remainder of his drink and laid his head on a greasy pillow. "She was beautiful, a beautiful whore. Those were great days." His voice wheezed heavily, his breathing was slow and deliberate. "Great days working for the Doctor, but all gone now." He took a deep breath. "All gone." He was silent.

Jake sat in the dark room alone at the table below the lighted portrait of Mae. A glance out the door showed a black sky thick with stars. The woods were a dark heaviness surrounding the cabin. Moyer was silent, and the only sound came from the dog who had crawled back in and slept under his bed.

"Moyer?" Jake called softly. "You awake?" No answer. He took the lantern from under Mae and held it over Moyer. The face looked old and drawn in the slanting artificial light. His eyes were deep sockets, the beard ragged, and his skin gray and creased. Jake held the lantern until he saw the chest rise sharply and settle again. He was asleep.

The last of the whiskey filled Jake's glass, and he silently toasted Mae's portrait, then made himself a bed on the floor.

Toward midnight a line of thunderstorms moved through the area. Jake woke briefly, thought he saw Moyer move around the room and dropped again into deep sleep. He awoke before dawn when one of the Indians entered with an armful of firewood and, without a word, built a fire, set a coffeepot on the stove and left.

Despite the hard floor, Jake had slept well and felt good. After washing in the cold brook behind the water tank he was anxious for breakfast and his flight south with the first load of bootleg whiskey.

Moyer had bacon frying on the stove and was mixing eggs in a bowl when Jake returned. "Hungry?" he asked.

"Very," Jake said. "What time is it?"

"Almost six; sit down." Jake sat. The pistol was wrapped in a cloth beside his plate. He pushed it aside, but Moyer stopped him. "Take that with you; it may come in handy." That was all he said. Jake put the heavy lump into his bag and fell to the mound of eggs and bacon.

They heard the first blare of the steam locomotive's whistle long before it appeared through the trees. Through cracks in the forest the Indians began to appear to meet the train. The two he had met the day before stood beside the tracks with what he assumed were their families, all silent with six pack animals tethered to trees. Jake counted eight small children, some swathed in blankets, the others in whiteman's pants and hats. All were silent, occasionally swatting at the many flies that peppered the morning air.

Jake had been to the airplane several times already that morning. He had checked the oil and added a quart. He checked the water and checked it again after pulling the propeller through a half dozen times. The new engine was tight with compression, something his own wrecked Standard lacked.

The Indians would glance at him without interest and mutter something among themselves, every now and then pointing at him. Put off by them, Jake kept to himself. "It was their country," Moyer had told him, and he truly felt it.

84

The sun warmed the damp earth, and he wondered if Susan had made her morning run to the train station. Was she in bed now? He remembered the feel of her soft body against his.

With an angry hiss of steam from under the locomotive, the train rolled to a stop. It dominated life at the watering stop as it clanked and rumbled while crew and Indians hurried about it loading, unloading, pointing and shouting. Moyer greeted the engineer with a familiar wave. Two Indians and a fireman busied themselves with the waterspout filling the boilers.

Jake stood well out of everyone's way, especially Moyer's. The giant with the red beard ran the stop, and everyone recognized his authority.

A boxcar door slid open, and Jake saw the whiskey stacked in wooden crates beside a crude metal tank he guessed would fit into the forward cockpit to carry the liquor. He confronted the physical evidence of his life stepping beyond the law.

"Give a hand," Moyer called.

Moyer slid two cases over to the door, and Jake lifted one at a time to the ground. "Why isn't this in barrels? Wouldn't that be easier to handle when it's loaded in the tank?"

"Your volume's too small to warrant buying in bulk." Moyer said. "When your demand increases, you'll get the discounts for bulk. Doc knows this. Meanwhile you get it bottled."

This is going to be a pain in the ass, Jake thought, and he watched the metal whiskey tank drop to the ground with a tinny bong. It was a cheaply welded square box with a cap on top and a drain plug on the bottom. There was an eyehook on top for lifting.

"Quite a load you've got there," the conductor walked up behind Jake and counted the boxes checking off numbers on a clipboard.

"That's the lot," Moyer said and jumped down. No one mentioned where the whiskey was headed, but now Jake felt like the sneak from across the border involved in a sleazy venture. Moyer signed for the whiskey.

"Your gasoline's off-loaded further up," the conductor said and pointed to several red barrels beside the tracks. He left, and the fireman climbed down from the locomotive after swinging the water spout back to the Indians.

The locomotive puffed out a great cloud of black smoke as steam blew out the sides and the train labored slowly away.

The two Indians carried the whiskey across the tracks while Jake untied the biplane. The train's whistle blasted twice, far down the tracks, and the forest was quiet again except for the birds along the wires and the countless insects chattering in the pines.

Jake pulled the canvas tarpaulin off the front cockpit, and water rolled from the cloth and collected in a puddle on the muddy ground. His shoes were caked with the mud from the short walk to the plane. Moyer walked up with the whiskey tank on his back and dropped it at Jake's feet.

"Looks too big," Jake said, his eyes darting from the cockpit to the tank.

"Arthur sent the dimensions; should slip right in," Moyer said and hoisted it onto the fuselage smearing mud on the fabric. Jake climbed into the rear cockpit and guided the container into the front. It squeaked against the maroon leather and plywood sides as it slid in, Jake and Moyer pressing to make it fit.

"There," Moyer said. He reached into a case at his feet, took out a bottle and uncorked it then poured the contents into the tank and splashed some across the wings before he tossed the empty aside.

"You are duly christened!" he said. "Grab a bottle."

Jake hopped from the plane, took a case and walked to the opposite side of the fuselage and set it in the mud. He and Moyer unloaded the entire load of whiskey four bottles at a time into the tank. When they were done, the empties laid scattered around the plane in two arcs. The tank was capped, and Jake buttoned himself loosely into his jacket and climbed into the cockpit.

"What time should I expect you back through here?" Moyer asked.

"Tomorrow around ten," Jake said.

Moyer looked at the ground momentarily then up at Jake. "Anybody tell you about Adair?"

"Doc mentioned her—said she was pretty ugly."

Moyer frowned. "Not ugly really, just scarred."

Jake stared.

"Cut up, actually...well, you'll see her." Moyer picked a clod of mud off the fuselage. "She's a leftover from the New York days. Her husband worked for Doc." He looked straight at Jake. "You just stay clear of her, that's my advice."

"Hadn't planned anything!" Jake said with a grin. "Anyway, she has a husband."

"And you don't want to go messing with him, either," Moyer said. "And don't stare at his face when you meet him, eh!"

"Why?"

"He doesn't have one!"

"None?"

"Well, he's got eyes, holes for a nose and one for a mouth, but no features like a regular face like you or me. Lost it in the war."

"How?"

"He was a forward observer up on this hilltop where he was supposed to watch for Germans, should they get the notion to push out. It was a nice job, quiet sector, and all he had to do was sit and watch. Every now and then he would fire off one of these flare guns." He mimed shooting a pistol into the air. "I guess that was a signal of some sort."

He combed his pockets until he found a bent cigarette and stuck it in his mouth. "So, one day, middle of December, wasn't a thing happening, and he was cooking soup or something, and he has to shoot off the noonday flare." Again he mimed shooting the gun.

"It must have been real cold, because the window was still closed, and he didn't look first, so this flare..." Moyer coughed a long rattling heave and spit on the ground.

"...This flare hits the window frame and flies back into the room right in his goddamn face!"

87

He found a match and struck it on the plane. "That stuff burns awful fast, and it—PHITT! Takes his face right off— melted it!"

Jake's own face had screwed into a cringing mask. Moyer lit his cigarette.

"They kicked him out of the Army, damn near court martialed him, too." Moyer slapped the fuselage. "So don't go near Adair, he's somewhat jealous of her on account of he knows how ugly he is."

Chapter
VIII

"It's easy!" Jake called from the rear cockpit. "You turn the propeller through—counterclockwise as you face it—and when you feel the compression build on one cylinder, you stop."

"Do you have to prime it?" Moyer asked.

"No, I've already done that. You just pull the prop..."

"And wait for the magic compression," Moyer said, and he grabbed the propeller.

"Now, don't touch anything until you hear me say, SWITCH OFF!" Jake called.

"Is it required to yell like that?"

"I don't want you to miss it; that prop can bite your head off if the magneto switch is on, and you go monkeying with it!"

"So, is the damn switch on or off?"

"It's off—SWITCH OFF!" Jake called.

"So, what's the big worry? The switch was off."

"Just pull it through now," Jake said, and Moyer took the propeller tip and spun it easily through one blade.

"That enough?" he asked.

"Give it a few more, you'll feel a good cylinder."

Moyer turned the engine over blade at a time, until Jake called, "That should do it! How's it feel?"

Moyer tugged at the prop tip and shrugged. "Feels like there's something in there."

"Fine, now stand back."

"You got a starter motor in that thing?"

"No, there's a booster coil in here..."

"Don't I have to swing my leg up like a ballerina while I spin the propeller. I saw them do it that way in "Wings.""

"This is a different ignition," Jake said. "I turn the magneto switch on, then I crank the booster coil in here," he pointed inside the cockpit. "And it acts like a telephone crank, sends a shower of sparks straight into the cylinder, where, hopefully, it fires. After that the engine runs on its own."

"Sounds clever. Does it work?"

"Mostly." Jake had little confidence in the Standard's ignition system. Starting was often a long process of priming, cranking, priming again only to flood the engine, then backing the propeller with the throttle wide open to clear it, and retry the whole procedure. He had seen an alternate method involving the use of a long elastic shock cord with a sock fitted to the end.

The sock was slipped over the propeller tip, and while one person held that, several others would stretch the cord out laterally to its limit. Then, on a signal from the pilot, seated at the controls, the one holding the prop tip would release. The cord would snap, and those holding it would fall to the ground, while, hopefully, the propeller spun with enough vigor to carry the engine through and run. From a distance, however, it merely looked as though Mack Sennet Studios were holding auditions.

"Here goes," Jake called and spun the booster coil handle in the cockpit. There was a brief grating sound, then the eight cylinder water cooled Hispano-Suiza engine belched orange flame and blue smoke from one stack and ran.

Jake looked up and smiled. Moyer waved, pleased with his part. Jake scanned the instruments for oil pressure and waited for the water to warm. Moyer ran around to the cockpit.

"See you tomorrow!" he called over the roar, his red beard leaning in the propwash.

"I might be back tonight, if all goes well!"

Moyer nodded and stepped clear. The Standard howled with power and slowly picked its way to the far end of the stump covered field. The plane felt heavy with its forward cockpit weighted with seven hundred pounds of whiskey.

The biplane turned into the slight breeze. The field was downhill from there, and Jake was relieved when he lifted and cleared the trees at the end with dozens of feet to spare.

Glancing at the 44 foot wing span, he climbed slowly over the treetops.

What lift, he thought. Imagine how much more this thing could carry, he thought. He knew there was much unused space in the forward cockpit. The Standard's potential slowly revealed itself to him. He turned southeast toward the shoreline of Lake of the Woods, he would follow that into the U.S. and then south from there toward Bimidji.

For two hours he flew over marshland and forest, occasionally spotting a settlement not marked on the map. He figured a rough heading toward Bimidji expecting to find any number of suitable landing spots near towns big enough to sell gas. A few he let slide past knowing he had plenty fuel aboard, but as he watched the gauge drop and no towns with easy access appeared he grew uneasy.

Finally, a small town took form off to his left, and without hesitation he flew over it. Less than a half dozen faces were there to watch him go over, and the nearest place he found to land was over a ridge and beyond a thick woods. He landed in a meadow of tall grass studded with yellow flowers.

The biplane rumbled to a stop, the grass whipping the lower wing. The air above had been colder than he expected, and he shivered and slowly pushed his way from the cockpit.

"Ahhh," he moaned slightly removing his helmet, while massaging the deep red creases around his eyes left by the goggles.

"Looks like I've got a bit of a walk ahead of me," he said.

Leaving the helmet on the joystick, he set out for town. After tearing his trousers on a wire fence, he headed into the woods where he spent almost 40 minutes wandering without direction.

The sun was directly overhead, and filtered warm through the leaves. The jacket soon became a burden, and regretting having brought it, he tucked it under his arm. At last he emerged onto a dusty road between the woods and a low soggy bog.

"Crap," he muttered. "The damn town's got to be over there!" he declared pointing across the bog. "I know it!"

Flies, attracted to his sweaty head, buzzed in pesky swarms around his face, catching in his hair. In mounting frustration, he boxed himself in the ear attacking a persistent bug.

He turned his face toward the flat blue sky. The wind was almost calm. He stopped to look again across the bog. "I know it's that way!" He pointed, and by now had convinced himself. He stepped off the road and into the soft damp grass. Immediately, his right boot sucked into the mud locking his foot.

"Shit," he swore. "Double shit!" He recoiled and pulled clear leaving his boot, only to fall backward and plant his bare foot in the same mire.

"Oh..." It was a sound of frustration more than anger. For several minutes he hopped as best he could on one foot, dangling the bare one above the ground, pondering his escape. Flies attacked him in swarms.

"Oh, phooey." He felt undeserving of stronger oaths and set his bare foot into the mud and reached for his boot.

Jake was seated beside the road, flies in his hair, when a wagon pulled by a single horse, crested the hill and drew toward him.

"Hey!" Jake shouted, and he jumped into the road waving his boot. "Hey!"

The wagon stopped and a squat boy in his early teens with red cheeks and small eyes stared at the man covered with dried mud jumping up and down in the middle of the road waving a boot and a leather jacket.

"You lost?" he asked.

"Not really," Jake said. "I just landed my airplane in a field over there." He pointed, and the boy's head followed the point. "And I need gas, so I was heading for town, over there." Again, he pointed, and again the boy's head followed, a frown always on his face. He held the reins tight and leaned way from Jake.

"I was hoping to buy some gas in town." Point. "Then get a ride back out, fuel up and take off."

The boy stared. Jake quickly looked at himself, the mud, the boot. He laughed, "Oh, I tried to take a short cut through the swamp." Point. "Rather than take the long way round on the road." He pointed behind the wagon. "You just coming from town?" No response. "I'll give you a dollar to take me in." Jake pointed, again, behind the wagon.

The boy looked slowly up the road where he had just been, then at Jake, and the swamp. A fly crawled across his nose, but he ignored it. "Town's that way," he said and nodded ahead, in the direction he was going.

Jake turned, and turned back.

"I'll give you a ride, but you can keep your dollar! You say you came in an airplane?"

"Yes," Jake said and climbed on the wagon. "It's over there." Jake pointed, but without conviction.

They both stared a moment into the woods. "You sure you had an airplane?"

Jake spent most of the afternoon hanging around the feed store beside the railroad tracks. Someone had told him that the one to see about his gasoline was a fellow named Brucker, Ernst Brucker, and he would be along anytime, now.

The train pulled in at 2 p.m., and, at 4, a second train came through in the opposite direction. The first dropped off bales of something brown. The second picked them up. At 5, the feed store closed, a thin man dressed in gray slacks and a shirt with no collar asked what he wanted.

"Gas," Jake said.

"Gas." The man thought a moment. "That would be Brucker; should be along anytime, now."

"Thank-you."

"What's the gas for?" The man looked around.

Jake hated to say it. "My airplane." He pointed, vaguely.

"Oh! Very good." The man in gray left. Twenty minutes later, Brucker drove up in a square old truck trailing a cloud of blue smoke. Three rusty barrels were strapped to the truck's bed.

He stuck his head through the window and called, "You the fella wants the gas?" A cold cigar was plugged into his wide round face. His nose was red and stubby, his breath stale. He smelled of beer.

"Yes," Jake said with a tired voice. "Is that high octane?"

"Yes, sure—high octane." Brucker nodded without looking at Jake.

Jake stepped onto the running board and opened the passenger door. Brucker said nothing, but a slobbering growl rose from the seat and Jake vaulted back. A small black dog sat coiled like a snake on the seat baring its yellow fangs.

"She don' bite none," Brucker said.

"I'll ride back here!" Jake climbed on with the barrels. Brucker shrugged and ground the truck into first gear.

The truck was old, its suspension tired. Jake bounced around on its bed as they drove through ruts and ditches. The gas barrels were lashed together with two lengths of dried rope tied into a confused knot on the side. With the constant vibration, one of the barrels wiggled free and slid toward Jake. About to see his fuel drop overboard, he lunged for the barrel as Ernst turned onto the field where the Standard was parked. He grabbed the barrel, lost his footing and hit the bed tearing his trousers on the boards.

With one foot hooked through the rails near the cab he pulled the barrel from the edge. He would have been safe had the other two remained stationary, but they, too, were loose and rumbled toward him. One crashed into his leg, and the other pinched his hand against the floor.

"WAAAH!" he cried and let go of the barrel as Ernst hit a rut and turned the wheel. Suddenly hovering in mid air, Jake saw the truck slide away, and down he came with a thud in the weeds, the barrels crashing around him, his knee smashing into a rock hidden in the grass.

"Ugggh," he moaned. The truck door slammed.

"Oh, I see you have already unloaded—very good!" Ernst said looking down on Jake. "We have work to do, now come! Up! Up!"

Jake crawled his way through the weeds and up the side of the biplane and onto the upper wing. Ernst handed him a cheap hose connected to a hand pump inserted into the first barrel.

"I pump; you run the gas through the chamois!" Ernst seemed to enjoy giving orders. He pumped steadily for twenty minutes stopping occasionally to light his cigar. Each time he would strike a match against the barrel and bring a brief red glow to the cigar butt.

The sun pressed down, and Jake soon had his jacket off and draped across the wing. His shirt and undershirt quickly followed. Gas fumes rose in shimmering waves from the tank and the barrels. Ernst pumped away on the truck clenching his cigar and wiping sweat from his face.

"Whoa! Whoa there!" Jake hollered, and fuel ran out the tank and down the wing.

"Enough?" Ernst called.

"Yes, enough." Jake handed the hose down and mopped at the spill. Ernst coiled the hose and poked a long calibrated stick into the barrels.

"How much I owe you?" Jake asked.

Ernst rummaged through the truck's cab until he found a pad and pencil, then scribbled some numbers while humming a martial tune through his nose.

"Twenty-five dollars."

"What?" Jake said. "That's kinda steep, don't you think? How much is it per gallon?"

"Fifty cents."

"That's twice normal!" Jake said. Brucker smiled. Jake started to count the money when he stopped suddenly. "At fifty cents a gallon...that comes to fifty gallons!"

"Yes, fifty gallons. That's correct. I measured it." Brucker held up the stick as evidence.

"Well, the damn airplane only holds forty-three gallons total!" Jake thrust his hands on his hips in a gesture meant to look threatening, but merely came off as feminine. "And I didn't fly in here empty!"

Brucker took a handkerchief from his pocket, wiped his forehead and blew his nose. "Maybe, I make a mistake, huh?" He took the notepad out again.

"Yea, maybe!"

"How many gallons you think? Maybe forty?"

95

"I guess—yea, about forty." Jake's head was throbbing from the sun and gas fumes.

Ernst tallied a new bill and said, "that comes to twenty-five, fifty."

Jake glowered.

"The price goes up, the longer you wait, my friend."

"Don't get much repeat business, I suspect." Jake counted out the money.

"Your kind of business I don' need!" Brucker took the money and stuffed it in his shirt. "You never come back this way—I know! Airplanes! Phitt! Who needs them?"

"Well, maybe I will come back this way, just to tell you I don't want any gas!" Jake felt he had to say something, even something stupid.

Ernst waved with his back to Jake, and soon the truck was a distant clatter going across the field. Jake pulled the propeller through until he felt compression build on one cylinder. He slipped his jacket into the front cockpit and pulled the leather helmet loosely onto his head.

"Too damn hot," he growled and pulled it off.

The engine started without trouble, but as soon as the biplane was airborne Jake knew he had paid more than double for the gas.

BRAP!....BLA...BRRAPP!

The Hisso motor coughed and sputtered sending interrupted surges of power through the machine.

"Son of a bitch!" Jake swore as the biplane struggled to climb just barely clearing a stand of trees at the crest of the meadow.

"Bad gas!" he shouted. BRAP-FRRAPP! The airplane barked and settled toward the ground. "That crap wasn't high octane! Wasn't worth more than ten cents a gallon!"

The engine was suddenly quiet, the propeller lurching to a halt. Jake's eyes were open as wide as they had ever been in his life, and only to watch another cluster of trees reach for the plane.

Too slow to maneuver, he pushed the nose forward for speed and awaited the impact.

BRAP! BROOM! The motor caught and over the trees the biplane struggled. Jake babied the throttle and found a setting short of wide open where the engine ran well enough to climb. At 2000 feet, the engine ran smoothly. Jake let the cool air dry the sweat from his body.

After ten minutes or so, he trusted the engine enough to climb to smoother air above the scattered layer of clouds. For two more hours he flew in the calm air, his fingertips making only the slightest of corrections to the plane. The warm sunshine still pressed on his head, but he wished he had kept the jacket handy for the cool air above the clouds.

Forest and marshland gave way to farmland, and he spotted the silo with the white cross painted on top. The sun was fast approaching the horizon digging deep shadows into the landscape. A large flat pasture, recently mown, stretched out beside the silo and barn. It was wide and inviting, big enough to handle approaches from any direction eliminating cross wind landings.

A face turned up at the sound of his motor, and someone stepped off the back porch of the house and climbed onto a horse-drawn wagon. Jake pulled the throttle back and immediately the Hisso engine began to misfire as the cheap gas ran through the idle jets.

"Ten cents a gallon, and even that's too much!" Jake complained to himself and slipped the Standard to a smooth landing over the house and past the silo.

The Hisso engine backfired viciously then roared with power. He jockeyed the throttle to swing the tail and taxi back. In a cacophony of engine farts and belching flame, the biplane made its way back toward the wagon. He clicked the magneto switch off. The propeller took several extra turns, hesitated, took another spin and lurched into reverse with a rattling death throe.

The odor of burnt oil and greasy exhaust swept back over the cockpit in a quick breeze. Someone stepped onto the wing and silently inspected the whiskey tank.

"Hello," Jake said.

"Is there a release to these straps holding the tank in?" The voice was a woman's, husky and frank. She was dressed in men's clothes, a loose collection of browns and greens. A brown felt hat hid her face. Her hair was also brown and short.

"There's two buckles near the bottom," Jake said. He dropped to the ground and pressing near her legs leaned into the cockpit and pushed the tarpaulin aside. "There."

She looked up quickly, only the faintest of smiles momentarily on her lips.

"You're Adair?"

Before she could reply, the ground shook with the pounding of hooves, and a chestnut stallion came galloping across the pasture toward them. Clods of turf hung suspended in the air behind it. The rider, a man, leaned slightly forward in the saddle seeming more an extension of the animal's back than a passenger. He wore high brown boots, tan pants and a white shirt open at the collar. A military Sam Browne belt crossed his broad chest and encircled the tight waist. A large revolver hung from the belt, its lanyard flapping as he rode.

He slid from the stallion's back before it had stopped, and with a vicious tug on the reins pulled the horse's face toward his. Then with a tenderness Jake found discomforting he stroked the animal's nose between the two flaring nostrils.

"There, now. There, now." The horse seemed to calm, and the rider lashed the reins to a wing strut.

"That's not a good place to..."

Jake was cut off when the man turned toward him.

"No face," Moyer had said. "He has no face!" And it was so. The rider with the broad shoulders, trim waist and powerful arms lacked a face—lacked identity.

Jake gave little thought to his own self. He was confident of his presence, pleased in a quiet way with his own features. His own arms were strong, shoulders wide enough, legs thin and well-muscled. He lived on the move and had the ragged look of one who rarely shaved over the same sink twice.

Looking now at the rider, he pictured his own face—long-jawed, pale blue eyes over a sharp nose and thin lips, but most importantly, his face had expression, life. The rider's was dead.

The rider approached. "I'm Vincent, in case you were wondering." He tossed an envelope to Jake. "Count it! I don't want Doc accusing me of cheating." He turned away and leaned past Adair to look at the whiskey tank.

Jake thumbed through the bills—mostly twenties. "It's all here," he said.

"Sign this," Vincent said and held a small receipt and pen in front of Jake's nose. Jake leaned against the fuselage and signed his first name only to the receipt and handed it back.

"First name only!" Vincent laughed. "That's good. Clever. No way to trace you, eh?"

"Something like that," Jake said. "Maybe, I don't like my last name."

"Good, I don't like mine, either—causes too much trouble."

Jake took the opportunity to stare at Vincent, and the man seemed to realize this, but said nothing.

Two gray eyes, a lump with holes for a nose and a tight opening over straight white teeth marked his mouth. There were no real lips, no whiskers, eyebrows or creases, only a pink smooth skin pulled taut across bone. Long white wrinkles led off the pink skin toward normal ears and hairline. The white wrinkles secured the mask in place. He made a slight whistling noise when he breathed, like someone with a cold.

"Pull the wagon closer," Adair called. "We'll have to off-load this thing."

While Vincent nudged the horse to back the wagon to the airplane Jake said, "Don't you have a barrel or something, so we can just drain it here. This is going to take too long."

"You in a hurry?" Vincent asked. The wagon bumped the fuselage.

"My time's valuable," Jake said.

"It's getting dark," Vincent said as he climbed onto the wagon and swung a winch over the cockpit. "Can't fly in the dark, can you?"

"Sure, why not?" The Standard had no lights, although some could be added, but Jake had no desire anyhow to navigate his way back to Canada after sunset.

"You'll spend the night," Vincent said. Adair hooked a cable onto the tank.

"It still takes too long the way you're doing it," Jake said knowing he had no control over the situation. "Next time have something here to drain the whiskey into!" He turned away.

Adair and Vincent stared briefly at each other. "Bossy type; isn't he?" Adair said.

The tank was soon loaded onto the wagon and headed for the house. Jake was alone with the airplane, and he took the opportunity to pull the sparkplugs. Bound to be fouled from that cheap gas, he thought.

Standing on the plane's right tire he tinkered with the still warm engine, but his thoughts ran in a confused circle to Susan, Moyer, and the woman he had recently met—Adair.

Doc said she would be ugly—"as river mud," he had said. Jake disagreed. Hidden beneath the men's clothing, Jake had detected the figure of a woman. The details were hazy, but he liked what he saw. Her face, too, was pleasant—trim with refined features and bright green eyes.

There was one glaring distraction, however. A scar, deep and vicious, bisected her face. Beginning above her right eyebrow, it split that and jumped over her eye and down the bridge of her nose slicing the left nostril before it ran on deep into her left cheek and tapered out near her neck. It was a white bloodless line, like a gash across a portrait, the beauty of the subject still visible beneath.

"Are you hungry?"

Jake was absorbed in wiping oil from the engine's cowling and watching the sunset below the wing when Adair spoke.

"What? Oh, ah, yes, I suppose," he said. He lay on his back in the warm grass below the cowling and squirmed free to face her. The rose glow of the sunset washed over her muting the scar. She wore a dress now, the blouse loose and soft. Her arms were folded across her chest pushing her breasts up.

"I've set out a few sandwiches. Vincent's almost done with the whiskey." She crossed her legs and briefly lost her footing. It was then Jake realized she was drunk. She turned, the skirt momentarily hugging between her thighs.

"Come up to the house anytime." She walked away, swaying gently. Jake sat up and clasped the lower wing.

"I will, thank-you."

As he approached the house, a tall thick man with a shotgun straddled across his lap sprang from the porch steps.

"Hello," Jake called with a wave. The shotgun was now pointed square at his face.

"Good evening, Mr. Jake," Vincent called from his seat on the lower step. The whiskey tank rested on the back of the wagon, a hose ran from the spigot near the bottom, and he filled bottles, one at a time, empties on his right, full on his left. He, too, was drunk.

"Friend of yours?" Jake asked and pointed toward the man with the shotgun.

"Oh, him? That's Marv. Say, hello, Marv."

Marv nodded and kept the gun trained on Jake.

"Say hello to Marv, Jake."

"Hello, Marv," Jake said. "Nice gun; ever use it?"

Marvin stared. "It's okay, Marv. He's a friend." Vincent spoke to him in the same tone he had used earlier on the stallion and with the same result—the beast was calmed.

"Can I offer you a drink?" Vincent held up a half full bottle.

"Thank-you," Jake said and approached, but kept one eye on Marv.

"This is excellent whiskey you've brought us." Vincent spoke easily, in a friendly tone, although his face never moved. It was like hearing a statue speak.

"Worth all the money you paid?"

"Ahha, the pilot becomes the business man, eh?" He filled another bottle and set it at his feet. "Yes, as a matter of fact, it is."

"I had no idea there was such a demand for this stuff in this part of the country." Jake took a swallow.

"Too many Bibles and farmers about, eh? Well, I'll tell you something, Mr. Jake..."

"Just Jake."

"All right, Just Jake, the demand out here is bigger than you can imagine, and the supply is limited. We can command any price we want."

"But can't this be manufactured locally, and cheaper than flying it in?"

"Not this quality!" Vincent filled another bottle. The screen door opened with a flimsy yawn and slammed. Adair stepped through. She carried a plate stacked with sandwiches.

"You've made it, good." She held the plate to Marvin who took one and smiled. Then she offered one to Jake and finally to her husband. He held a bottle for her and she drank.

Vincent spoke. "We sell this uncut in Minneapolis and it'll bring twice maybe three times what it would bring in New York! Locally, we cut it in half, even down to a quarter strength, and make even more money.

"Mr. Just Jake, we can peddle everything you bring."

Jake slowly chewed on the sandwich.

"How many other pilots does Doc have?"

"Only me, so far."

"There'll be more—you watch. My guess is Doc will clear a couple hundred, easy, on each flight—each airplane. That might not seem like much, but you think about it. If he gets, say, a dozen planes, or so, and runs them back and forth across the border, he'll make out. You watch."

The evening slipped on, and soon the tank was empty, and Jake helped them carry the bottles into the house where they were stacked in crates.

Adair fell asleep on a worn out couch near the stairway and Vincent slowly drank himself into a stupor. When he was asleep Marvin walked over and gently lifted him into his arms and carried him up the stairs. Jake watched them disappear into one of the bedrooms.

Minutes later he returned and took up the recumbent Adair and carried her to a separate room.

"You sleep in the barn," he said to Jake as he walked past him and out the door. "Come."

"What? I don't get carried?"

Marvin turned, and with a confused look half started to reach for him, but Jake added, "Lead on; lead on."

Marvin lived in a small room built into the southeast corner of the barn. The door was open, and Jake followed him inside. A bed, nightstand with lamp and small Bible were against one wall. Another table with wash basin and pitcher stood against the far wall. The floor was rough board except for a braided rug beside the bed. A cast iron stove stood in the center of the room, a pair of shoes on top. One wall had a small window and the opposite wall a portrait of a fair-skinned Christ with blue eyes. Marvin hung his shotgun on two pegs below this.

"Nice," Jake said.

"You sleep outside—over there." Marvin pointed to a pile of stale hay in an unused stall. The farm had been idle for many years, and many of the stalls were empty. Only the chestnut stallion, whose breathing rose in deep regular drafts from the dark, and the dray horse for the wagon lived there now.

Marvin handed Jake a thin blanket, said, "Good night," and closed the door. Jake stood alone in the dark listening to the stallion and tried to calculate the profits from the whiskey trade. Shortly, he gave up and threw the blanket on the straw. When he pressed his face to the blanket his thoughts maneuvered toward Adair and her reclining figure on the couch. He tried to picture Susan, and wondered what her face would be like with a scar across it. He slept.

The few clouds that hovered at sunset disappeared after dark as the earth cooled and the wind grew still. A thick dew covered the ground before dawn, and Adair's bare feet were wet after the walk across the pasture from the house to the barn.

"Whass-that?" Jake started and felt Adair's warm form sink into the hay beside him. She was wrapped in a soft cotton blanket, her skin cool and smooth. She carried the fresh aroma of cut grass from the field.

"Sshhh," she hissed softly, and the blanket slipped from her body and covered him.

"Susan...Adair?" Jake shot upright in the stall, his head snapped left and right from the door near the stallion to Marvin's room. His eyes adjusted to the darkness, and he saw stars peek through a hole in the roof.

"Lie back," she whispered. "And don't worry."

"Your husband..." he said unsure what should follow.

"Is asleep, and does not know, nor care."

"Marvin?"

"If you keep calling him I'm sure he'll join us, but I don't think we want that, do we?" Her leg slid across his thighs bumping his sore knee. He groaned. "Now, that's better," she said.

Before the sun rose, Adair's feet were once again wet as she climbed the steps to her room. The blanket was draped over her shoulders, and light escaped under the door to Vincent's room. She opened it. He was propped upright in bed, a pistol disassembled on the covers stretched across his legs.

"Nice walk?" he asked, his eyes blank and staring, his tone flat, as expressionless as his dead face.

"Nothing special," she answered. "Can't sleep, again?"

"As usual," he said and dropped the cylinder drum onto his pillow. "Sinuses clog the minute I lie down."

She stared at her husband, alone in his room, grotesque in his isolation. Yet, she stood resolutely outside the door, her blanket clutched tightly around her body, low across her small breasts.

"Well." She hesitated; there was nothing. No words could handle the immense crevice that had separated their lives since the war. "Try to sleep," she said. She left.

He continued to stare at the closed door after she had gone. He wiped the revolver with an oily rag, reassembled the pieces and took six bullets from the table at his side. One by one he pressed them into their chambers turning the drum with each round. The pistol snapped shut with a dry click, and he pressed the barrel to his temple. He cocked the hammer.

"Ka-Pow!" he said, and he lowered the gun slowly to his side.

The single window in his room overlooked the pasture to the east where the purple trace of dawn grew. The outline of Jake's biplane was discernible in the faint light.

Marvin snatched the blanket from Jake's body, and left him curled in the hay. Jake stayed in that position for a moment, then woke with a start. He heard the barn door squeak as the dray horse brushed past it on its way out. Jake remembered Adair's visit.

"Oh, Christ!" he stammered and pushed to his knees and gathered his clothes. Sunlight poured in through the windows and cracks in dusty slants. He pulled his trousers over his sore legs and quickly found his boots.

"Oh, double Christ!" He peered over the window ledge above him at the quiet house and pasture. The heavy dew rolled off the Standard's upper surfaces and drummed the lower wing. Jake hurried across to the plane and reached for the tarpaulin over the cockpits. They were already off; the whiskey tank replaced.

Jake pulled his boots over bare feet and noticed the fuel gauge read full. He grabbed the propeller and with short strokes pulled it through. The clank of the cold Hisso engine blended with his muffled grunts to make the only sounds of the morning. Gasoline dripped from the lower cowling, and he set the propeller in a position to fire.

"Switch on, fuel on and brakes," he said to himself, once inside the cockpit, and was about to crank the booster coil when over the ridge Vincent appeared astride the chestnut stallion bound toward him. The low sun created a long shadow off to his right. Jake froze.

The dull thud of the hooves was now the only sound. Jake contemplated several options, running, fighting, explaining. He suddenly remembered the gun Moyer had given him, but did nothing except sit and watch the tall rider, with the pistol flapping at his hip, approach.

Instead of dismounting, Vincent pulled alongside the fuselage, until Jake could smell the sweat from the horse's flanks.

"Not staying for breakfast?" Vincent asked, his voice cheerful—toying, Jake thought.

"There's money to be made," Jake answered and knew his voice was strained, his mouth dry.

Vincent shifted in the saddle and tugged the reins with a snap as the horse lowered its head to nibble at the wing. His right hand rested on the pistol's grip. "You'll notice I fueled you already."

"I noticed—thank-you."

A fat pause elapsed while Vincent gazed around the farm and at the sky. The air was dead still. Dew still tapped onto the lower wing, the sound amplified in Jake's mind.

DRIP...DRIP...DRIP

Vincent turned his expressionless stare on Jake. Although the face was immobile, Jake detected something—an emotion, an effort—but it was vague. "I'll tell Adair you've left."

Before Jake could speak, Vincent yanked the stallion's reins and was away at full gallop across the meadow. Mercifully, the engine started with little effort, and Jake was airborne before he realized his seat belt was unbuckled and his shirt was still off.

Chapter
IX

Normally, Susan would return home after delivering the milk cans to the train station in Luther, but today she drove to Boone instead. It took almost an hour in the old Ford truck over roads often impassable with mud from springtime rains. The countryside was deep green, the air ripe with the smells of the earth.

The pilot, Jake, had been gone for several days, and life around the farm had resumed its quiet routine. Her father pursued his beer-making to the almost total exclusion of everything else around him. A wonderfully kind man, she thought, he was nonetheless lazy, detached.

She loved her father, and had grown increasingly close to him since her mother's death three years earlier. With the death, he had eased into the comfortable yet irresponsible position of permanent mourner. She adjusted to the new role and managed the house as best she could watching it slide into disrepair. She poured devotion on him until he was void of motivation, an old bull without purpose.

Beer-making was his one enterprise, and it had started as a curiosity but soon offered an illicit prospect of income to be exploited. Had it not been for Susan, however, the beer would have vanished inside himself. She spread the word, discreetly of course, and soon she was overseer of a small cottage enterprise. John was content to dwell in the cellar with his tanks, bottles and cappers.

She was self conscious of the muddy truck when she pulled into Boone, the tires kicking clods of dried mud into the air behind her. She never once turned a head. Boone was every bit the farm town as Luther, only larger.

She parked a block away from the municipal library partly to hide from the shame of the poor old truck, but mainly she enjoyed the walk through the square, past the three-story county building with its spires and battlements like some medieval castle. Off the square, the streets ran in parallel rows of substantial houses with broad porches and shade trees. The Carnegie Library was under two chestnut trees amid the houses.

Inside, the air was quiet and stuffy barely stirred by the woman dressed in gray behind the circulation desk. She wore a small brooch at her throat and smiled kindly, but suspiciously,when Susan brushed past toward the card catalogue.

Susan was familiar with the catalogue and pulled out the drawer labeled, 'Fa-Fz'. She looked for FLYING but found nothing about airplanes only a card that said, SEE BIRDS. She checked under AVIATION but nothing. Finally, she found a card under, AERONAUTICS, and a book entitled, YOU TOO CAN FLY, by Edgar Blythe, Capt., US Army Air Corps (ret.).

She hurried back along the streets past the houses and courthouse without giving them so much as a glance. Her fingers flipped through the pages of Capt. Blythe's book exposing photographs of Curtiss JN-3's and JN-4's, of flying boats and large twin-motored transports capable of carrying as many as ten passengers at once.

Biplanes! Three-winged machines! She turned through the pages as quickly as her eyes could absorb the images. A picture of the White House, in Washington, D.C., taken from the air, spread across two pages. She giggled at the thought of simply flying over something as important as the White House.

She stumbled over a curb and struck her knee on the sidewalk outside a drugstore.

"Careful there, now!" A large hand reached from above and pulled her to her feet.

"Thank..." Susan looked quickly around. The book had slipped from her hands.

"I don't think there's any damage." The man who had helped her to her feet picked the book off the pavement, and with a glance at the title his thick eyebrows shot up. "Becoming a flyer, eh?" He laughed thinking himself witty.

"No!" Susan shot back. She reached for the book. YOU TOO CAN FLY, the title read. "I mean, yes," she said and looked at the man with the thick eyebrows, his face a rosy mask of polite condescension. "Why-the-hell-not?" she said and strode to the truck, and with a vicious crank of the handle started the motor and left.

With the book open on the seat beside her, she drove home, twice becoming stuck in the mud. The second time required her to dig the rear wheels out with a shovel she carried expressly for that purpose.

When she arrived home it was well after noon, and the house was quiet except for the clink and rattle from the cellar—the sounds of beer-making.

"Hello, Daddy. I'm home," she called from the head of the stairs.

He was singing a Christmas carol and stopped at the sound of her voice. His face, bright and red, appeared around a corner, and he smiled. She often felt he smiled like that, because he thought he was expected to, not because he was happy. It was an empty smile—confused.

"Hello, Susie! Meet the train okay?"

"Yes. Had lunch, yet?" She glanced into the kitchen at the unwashed breakfast dishes on the counter.

"No, I've been busy. There's two batches ready to prime. What time is it?"

"One-thirty."

"Oh," he said. "Where has the day gone? I was hoping to get some work done on the barn, you know." She knew this was an empty statement. The barn had been a pending item for years, something planned before her mother died, begun in a feverish moment after her death and since abandoned. It stood now merely as three stone foundation walls and a scattering of lumber. Jake's crumpled biplane was parked within the walls.

"Let me know if you want something," she said. He stared at her with his blank smile, anxious to return to the beer. She walked into the kitchen and spread the book open on the table.

Below, her father's voice drifted into the sad Christmas carol. She started to read, "The Aviator—His World, Chapter One".

POW! POP! POW, KA-POW!...KA-POW!

She was half way through the chapter on flight maneuvers and trying to envision something called the Immelmann turn when explosions ripped through the cellar punctuated by her father's voice.

"God Blammed Son of a...!"

She ran to the top of the stairs and looked down on her father wading through foam kicking broken glass and swearing.

"Overprimed another batch?" she asked.

"I think so, Susie," he said and looked up with a sheepish grin.

POW! Another bottle blew.

"And that batch was ready for delivery, too," he said.

"Oh well, come on up; let it settle." The yeasty smell rose up the stairs. "You want some lunch, now?"

POP! A final bottle went, and the cap ricocheted off the walls.

"I think so," he said.

*****

"Watch your airspeed! Watch your airspeed!" the instructor screamed from the front seat of the Standard J-1 biplane, but his voice was lost in the slipstream.

Henry was preoccupied in the rear seat, slamming the controls from stop to stop, his face twisted into a tight grimace behind loose goggles.

"You're too slow! Too damn slow!" The instructor poked the stick forward, and the biplane picked up speed diving steeply toward the ground. Henry, stronger than the instructor by a wide margin, pulled back on the stick, and the plane stalled inches above the grass dropping tailskid first onto the turf. The main wheels followed with a solid thud.

"That one didn't look too bad," Mr. Howard said to Art. The two men sat on a bench outside the factory hangar while

Henry completed his third flying lesson in the factory's trainer.

Each morning Henry and Art would leave the hotel, ostensibly for breakfast, but, instead, they would hurry to the airfield where Henry took his lessons. Doc never let on that Art kept him informed about Henry's progress, and even with Art he pretended his silence was merely tolerance and not approval.

The biplane's wheels cut a twin path across the dew-covered grass, and the engine shuddered to a stop just as they pulled in front of the two men. Pushing his goggles off his face, Henry waved to Art who flashed back a proud thumbs up and smiled. The instructor, a young man with less than three hundred hours logged, shook his head with the practiced impatience he considered essential to his office.

He spoke. "This is one terrific flying machine." He thumped the cockpit's rim, half turned in his seat looking at Henry. "A very forgiving machine. But! A machine nonetheless! You have to fly it. You, Henry, must pay attention. The machine will not fly without airspeed. Do you understand that?"

Henry shifted nervously in his seat like a schoolboy chastised. He looked at the floorboards.

"Yes, I know. You keep saying, airspeed—airspeed! But I got too many things to do at once..."

"Well, airspeed is important, Henry," the instructor said carefully to his student. "Without air flowing across those wings—airspeed—you don't fly!" His hands flashed in a sequence of stalls and spins as he spoke.

"I understand, but..."

"Have you been reading that book I gave you—YOU TOO CAN FLY?"

Henry shifted in his seat and toyed with the joystick.

"Well?" he asked. Henry was silent.

"Henry, you go home and read chapter three, and read it good. I want that done before you come out tomorrow. Promise?"

"I will," Henry said. "When do you think I'll solo?"

"Not until you've read chapter three, and I mean read it so you understand it!" The instructor swung down from the plane and walked toward the hangar stopping briefly to talk with Art. Henry stayed in the cockpit and caught a finger being pointed his way with a quick shake of the instructor's head. They spoke in low voices, only the connotation drifted his way.

Henry climbed down after the instructor was gone and said to Art, "Well, what did you think, huh? Pretty good? He says I'll be soloing real soon. I think I got the knack of it even now..."

"What's this about chapter three?" Art asked deflating Henry's enthusiasm.

"Oh, he's all concerned about bookwork, and I say I can fly a plane without it. Hell, the airplane doesn't know if I can read a book or not!" He looked around for something to throw.

"We'll look into it together, Henry. Now, let's get back to the hotel. Doc will wonder where the hell we've gone."

Henry led Art to the Packard.

"I appreciate your helping me, Art. I do." He closed the door and they drove off. Behind them, a de Haviland D.H.4 biplane, with US MAIL stenciled on the side, landed.

Chapter
X

The team of horses strained at the harness; one Indian guided them by the bridle while the second chopped at the roots with an axe. Moyer left the cabin near the tracks and walked over to the airstrip. Most of the stumps were pulled from a narrow strip parallel to the tracks.

"Amazing what those lads can do when you get a hold of 'em," he said to himself reviewing their work.

He dug his heel into the ground. There had been no rain for over twenty-four hours, and the ground had begun to dry. The sky was cloudless, the wind strong out of the northwest.

"Couple more dry days like this, and we'll have an airfield," he called to the Indians who waved politely.

The Indians, both men and women, worked in small groups. Two drove the team of horses pulling stumps while a second group of about six manhandled the stumps into a pile at the edge of the woods. The women walked along the cleared earth with shovels and hoes removing stones and filling the holes. No one spoke other than to shout at the horses, and even that was done softly.

Moyer put a cigarette out and reached for another when over the trees he heard an airplane motor groan, muffled by the woods. He glanced at the shack where the Canadian mapleleaf fluttered in the wind. A cross wind, he thought. He wondered if a second runway could be carved from the forest to accommodate the wind; not now, of course, but for the future.

Jake's biplane roared low over the fir trees, banked and returned along the tracks. The Indians pulled one last stump from the strip before driving their horses aside. Moyer shaded his eyes. The biplane banked again over the trees and dropped to the runway. The landing appeared simple, the pilot easily following the narrow path between the stumps. Jake turned at the end, and taxied back, the tailskid etching a neat furrow in the soft earth. Gradually, the Indian crew returned to work.

"Hello," Moyer called through cupped hands, the unlit cigarette protruding between two fingers.

The propeller stopped. "Good afternoon," Jake answered. He stood in the cockpit and examined the Indians at work. He waved, but only one waved back.

"Have a good flight?" Moyer asked.

"Very," Jake said. "You've been busy since I left. You keep this up, and you'll have a first rate airfield on your hands."

Moyer was pleased. "You think so, eh?" he asked. "I think so. You know, Jake, there's a future in this airplane stuff. I mean other than your clandestine enterprise with Doc.

"Just think of it. Someone in Minnesota, say, could hop in your machine, fly up to this airfield, meet the train and be in Saskatoon before breakfast the next day."

Jake strapped the joystick back with the seat belt then jumped to the ground. "Do you really think anyone would want to go from Minnesota to Saskatoon?"

"Why not? People travel to the damnedest places. You never know. Anyhow, might as well get started on the future right now."

"A fine idea," Jake said, his voice amused. "Where's my next load of whiskey, there's a lot of hours of sunlight left ahead of me yet."

"You must be hungry, eh," Moyer said.

"That too."

"We can fuel your machine after you eat. There's plenty in those barrels, and I've ordered a stand so we can hoist them up above the wing. That way gravity will do all the work for us, saves pumping."

"That's nice," Jake said. "Where's the whiskey?"

"Be in on the afternoon train—six o'clock."

Jake looked at Moyer, annoyed, awaiting an explanation. Moyer put the cigarette in his mouth and lit it curling the smoke through his head and out his nose.

"Gasoline I can get easy; whiskey I need to see cash up front. Doc knows that. As soon as you've established a good reputation we can have it here waiting for you."

Jake reached into his jacket and removed the envelope with the money. In a show of counting out Moyer's fee, he slowly fanned the bills into an arc and held them out.

"Like I say," Moyer said and took the money, "before long, I expect Doc'll have a dozen planes running through here regular as a train station. When that happens the whiskey will be waiting, and all you pilots will have to do is pull in and shout, FILL 'ER UP! Just like a filling station, and someone will stick a hose in the gas tank while someone tops off the whiskey hopper, check the oil and wash your windshield. Why we'll have you turned around in 15 minutes—long enough to stretch your legs and take a leak. Meanwhile, we start slow, and you wait for the six o'clock train, eh."

Jake shook his head, the easy hop to Canada and back for two hundred dollars was dragging out longer than he had planned.

He ran his hand through his hair and sighed, but the effect was lost on Moyer who calmly smoked his cigarette.

"Bum a smoke off you?" Jake said at last.

Moyer held out the pack. "Still hungry?"

"You cooking?"

"Come on," Moyer said with a wheezing laugh and clasped Jake's shoulder. He led him across the tracks to the shack.

Stepping inside the shack, Jake saw that Moyer had been busy here as well. The room was clean, everything neatly stacked on shelves. The dirty clothes had been picked up, and the bed made. The cabin smelled of disinfectant, and the window was open wide allowing air to circulate.

"There's something different here...Wait! Don't tell me! You've convinced the dog to shit outside!"

Moyer cast a sideways glance.

"The dog convinced you to shit outside?"

"The maid came back," Moyer said.

"And she promised not to shit inside anymore. Now, I understand!" Jake said.

Moyer waved a fist weakly. "I'll find us something to eat."

"Where's the picture of Mae?" Jake asked.

"Took her down," Moyer said quickly. "Did you meet Adair?"

"Yes."

"How is she these days?" Moyer placed a dry salami on the table and hacked at it with a long knife. He slipped a chunk into his mouth and tossed a piece to Jake. "I haven't seen her since the summer of '26. We were all out in California for Mae's wedding to that Prince David M-something. Did I ever mention him? Worthless little turd. Nobody knew just what he was supposed to be prince of."

"You mentioned him. Did she have that scar across her face then?"

"Adair? No." Moyer looked quickly around the room and opened a trunk in the corner. He rooted through a collection of books and papers and pulled out the picture of Mae that had been on the wall.

"Thought I'd keep it a while," he said with a grin and continued to rummage through the trunk.

Jake took a piece of salami from the table and chewed it slowly. He noticed a group portrait of several men in uniform. The photograph was in a cheap metal frame, the glass cracked diagonally across the face.

"Can you spot me?" Moyer asked.

Jake held the photo and instantly recognized Moyer, third from the left, partially hidden by a short fat sergeant with a pipe. They all wore helmets and combat kits. They all carried rifles except one man, apparently an officer. He wore a Sam Browne belt across his chest and a pistol in a military holster strapped to his hip.

"Third from the left," Jake said. "Although, I must say, you look younger here."

"That's Adair's husband, Vincent, on the end."

Jake stared at the young officer with the proud chest crossed with the Sam Browne belt and the pistol at his hip. He remembered Vincent, wearing an identical holster, riding the stallion. He looked into the man's face for the first time. The face of the man he had, and the only word he could think of was: cuckolded. He stared at the face of the man with no face.

"Was he married then...to Adair?" It was important that he know.

"Yes, she was at the train station when we left for France. I only saw her from a distance. Someone whispered something about the lieutenant's wife. We all looked. I didn't see her again until after the war when we all worked for Doc in New York."

"So, she's Canadian. Vince, too?"

"You make it sound like a handicap," Moyer answered. "Yes, Vince led our company at first, then was transferred to an observation post...the rest you know."

"How did you manage to keep in touch?" Jake asked.

"I was back in Canada at the training camp when I heard about his discharge. Well, I looked him up, saw his face all burnt away, and, well...I just kept in touch. When I hooked up with Doc I found an opening for Vince. He turned out to be a brilliant organization man—all that officer experience, I suppose. He ran the trucks; kept the drivers happy, did well."

"And Adair?"

"Adair...?" Moyer rolled his eyes at the ceiling. "Adair enjoyed New York. Adair met Mae, and the two became inseparable friends, always running back and forth to the West Coast, party hopping, generally enjoying life. Adair was very young; younger than Vince."

"So what happened to her face—the scar?" Jake persisted.

Moyer handed Jake another photograph, this one loose, without a frame. It was another group portrait. Jake recognized Mae Murray dressed in a short fur-trimmed gown, her back arched in her trademark seductive pose. She stood beside a dreamy, curly-headed man Jake guessed to be Prince David.

"Mae?" Jake asked.

"Yes," Moyer answered. "And that's His Royal Worthlessness, Prince David on her left. I forget who the Dago is beside her, but that's Agnes Ayres on his left."

117

"Pretty," Jake said. "She's another movie star I assume." Moyer threw him a disapproving look. "Hey, I don't get to the movies much. Sorry," Jake said.

"Yes, she's another star," Moyer said. "And that woman under the floppy hat..." He pointed.

"Hey, that's Adair!" Jake said. "I didn't recognize her."

Adair stood in a brassy stance at the edge of the picture, arms akimbo, her sharp chin jutting forward in a defiant pose. Jake ran his finger lightly across the photo, across her face.

"She got that scar right after we returned to New York."

"You were there, too?"

"Who do you think took the picture?"

"How did she get the scar?" Jake asked again, his voice serious. Moyer took the photograph and pressed it into a large book and slid it back into the trunk.

"Car accident. Went through the windshield," Moyer said. Jake said nothing but thought, is that all? Moyer looked at him and closed the trunk lid and sat on it. He lit another cigarette offering one to Jake. Jake lit his off Moyer's.

"It wasn't a car accident, was it?" Jake asked.

"Vincent did it. I've never had any proof, but I know he did it." Moyer calmly sucked on the cigarette, his eyes half closed.

Jake hated the man with no face.

Moyer spoke. "It was such a waste. She was a pretty girl." He drew on the cigarette, his voice flat, impersonal. "But it was none of my business. It's just not smart to stick one's nose into another man's marriage." He exhaled a blue cloud.

Jake felt his jaw tighten with the hate and disgust that welled inside for Vincent and now for Moyer as well. Cowards, he thought, and wondered how Moyer could be so detached. He said nothing, and decided to leave as soon as he could get the whiskey and refuel.

A wave of pity, and longing, for Adair flooded across him, and he hated Moyer and Vince all the more.

Moyer finally spoke again. "He was jealous, I suppose." He looked at the red ember of the cigarette. "He hated to see

118

her slip away, I suppose..." He looked at Jake. "You hungry?"

"No," Jake answered and left the room. Moyer stayed behind.

Outside, the Indians had stopped work to eat. They sat loosely around a blanket covered with dull looking food. There was much talking, and a small band of children ran about jumping from stumps and throwing dirt clods at each other. One stray child walked to the biplane and hurled a clump of wet mud at the fuselage. It bounced off the ALMIN-CO lettering leaving a brown smear. The adults ran forward and took the child away with a quick swat on the rear.

"Don't worry about it!" Jake spoke slowly in the universal assumption that native populations harbored an innate ability to understand English if spoken slowly.

"It will come off," he said. "I will scrub it off," he added making a circular motion with his hands. "I clean mud; you eat!" Everyone smiled and nodded. When he turned his back the Indians shook their heads.

Jake kept a small knot of clothing in a bag inside the front cockpit. He rummaged through it until he found the pistol Moyer had given him on the last trip. It was at the bottom of the bag wrapped in an oily rag.

A large portion of a pilot's life is spent on the ground waiting, unproductive, but necessary. Only a fraction of the busiest aviator's time is in the air; once earthbound, the flyer is of little value.

Jake was a man of the sky, a man alone. The airplane, the wind, these were his contacts. Anything beyond was a burden, and this thing with Adair ate at his insides. Women he had had before—many times. In each case, it lasted the time it took him to wring what he could from the visit, then before anything could latch on to him he was gone. Never considering himself cruel, he treated women as equals. He figured they were like him, alone, self-motivated, and could just as easily abandon him as he would them. Susan, he reasoned, was like that. Mostly, though, he never thought about it.

Jake had to wait for the 6 p.m. train from Winnepeg, and until then he was a wingless creature without purpose. He took the pistol from the bag and headed into the forest.

"Think I'll kill me some time," he said to himself. The unchanneled anger from his discussion with Moyer brewed inside him, thinking about Adair and the vicious scar across her face. The image of the faceless husband astride the stallion replayed again and again in his brain. Stepping deeper into the woods, the vague anger mounted into obsession.

He unwrapped the pistol, stuffed the rag in his back pocket, and stepped over a small creek. He looked for something to shoot—something that needed to be shot. He clicked the hammer back and carried the gun loosely, the barrel pointed at the ground. Nothing stirred in the woods except the very tops of the fir trees in the wind. The only sounds came from insects in the branches and the songbirds who chased them. Jake sat on a decaying trunk.

Ahead, yellow wild flowers grew in a sunlit clearing beside a stand of younger trees. Jake looked across the flowers at a tall healthy tree near the edge of the clearing.

It had been marred, apparently, when a limb had broken away from one of the older trees ripping through its green flank. The new growth over the wound was ragged and grotesque, incongruous with the grace of the rest of the tree. Jake lifted the pistol, already cocked, and aimed.

Moyer heard the shot echo from the woods, a short flat pop, but he was busy installing a shower head on the water tank and merely waited for a second report. None came.

Jake lowered the pistol and stared at the gouged wound he had made. He listened to the silence watching the thin blue smoke from the shot dissipate in the air. He slid off the log and lay down on the spongy ground of pine needles and wild flowers.

For some reason, he almost expected the tree to topple over from the shot, but it stood, as before, healthy except for the torn branches and the new wound.

Without looking, he slid the pistol onto the log above his head and closed his eyes against the sun. The wind swished in whispers in the treetops and carried the distant sound of the Indians pulling stumps. He slept.

The train from Winnepeg was several miles out when he awoke to the faint shrill of its whistle. He woke slowly, the pine needles and wild flowers reluctant to give him up. His mouth was dry, his body hungry, but all he wanted to do was fly. The whistle sounded again, closer now, and he wrapped the pistol in its rag and stood. At first his legs moved stiffly, but soon he was charging through the trees, full of energy, and feeling the best he had in weeks.

The smoke from the locomotive was visible over the trees before any of the train. The Indians had stopped work and waited beside the tracks near the octagon water tower. The train rolled to a stop, its brakes screeching, the line of boxcars and flatbeds rattling and compressing into each other like a row of drunkards.

The whistle released a long blast of white steam, and Moyer strode from the cabin along the tracks, glanced once at Jake and greeted the engineer with a wave. The fireman climbed onto the tender and grabbed the water spout from the Indians. Once in place the chain was pulled and 20,000 gallons rushed through.

Jake climbed over the couplings between the cars while still holding the rag-covered pistol. A boxcar door slid open, and someone pushed a case of whiskey toward Moyer. Jake ran and took it.

"I'll take that for you," he said and hustled away toward the plane. Moyer was taken aback with his suddenly cheerful attitude.

"Thank-you," he answered.

The remainder of the whiskey was off-loaded before the water spout was turned away from the tender.

"That's the lot," someone called from inside the boxcar and handed Moyer a clipboard to sign. Moyer ignored it watching Jake run back and take two cases, one under each arm and trot back toward the plane. "Moyer?"

"Huh?"

"Was there anything else?" the voice inside the boxcar asked, the clipboard still thrust toward him.

"Ah? No," Moyer said, and he scribbled his name across the bill of lading.

"Thanks," the voice said. "Did you hear about Jim Huston, fellow that used to fire for us about a year ago?"

"No," Moyer answered absently, his attention diverted by Jake's energy.

"He was firing for that line up north, had the boiler blow 'bout a week ago. Blew that locomotive right off the tracks— water must have foamed, probably contaminated or something. Hope the water here's clean, eh. Anyhow, they only found the bottom half of old Jim..."

"Yea, that's swell...Listen, I gotta run," Moyer said and picked a case of whiskey, hesitated and took a second one then headed down the row of cars toward the airplane.

He wheezed heavily under the load and halfway across the field passed Jake headed back for more. He seemed to move faster, and Moyer picked up his pace to match. They passed each other again, Moyer wheezing, and Jake smiling until all the whiskey was stacked beside the plane.

The Indians came down from the water tank, and a cloud of black smoke billowed from the locomotive's stack. The whistle blared, and the train moved slowly away from the stop. Moyer sat heavily on the whiskey, a cigarette dangling from his mouth, his breath pumping in short choppy puffs like the locomotive straining to pull the row of lazy cars.

Jake scrambled into the Standard's rear cockpit setting a case on the fuselage before him. He uncorked a bottle and jabbed it into the tank's neck then uncorked another. He worked quickly and silently draining six bottles before Moyer spoke.

"You're not being very good company today."

"No?" Jake said and tossed an empty over his shoulder.

"No. And I think maybe I know why." Moyer stood.

"You do?" Jake said. He threw another bottle over his-shoulder hitting the Indian boy who had thrown the mud at his plane earlier.

"I think, maybe, it has something to do with Adair, eh."

Jake nodded slowly and smiled. He felt good after the sleep, somehow feeling above Moyer and Vincent and their treatment of Adair. He thought of Vincent astride the stallion; he thought of the bullet hole he had put in the pine tree. He smiled.

Moyer caught the smile. "See," he said. "I thought so. You should stay away from that one."

"What do you know?"

"I know Adair, and I know Adair has certain, oh, beguiling attributes, eh, and she has a tendency to advertise them."

Another bottle emptied and Jake hesitated, suppressing an urge to throw it hard at Moyer. Over his shoulder it went.

"You met her," Moyer said. "What do you think? Not bad to look at."

"That scar."

"Fuck the scar!" Moyer snapped. "You think she's pretty, I know!"

"I was there on business," Jake said.

Moyer glanced at the ground. "There was a time when I had business relations with her." He looked at Jake who let a bottle drain without opening another.

"Then how could you just do nothing?" Jake asked.

"About what? Vince?" Moyer yelled. Jake stared. "About him cutting her? Cutting that beautiful woman? Or do you mean how come I didn't do anything about Vince? About his beautiful face?"

"You know I mean Adair," Jake said quietly taking the empty bottle from the tank.

"There are certain things one has no control over. None!" Moyer said. "Maybe, she had it coming, I don't know."

Jake whipped the bottle as hard as he could at Moyer who saw it coming and caught it with both hands. He stared at Jake and neither spoke. One of the Indians yelled something at the

123

draft horse, and a stump lurched from the soft earth. Jake uncorked a bottle and emptied it into the tank.

"I don't understand how anyone could do such a thing," he said.

"Adair was a reminder, a constant reminder to Vince of what he was, what he had before the war. He survived the war, but for all purposes, he had been erased, his identity gone.

"Oh, sure, he still functioned, still had legs, arms, even his name, but he, the individual who loved and married Adair, was gone."

Moyer pushed a case of whiskey before him onto the biplane. He opened a bottle and stuck it in with Jake's in the tank.

"I think, when he first came back from France, Adair was just glad he was alive, but after a couple of years she had to have something—something human. She was young, Jake, a young vital woman. Vincent had ceased to be in 1917."

Moyer threw the empty away and opened another. "I wasn't the first," he said.

"It still wasn't right," Jake said.

"No, but it got the point across. I don't think anyone touched her after that. I was off to Canada by then running the shipments from there, and I never saw her again. Strange thing was, I saw him regularly, and I knew he knew about me and Adair, but he never said anything."

Jake and Moyer poured bottle after bottle into the tank in silence with each empty tossed into the weeds. The cases slowly emptied, and the tank was capped. Dozens of bottles were scattered around the area along with the wooden boxes. Jake took one, overturned it to stand on and added oil to the engine.

Moyer walked to the gas barrels and returned with two five gallon cans. He handed these to Jake.

"I ordered a three hundred gallon gas tank to go on that stand I told you about," Moyer said. "Should have it up next time you're in."

"Shows confidence in this scheme," Jake said.

"Just good business. I was thinking of installing a similar tank for the whiskey."

"Good idea so long as nobody confuses them," Jake said and dropped the empty five gallon can to the ground. Conversation was strained with both men trying to reestablish a relationship without reference to Adair.

"I could paint one red, one brown," Moyer said. He returned to the barrels for more gas.

Jake stuffed the pistol back into his bag then wrapped the tarpaulin over the whiskey tank and tied the ends. Moyer returned with two more gas cans, his cigarette dropping ashes on his beard. Jake envisioned his whole head suddenly engulfed in flames and only gasoline or whiskey around to snuff the fire.

"You really think this operation will make money?" Jake asked.

"Everyone's going to make money."

"Well, I do get two hundred bucks a flight," Jake said.

"Peanuts," Moyer said. "Why, I bet Doc will have a half dozen planes like yours operating within a month. Now, if each plane completes, oh, say only two round trip fights per week..."

"Only two?"

"I'm figuring conservative," Moyer said. "You have to figure weather delays, maintenance, wrecks, disease, famine, pestilence, whatever. Anyhow, he's not peddling mountain hooch, but good quality Canadian whiskey—high dollar stuff.

"So if each plane hauls about two hundred gallons per week, and he cuts it and sells it. You follow?"

"Yes," Jake feigned disinterest although the figures raced in his brain.

"At that rate, Doc pulls in a few thousand profit, after all expenses, per airplane, per week! And what does he do?" Moyer paused. "He sits on his duff and figures ways to haul more, and get you pilots to fly more."

"More flying—more money," Jake said.

"And with a half dozen airplanes Doc rakes in about twenty-five grand per week."

The figure shot before Jake's eyes.

"What does he pay you?" Moyer asked. "Two hundred a flight?"

Jake shrugged without answering. Moyer coughed violently inhaling smoke. "You know," he said, "I don't think these things are as good for you as they claim." He inhaled. "I'd say Doc's going to make a tidy profit."

Jake was having trouble visualizing $25,000. "What would you do with twenty-five grand?" he asked.

Moyer grinned. "Same thing I did with it last time—blow it!"

Several minutes later, Jake slid himself into the cockpit, buckled his seat belt, and pulled the helmet over his head leaving the ear flaps up. Moyer walked to the propeller.

"Switch is off; fuel's on!" Jake called.

"Yea," Moyer mumbled and pulled the propeller through until it was set to fire.

"Give it a bump," Jake called and reached for the booster coil. "Hey, Moyer! Did you ever have business relations with Mae Murray?"

"Go to hell!"

Jake cranked the coil. Moyer tapped the propeller. The engine burped and fired.

Jake was soon a distant murmur beyond the trees. Moyer walked into the cabin, let the rust-colored dog in for the evening and took a small .32 caliber automatic from his jacket and placed it on the shelf.

He lifted the earpiece off the wall phone and cranked the handle.

"Hello, Magalar? Can you hear me? Yes, this is Moyer, I need to send a telegram, eh. That's right. No, I'm fine...Huh? No, I don't owe you any money! No, I didn't hear anything about a boiler blowing up north. Look, send the telegram to Mr. Arthur Reynolds, Hotel Lindell, two-four-five South Thirteenth, Lincoln, Nebraska, U.S.A. Yes, the U.S. You think Nebraska's up here, eh? What? I don't know, I think it's next to Ohio or something. Send this: Cargo left six-forty p.m.; full up. Moyer. Just send it like that. What? None of your damn business; I'll send the money tomorrow on the train. Thanks, Mag. Bye."

Doc sat at a table in his hotel room in Lincoln finishing dinner. Long slants of orange sunlight crossed the room. Art answered a knock at the door. A bellboy in a wrinkled uniform handed him a telegram and waited while Art dug fifteen cents from his pocket, handed it to him and closed the door.

"Who was it?" Doc asked.

"Bellhop; a telegram." Art ripped the envelope open and read the message. He dropped the telegram on the table in front of Doc and picked a glass from a sideboard and filled it with liquor.

"Ah-huh," Doc muttered and wiped his mouth. "Is Henry back from the airfield, yet?"

"Not yet," Art answered. He made a note of the fifteen cent tip in his ledger.

"When he returns, tell him to get the car washed and be ready to drive out to the university. There's a fraternity house we need to visit, several in fact. Were you ever a fraternity man, Arthur?"

"No," Art scowled. "Silly collection of snots."

"How is Henry doing with his lessons, anyhow?"

"I helped him through the bookwork last night. He should be ready to solo soon, but then what the hell do I know about flying?"

"He likes it, does he?"

"Loves it."

"Airspeed, Henry! Watch your goddamn airspeed!" the instructor shouted over the noise of the wind and the airplane's motor. Henry struggled with the stick and rudder against the instructor's efforts. His goggles were loose and slipped off his nose blocking his vision. He flew with his tongue sticking out.

"Power, Henry! POWER!" the instructor screamed. The biplane fluttered in slow flight twenty feet above the earth, shuttered on the verge of a stall then dropped with a teeth jarring thud.

"KEERIST! Henry, what the..." His words were lost as Henry added power, and the biplane staggered back into the sky.

"Am I ready to solo?" Henry called. The goggles had progressed over his mouth and muffled his voice. "I read chapter three!" The plane aimed straight at a tree at the field's edge.

"HENRY! POWER, HENRY! AIRSPEED!"

Chapter
XI

Jake looked at his watch—9 p.m.. The sunlight would linger for at least another half hour, he thought. Enough time to find a town somewhere along the Minnesota River big enough to have fuel but small enough to avoid trouble. Except for a few isolated thunderstorms, the flight south had been a dream—blue skies and a northeast breeze to push him along. He enjoyed the flight and would have even if not being paid. However, he decided, being paid was better than not.

The last forests were well behind him, and he stared across the vast prairies of western Minnesota and the Dakotas. The low sun washed everything in pale red. He rocked the wings to bob the fuel gauge. As long as there's motion, he reasoned, there had to be fuel. He had yet to find a gauge he trusted.

"Another approach on faith and fumes," he said aloud to the passenger who always flew with him. He knew other pilots who admitted to speaking to that same passenger, unseen, but whose presence was unmistakable. Only the low time pilots called it a deity; to Jake it was just the presence that rode with him, and unlike a god, would follow him straight into the side of a mountain or burn up with him in a heap somewhere. These, however, were thoughts he seldom shared.

The Minnesota River ran strong, fortified by the heavy spring rainstorms. He dropped low enough to distinguish landmarks—towns, river forks and railroads. A glance at the map then at his watch, and he decided he could reach Ortonville, just across the river from South Dakota. He picked out the town's outline ahead.

The wind near the ground was as gentle as it was soft, a rare combination. Jake made one pass over the town to attract attention and closed the throttle gliding to a smooth landing on a hilly field a mile away. The grass was longer than he expected, but the twenty-six inch wheels cut through it with ease; the tailskid etching a thin strip along his path.

While the tires rolled through the tall grass, the elastic shock absorbers dissipated the beating from the rough surface. The landing was normal, almost dull, and Jake felt the speed bleed off almost to the point when he could begin his turn to taxi back. He was congratulating himself on another flawless landing, when a sharp crack shook the tail and was transmitted into the joystick in his hand.

"Blasted!" he said. "What the hell was that?" The tail bounced slightly and settled back onto the field, lower than usual. The biplane stopped, and despite his efforts to swing the tail around with power, it anchored fast in the grass. He throttled back to idle and sat. The engine ticked over at 800 rpm, the fuel gauge sat near empty, and the water temperature showed 120 degrees—normal.

Motor's okay, he thought. A quick glance around showed no damage to wings or mains. He killed the engine and unbuckled from the seat and climbed down. Once on his hands and knees, he combed through the grass under the tail until he found the steel toe from the skid pointed awkwardly at the empennage with grass and mud smashed into a break half way up the shaft. A quick inspection of the underside showed no other damage, so he pulled weeds from the area to make room to work.

Jake had the airplane's tool kit spread open on the grass when two men astride mules rode up. One was about sixty, Jake guessed; the other a teenager, ripe with acne and dressed in ill-fitting overalls, brown shoes and a plaid cap.

"Heard you go over town," the older man called still far off. "You lost?"

"No, just need some gas and maybe a car jack." Jake pointed. "I need to lift the tail to replace the skid."

The man nodded climbing off the mule. "The gasoline's no problem. What kind you need?"

"Hi-test. Sinclair HC if you can find it."

"Yeah, we can get that, and I'm sure we can dig up a jack..." He turned to the boy still on the mule. "Say, you think Harold has one in his shed, you know, alongside the stove?"

"Uh-huh," was the answer.

"What parts you need?" The older man stepped away from the mule approaching Jake and held out his hand. "I'm Ronald, that's my nephew, Eldon."

"Pleased to meet you," Jake said and gave his name. "It's the tailskid; it's made of hickory and steel, but all I need is a piece of hardwood, oh, about this long." He indicated the length with his hands.

"Think you can find that for the gentleman, Eldon?"

"Uh-huh," Eldon grunted.

"Good, now get him one, and get that jack out of Harold's shed." He turned back to Jake. "How much gasoline you going to need?"

"About thirty-five, forty gallons."

"I've got a five gallon can, use it for kerosene, we can empty that out, should be all right. Eldon, look around Harold's shed for a can of some sort—something for gasoline."

"Uh-huh," Eldon grunted and rode away. He passed a truck making its way across the field toward the airplane. It spewed a blue cloud of exhaust behind.

"Looks like Mr. Grover's truck," Ronald said. "He's a curious sort; be asking you all sorts of questions; you just let me handle him. By the way, where you from?"

"Ah, Lincoln, Nebraska," Jake said and decided he had said too much.

"What brings you up here?"

"Airline business; scouting new routes."

"What's that ALMINCO?"

"That's the name of the company: Airline Minnesota...Company!" Jake struggled to make the name sound plausible.

"Never heard of it. Anyhow, I don't think there's anyone around here wants to take an airplane ride. Maybe you should try Minneapolis." He pointed east.

"I'll do that. Thank-you," Jake said.

Mr. Grover moved slowly out of the truck and with short careful steps made his way to the biplane. He was at least 85, Jake guessed. He seemed to be swearing to himself.

131

"Mr. Grover," Ronald shouted in man's ear. "This man's from the airlines. Says they're planning to run an airline through here, from Minneapolis." Mr. Grover merely nodded and walked around Jake's plane poking at the fabric, carefully, as if it would rip.

"What's in here?" he called, his voice high and raspy. He tugged at the tarpaulin over the whiskey tank.

"Weight!" Jake said running to the cockpit to tighten the ropes. "We place weight in the front seats on these trial runs...to, ah, measure the effects of weight on...ah, on flight!" It sounded ridiculous to his own ears, but Ronald nodded approval, while Mr. Grover crawled under the tail to inspect the broken skid. The sun was almost below the horizon, and the deep red glow flooded the underside of the airplane.

"That's busted!" Mr. Grover announced, and he shook the skid.

It was dark when Eldon returned from town with the jack and two cans of gasoline strapped to the mule.

"Did you get a piece of wood for the airplane?" Ronald asked as he helped him unload the fuel. Eldon nodded and handed him a crude piece of hickory. Ronald passed it to Jake.

"Look, it's hickory," Ronald said. "We can whittle it for you, just show me the shape."

The biplane was illuminated by the lights from Mr. Grover's truck. The old man was asleep in the cab shrouded in exhaust fumes. Jake took a pencil from the cockpit and traced the outline of a tailskid on the wood.

"Think you can handle that?" he asked. "It doesn't have to be fancy, just good enough to get me home. I'll replace it with a real one later."

Ronald studied the diagram and turned to Eldon. "Take this into town. Use the saw in Harold's shed and cut this down as close as you can to the lines here. Then take your knife and carve it best you can. Got that?"

"Uh huh, you want more gas?" Eldon spoke something close to a complete sentence.

"Good point!" Ronald said. "Tell you what, you run into town and take care of that tail thing like I told you, and we'll

132

finish up here and take the cans back for more. Now, be careful with that saw!"

Eldon rolled his eyes and turned. Only Jake caught the expression as Ronald never seemed to look directly at the boy. He climbed onto the wing and was handed the fuel. In the yellow lights from Grover's truck he managed to pour most of it into the tank; the rest rolled off the wing or soaked into his pants.

"You hungry?" Ronald asked taking the empty can and handing back the full one.

"Yes, very."

"Well, when I run back to town, I'll bring something back for you. You like chicken and potatoes?"

"Hard not to like it," Jake said and felt hunger twist at his stomach. Aside from the salami, he had had nothing all day.

Ronald strapped the empty cans to the mule and left. Jake placed the jack under the tail, but when he began to pump it simply pressed itself into the soft earth. He walked over to the truck.

"I need a piece of wood, or something, to set the jack on," he said to the old man on the front seat. "Got a board back here?"

No answer.

"Anything to keep the jack from settling."

Silence. Gray exhaust swirled through the cab around the old man with his eyes half closed.

"Mr. Grover?" Jake said. "Hello, you awake?" He reached in and shook him. The old man slumped over.

"Oh, Christ," Jake said. He pulled the door open, grabbed him under the arms and dragged him clear. The limp body weighed little and slid easily off the seat, the feet knocked sharply on the running board as he pulled him from the gasses and toward the light.

"What the hell!" the old man suddenly growled.

Jake stopped. "You all right?"

"Let go of me, you...of course I'm fine. What'd you go and wake me up for? Damn fool!" Jake helped him to his feet, and he staggered back to the truck and crawled inside.

133

He dropped off asleep again, and Jake helped himself to a board from the rear.

He had the empennage jacked up and the broken skid removed when Ronald returned with the gas. There was a passenger with him, a pretty girl with long red hair, about sixteen years old. She wore a plain green dress, black stockings and her hair was pulled back and tied in a long tail down her back. Jake thought she had outgrown the dress as it fit snuggly around her hips, the cloth tugging at the buttons between her breasts. She handed him a basket.

"Thank-you," he said and inhaled the aroma of fresh bread, and baked chicken.

"That's my daughter Neili," Ronald announced. "Pretty isn't she." He laughed. Embarrassed, she turned her head, lifted her eyes and grinned at Jake who bit greedily into a thigh.

Out of the darkness Eldon appeared with two more gas cans strapped to the mule, and the tailskid protruding from his pocket. He set the cans near the others and handed the skid to Jake.

"This is excellent," Jake said amazed with the craftsman-ship. "I may not even have to replace this after all. Thank-you."

Eldon nodded and retreated into the shadows where he could watch without being seen. They all waited while Jake finished his meal. This done, Neili gathered the remains into the basket and joined Eldon where she, too, could gaze, only she made certain to keep herself at least partially in the light—where Jake could see.

"Hand me a knife," Jake said.

Ronald did and Jake shaved tiny slivers from the skid to fit it into the bracket.

"What holds it in?"

"Couple of bolts," Jake said. "I can carve some holes with the knife. Hand me that wrench...no, the other. Thanks."

Within an hour the skid was on, and they slowly lowered the tail. The biplane groaned as the weight settled onto the new piece. Jake reached under and shook it violently.

"That's not coming off. You do good work," he said to Eldon who still hid in the shadows.

"Umm," was the reply.

"Will it work?" Neili's voice chirped close to Jake's ear. She had moved closer without him noticing. She leaned close to his face and kept a serious frown on her brow. She smelled of soap, and Jake decided the dress was definitely too small enveloping her hips seductively.

"It'll hold."

"Oh good," she said and nodded as if any of it really mattered to her.

Mr. Grover emerged from the truck, and pushing by Neili he squatted beside Jake and tapped lightly at the tailskid.

"It's fixed," he announced. "I'm going home." He shuffled back to the truck and drove away. The field was once again silent without the clacking of the truck's motor. A three-quarter moon rose providing enough light to make out figures and finish the fueling.

"I could use another can, but I suppose I can get along without it," Jake said.

"I'll send Eldon back with the rest," Ronald said. "I'll take Neili back with me now, time for bed. Where are you planning to sleep tonight, or can you fly at night?" He mounted the mule. Eldon climbed onto his pulling his cousin behind him.

"I'll get a couple hours sleep here, then depart before dawn. Don't forget your jack." He handed it to Ronald.

"Eldon will bring the gas. Good-night. Stop in again. Of course, if the airlines are planning to run through here I guess I'll see more of you. Good-night."

"That could be," Jake said knowing he would quickly forget this place. "Thank-you."

"Bye," Neili's voice popped quickly. He saw her vaguely in the moonlight, her arms wrapped around her cousin's waist.

They were gone, and the night was heavy with damp silence. The moon hung three hands above the horizon burying the field in deep shadows. A dog barked in town, and Jake threw a blanket onto the ground beneath the biplane.

135

Eldon returned 30 minutes later with the gas, and Jake poured it into the tank through the chamois. He paid for the gas and included a five dollar tip for Eldon

Eldon stared at the bill. "That's a lot of money." He kept it.

Jake waited for the boy to leave, but he stood there glancing from the five dollar bill to the biplane. Finally, he mustered the courage to ask, "Can I watch you start it?"

"I don't plan to leave until about 4 a.m., maybe a little earlier."

"I don't mean now; how 'bout later, when you leave. Can I come back and watch?"

Jake had been thinking of only getting to Lincoln to deliver the goods and had forgotten what it was like to hop rides, see honest excited faces in a crowd possibly looking upon their first airplane. Eldon, he gathered, had never been this close to an airplane, and it had taken all his courage to ask merely to watch Jake start the motor.

"I could use help starting it," Jake said. "Be here before four."

Eldon's eyes opened wide. "Thanks."

Jake expected he would leave now, but he stood, rooted to the spot looking at the airplane.

Finally, he asked, "Do you like being a pilot?"

"There's nothing else for me," Jake said. "I couldn't stand being stuck somewhere in a real job."

"Or in a town like this?" Eldon seemed more at ease without his uncle around.

"Looks like a nice town..."

"I'll bet you've been to some swell places. Ever been to Chicago?"

"Many times. As a matter of fact, if we took off right now we could be in Chicago by breakfast time."

"That's fantastic!" Eldon said. "You can just go anywhere, anytime."

"Well, it's not that easy..."

"What's a person have to do to become a pilot?"

"You have to want to be one," Jake answered.

"Then what?"

"Then be one."

Eldon grinned, he knew nothing was that easy. Flying was impossible for him. Never, he thought. Just be one? No, there had to be more. He knew there was no place for a kid from Ortonville, Minnesota in aviation.

"You don't believe me," Jake said. Eldon stared. "If you want to be a pilot—you'll be one. Just go do it."

"I'll see you later," Eldon said and turned.

"Hey," Jake called. "Tell Harold, thanks, for the tools."

"Can't do that. Harold's dead. He was my father. Anyhow, everybody just borrows his stuff like nobody owned it." He left.

Jake wrapped himself in the blanket, although the air was warm enough to do without. He stayed awake for a while staring at the flat black sky softened by the moon. He casually wondered how much younger Neili was than Susan, then he tried to calculate how much older Adair was than himself. He slept.

Eldon propped himself on one elbow in bed to stare out the window at the moon as it crept across the sky. Sleep was impossible, his thoughts held captive by the airplane in the field beyond the rooftops.

"Chicago by breakfast," he said aloud. "Then Ohio by lunch." He sat up. A warm breeze rustled the curtains in a lazy snap. "Then Pennsylvania by supper and the Atlantic Ocean the following day!"

He heard his voice bounce off the ceiling and froze. His cousin already thought him weird, and if she heard him talking with himself in the middle of the night she would never let it rest. He slid onto his elbow again.

"Chicago," he said. "Kansas City!" He pictured himself togged in leather with a long silk scarf casually announcing the departure of his flight to, "New York; Miami; Havana." He pronounced the names slowly, savoring each one. "Buenos Aires!" he called softly and sank into darkness.

Down the hall from Eldon's room was Neili's. She, too, lie awake staring through the window at the same moon and for almost the same reasons.

In her room, the curtains fluttered in the night air, but her thoughts were less of the machine in the field beyond, than of the man presently sleeping somewhere beside that machine.

"He would never live here," she said. "Never in a place like this." She looked at the cracked plaster ceiling, stained brown from a leak in the roof.

Somewhere a dog barked, and she considered getting dressed and sneaking out to the field. The dog barked again, and she felt the darkness tell her it was impossible. Lightly, she ran her fingertips across her thigh, feeling that same warmth that came over her whenever Roger Whelm held her and kissed her. Only now she was warmer. The curtains pulsated in the breeze. She closed her eyes and thought she heard her cousin mutter something down the hall.

Jake woke. The moon was on the far side of the sky now, and the grass around him was wet with dew. He rose stiffly, thinking he should be hungover, but realized it was just the usual discomfort that came from too many nights sleeping on the ground under an airplane.

"Ohh, crap," he moaned as he pressed to his feet grasping the damp wing above him. The air had turned cooler, and he shook the blanket off and reached for his shirt strung across a flying wire between the wings. It, too, was damp. He found his boots and with much complaining slid them on.

"Hmmph," he grunted and looked around at the empty field. "I thought that kid might show up. Gotta stop talking to myself like this.

"Good idea.

"Thanks.

"You're welcome."

He climbed onto a tire and popped the cowling open to check the oil. The stick was hard to read in the dim light, and he added a quart just to be safe. He stuck his finger in the radiator. "Yup, it's wet.

"Gotta stop talking to myself.

138

"Okay."

With his blanket stuffed into the forward cockpit, and the engine primed, Jake gave one quick look around the dark field.

"Thought he might show," he said. "Funny how people are. You never can tell." He swung a leg into the cockpit, then the other, sat and pulled his helmet on.

"Wake up Ortonville; its four a.m.!" He cranked the booster coil, and the silence was punched away by the healthy roar of the Hispano-Suiza.

While the motor warmed, he surveyed the field visible in the moonglow. Flying was like shooting pool—nothing in writing, no books, graphs or figures, just gut feeling and a good eye. A downhill take-off looked best from right where he was, so, as the water temperature rose, he opened the throttle, and the heavy biplane bounced across the grass and lifted into the night.

He found the few lights of Ortonville, and turned over the town to pick up the river. In the fields to his left he could see a farmer with a lantern head toward a barn. He found the river and turned south, climbing slowly in the smooth air. Breakfast in Lincoln, he thought and forgot about Ortonville as he had forgotten almost every little town where his flying had touched someone.

Neili heard it first, pushed herself from bed and ran to the window. Eldon woke a second later and did the same. They both strained to see and at the same instant caught the outline of Jake's biplane against the moon as it crossed the houses and disappeared over the river.

Eldon felt his stomach twist and tears of rage welled in his eyes realizing he had fallen asleep and missed his chance. He pounded the window sill and cursed through clenched teeth.

Neili smiled and waved, felt a little sad, but went back to bed. She would meet Roger Whelm by the river that night.

In the room below Eldon's, his uncle woke to the pounding of his nephew's fists on the window sill.

"What the..."

He heard the deep rumble of the biplane's motor cross town then turn south.

"Damn fool's never going to find Minneapolis that way," he mumbled and pulled the covers away from his wife and fell asleep.

Chapter
XII

Nicholas Barchek was the only person in the office on Friday afternoon when the last mail was delivered. The bulk of it was ordinary correspondence from other Treasury Department branches as well as the usual ignorable memoranda from the head office in Washington.

One letter, however, stood out from the rest. It was addressed simply to the office in Manhattan and bore no return address; the postmark was smudged. Nick stood beside an open window five stories above the street and slit it apart with his finger.

It contained a single sheet of plain paper with the missive: THE DOCTOR IS ALIVE AND IN LINCOLN, NEBRASKA.

There was no signature, no name, and the note was typed without betraying any source. Nick had no idea who The Doctor was, and as there was no one left in the office—there never was after three on Fridays—he folded the note back into its envelope and threw it into his desk. He decided to ask someone Monday if there were any doctors of interest in Nebraska. At 3:55 p.m., he locked the file cabinets and the office door and left for the weekend.

"Nebraska?" he muttered riding the elevator down. "Is that above Kansas or below?"

*****

"We have cash deposits from two of the bigger fraternities; Clara's, of course, and that kid, ah, what's his name. Yes, here it is, Sam Nyberg at Randall's garage. He alone should take up half the load as soon as we can bottle it." Art read mechanically from under a reading lamp in the rear of the Packard.

"Nyberg says there's an alky cooker about twenty miles from here. We can cut with that for less than a buck a gallon."

"Is that what Randall used?" Doc asked.

"Yeah, Nyberg says no one's ever gone blind—that he knows of."

Doc shrugged. "We should get it for less than seventy cents per gallon. I know for a fact those dagos in Chicago, the Genna brothers, only pay forty cents, and they got stills going in every Sicilian kitchen in town."

"Maybe," Art said. "But this isn't Chicago."

"Or New York, but we can make money anywhere; it's all the same product, same rules."

Silently, Art reviewed his books without looking up. He recognized one of Doc's depressing mood swings. He had seen it before in Lake Placid each time Doc was about to expand business or branch out into new territory. Doc had the ability to make all the right decisions when the pressure was on, but it took a toll in the depression it wrought until the money actually began to flow.

"How are we set for actual cash on hand?" Doc asked. He stared ahead past Henry into the predawn darkness and spoke without moving his head. Art imagined the brain inside calculating margins, exploring angles.

"Rather well, actually," Art said adding a cheerful note that sounded forced. "Excluding the money Hollow should have..."

"Has!"

Art glanced over his reading glasses. "Has. Even without that we have just over 8,000 dollars."

"And after we deliver?"

"Right around sixteen," Art said.

Doc smiled and glanced briefly at Art; he liked to hear the numbers played out. He leaned forward clasping Henry's shoulder.

"Keeping awake, Henry?" he asked.

"Yes, sir!" Henry flashed his wide smile only briefly taking his eyes from the road. The Packard glided smoothly over the gravel surface, the sound of pebbles striking the chassis added to the sensation of speed he so enjoyed.

His mind, however, was less on the automobile than it was on flying. Over and over he replayed each landing, each take-off in his mind until he had analyzed every mistake, every botched approach to the finest detail.

"I'm ready," he said quietly. "I can solo—that bastard's just stringing me along for the money. I don't need any goddamned instructor!"

"You say something, Henry?" Doc asked.

"No, sir."

Art spoke. "That colored guy from the restaurant, James—I think that's his name—I spoke with him yesterday, and he says he's got the bottles all ready, and with the help he's lined up we should have the final product ready to move by noon." He massaged his knees as he spoke, the ledger balanced on his lap. "Assuming Hollow makes it in on time."

Doc glowered. "He'll be here, Art; he'll be here." He looked straight at Art. "We're on our way back, I know it!"

"I believe you," Art said and he did.

Henry turned onto O street, past the entrance to the Standard Aircraft Factory and continued along a gravel road to the airfield. The entire area was dark except for the double string of runway lights left on for Jake. Mr. Howard had arranged that under protest.

"You can't land here!"

"Why not?"

"Someone will see."

"Not at that hour—besides, that way you'll get your money sooner and can keep track of your collateral."

"I wish I had never met you!"

"But you have."

Henry stopped the car about a hundred yards from the runway. He switched off the motor, and the three sat in darkness awaiting Jake and the sunrise.

They heard the murmur of the Hispano-Suiza engine before they saw the plane, at that hour only a silhouette against the weak dawn sky. The sun peeked orange behind the hangars. The biplane moved like a shadow, turned over the field and banked toward the runway. A faint rumble of tires on grass was all they heard before it rolled to a stop at the end of the field.

"There they are," Henry said pointing.

"Where?" Doc asked, his head poking through the window.

"Just past those trees; see the truck?"

"Oh yes."

They watched a truck move quickly to the aircraft. The sun moved slowly above the hangars. The pilot climbed from the plane and met the driver. Voices carried across the field but without modulation. The pilot waved and the truck backed to the plane.

Doc stepped from the car and paced from bumper to bumper. He could only see the occasional figure move between the plane and the truck. Someone pointed; someone shouted. The sun rose higher. Someone climbed onto the back of the truck and more voices carried.

"They're taking too long," Doc said and turned toward the sun, now full above the lowest hangar. A car drove past on the road.

"They're not waiting for us, are they?" Henry asked.

"Damn well better not be!" Doc snapped. "I specifically made it clear that we would never be around to meet shipments. That's what they get paid for!" He nervously checked the road and the factory. No one was there.

"What the hell could they be doing?" he growled.

Jake Hollow jumped from the cockpit and greeted James, the black dishwasher he had seen at Clara's. "Morning! Hope you haven't been waiting too long."

"No, just pulled up," James lied. He had been there with his cousin since 3 a.m.. "That's Herbert, my cousin. It's his truck. Those are my water barrels." He pointed to four barrels on the truck's bed. It was then Jake noticed the truck had no winch, no way to remove the whiskey hopper from the cockpit.

"What are the barrels for?" he asked cautiously.

"I had supposed for your whiskey!" James said with the same caution and tinge of alarm. "Or are we talkin' 'bout different subjects here?"

144

"No, I've got the booze, but I think we have a slight logistical problem," he said and led James to the cargo and undid the tarpaulin.

"What's that supposed to be?" James asked staring at the tank.

"Whiskey. One big can of it! You're supposed to have something to lift it out with!"

"That's not the way I heard! You're supposed to have some way for me to put it in the barrels," James said. The two men stared at the barrels then at the whiskey hopper. "Think the three of us could lift it out?"

"No. Damn thing weighs over 700 pounds." Jake walked around the plane trying to think up an alternate plan.

"I got a hose," Herbert said. "We can siphon it."

"Good idea!" James said.

"Better yet," Jake said and climbed onto the plane. He stuck his arm deep inside the cockpit feeling around the base of the tank. "Yea, there it is."

"What?"

"There's a drain plug on the bottom. That's where they stick the tap in to fill bottles. If you've got a hose we could open that and run it straight into the barrel."

"Think it'll work?"

"What choice do we have?"

"Here." Herbert pulled a short length of rubber hose from behind the truck's seat and handed it to Jake. Jake took it sniffing at the end.

"Wheww!" he said. "This thing's been used for gas!"

"And kerosene," Herbert added. "Don't worry, I don't think you'll hurt it."

Jake shrugged. "Have you got a wrench; a pipe wrench or something?"

Again, Herbert rooted behind the seat and produced a small pipe wrench and handed it to Jake. Jake dropped the hose over the rim of the cockpit and fumbled with it for a minute before he stood. "Won't work. It's too damn tight a fit; I can't get the hose on and make it bend back out. It pinches."

James, Herbert and Jake looked at each other.

145

"It's got to come out the side of the fuselage, somehow," Jake said, pointing.

"No problem," James said and took his pocket knife out; Jake hesitated. "You got a better plan?"

Jake thrust the blade into the airplane striking the plywood beneath. With the aid of claw hammer they managed to hack a crude hole through the side and soon had the hose attached.

"I hope this Al Minco fella don't mind you puttin' holes in his airplane," Herbert said.

The whiskey flowed from the tank into the barrels while the sun burned flat and yellow well above the trees. One by one they manhandled the barrels back onto the truck.

"What the hell are they doing now?" Doc asked. He looked over the car where the first employees were starting to arrive at the factory.

"Christ! That idiot's going to blow this whole thing!" He jumped into the car, and Henry took his seat. "Someone's going to see them...Ah, blast!" He kicked open the door again. "Let's go see what the hell they're doing!" He jumped out.

Art scrambled as best he could across the seat to follow.

"Those idiots!" Doc spat, livid with rage, and slammed the door behind him.

"Whoomph!" Art moaned as the walnut door panel struck his left knee square on the cap. He rolled back on the seat groaning and clutching his knee, his eyes clamped shut in pain. Doc was several yards away storming toward the airplane when Henry jumped from the car, took a glance at Art coiled on the seat and ran after him.

"Wait!" he called.

Doc spun around at the command, anger bright on his face, but he froze at Henry's voice.

"You look what you did!" Henry shouted and was amazed at his own boldness. "You hurt him, and I think you hurt him bad."

Doc moved slowly, cautiously, a half step then a full stride toward the car. He saw in Henry's face more than the usual bovine subservience he had grown to expect. Henry stood a full two heads taller and easily topped him in weight, youth and raw power; hence his value to the enterprise.

Doc felt his tenuous authority slide momentarily. "He isn't hurt too badly."

"You haven't even looked," Henry answered. They opened the door where Art rocked from side to side clutching his knee, fat tears in his eyes. "He's hurt," Henry said and brushed past Doc.

He took Art by the shoulder enveloping him in his huge arms. The accountant pressed his face into Henry's chest, and Doc heard the muffled sobs of a man in pain.

"Art, I'm sorry," he said. "I wasn't looking..." Art's fingers tightened around the battered knee. "You all right, Art?"

Doc continued to watch the Standard being unloaded even as he spoke. The barrels were lifted back onto the truck and covered with a tarp. He saw Jake cover the forward cockpit and shake hands with the other two.

"Everything's going to be fine," Doc said.

The truck lurched away. "Let's get him to a doctor, Henry!" Doc commanded. He slid onto the seat beside Art, but avoided touching him. Art pointed toward the glove box where Henry found a bottle.

"I'll be fine," he said wiping tears from his face. He forced a smile and winced.

"Nonsense!" Doc declared. "Henry, find Art a doctor. Let's go!" Henry closed the door and hopped into the driver's seat. As the motor fired, Doc muttered, "I'm sorry, Arthur," but the words were covered by the motor.

The Packard turned across the damp grass and onto the gravel when someone ran toward the car waving anxiously.

"Oh Christ-in-a-cathouse," Doc mumbled seeing Mr. Howard, the salesman, waving his arms for them to stop. Henry slowed the Packard without stopping, so Howard was forced to trot alongside as he spoke.

"Mr. Howard!" Doc called pleasantly, barely looking his way. "Up early."

"The plane's back!" Howard said and pointed.

147

Doc made a show of looking where he pointed.

"Eh? Well, so it is." The Packard crept along, Henry angling it over to drive through a mud puddle.

"Where's the money?"

"Mr. Howard! Business is not conducted like this. We had an arrangement, you supply the airplane, and we shall pay— this afternoon...or tomorrow."

"But it's back now!" Howard shook mud from his shoe and tried to keep up with the car, finally hopping onto the running board; his head thrust through the window. Doc cranked the window handle until the glass pressed against his chin.

"Everything is exactly as planned, Mr. Howard. You just get that second airplane ready for us—same arrangement."

"No!" Howard shouted.

Doc tapped Henry's shoulder, and the Packard stopped.

"I'm out," Howard said. He pulled his head from the window and dropped off the running board. "You pay cash!"

Doc pressed a cold stare on him as a faint smile spread across his mouth. "No, Mr. Howard. You are very much in." He let that sink in, until he saw Howard's shoulders droop in nervous despair. He waved a hand at the factory while shaking his head.

"This factory hasn't sold an aircraft in months, Mr. Howard. We have the data." He waved Art's ledger to support his case.

"Why, I even believe the factory has been sold, am I right?"

Howard looked at the ground and nodded.

"Refurbishing old biplanes is not the way of the future, Mr. Howard. This," he waved his hand again, "will be gone very soon, so you might as well take what you can. Plus, Mr. Howard, we hate to lose friends." With this, Henry turned a stone-hard glare at the salesman.

"We have others things to do, Mr. Howard. I suggest you get to work." He started to roll the window all the way up, then reversed himself calling, "And, Mr. Howard. Those prices are much too high; be thinking of discounts, eh?" He tapped Henry's shoulder, and the Packard growled and pulled away.

"How did you know the factory's been sold?" Art asked through the pain.

"I didn't, but I do now."

Jake spotted the Packard as it left, then caught a ride into town on the back of a truck, had breakfast at Clara's and found a room at the Hotel Cornhusker, a block away from the Lindell where Doc stayed.

He awoke around noon and made his way to the Lindell and up to Doc's room on the top floor. Henry was stretched across a small couch, eyes closed, his feet dangling shoeless off the edge. Art sat in a stuffed winged chair, his left leg propped on a pillow atop another chair. He appeared drunk.

"Mr. Hollow, how's our aviator?" Doc strode across the room to take his hand. "I trust you had no difficulties."

"None to speak of," Jake said. "What happened to you?" he asked Art.

"Nothing," Doc interrupted. "Just an old injury acting up."

Henry stirred on the couch, one eye taking in the room.

Doc spoke. "He'll wear that bandage for a few days, then be good as new." Henry grunted turning his head away. "You've had time to settle in?"

"Got a room at the Hotel...Cornholer, or something. Down the street—I started a tab in your name."

Henry chuckled with his back still to them, and Doc kept his smile rigid. "Fine, but we may want to move you. Do you have the money?"

Without a word, Jake reached into his jacket and produced a crude wad of bills and lint dropping it on the coffee table in front of him. Doc pushed it toward Art who leaned painfully forward and took it. No one spoke as he sorted the bills, turning all the faces in the same direction. Once in neat piles he licked his fingers, took a swallow off the bottle at his side and counted, occasionally jotting a number in his book.

"It's all here, less what I assume was spent on fuel," Art said. "I'll need receipts from now on, Jake."

"Sure."

Art counted 200 dollars from the stack, made a notation in his ledger, and with a glance at Doc handed the money to Jake.

"Quite a bit left over," Jake commented stuffing the bills in his pocket.

"Thanks in part to you," Doc said and with a flourish picked a fifty off the pile. He held it before Jake's face.

"Go ahead, take it," he said. "You deserve it."

Jake glanced from the fifty to the pile on the table. The room fell silent except for Henry who sat upright on the couch searching for his shoes. Jake made no effort to take the fifty, so Doc pressed it into his pocket.

"No need to be rude, Mr. Hollow. Someone offers you a bonus, you should take it; be grateful and maybe say, thank-you!" His voice was icy.

Jake chuckled. "Looks to me like you did pretty well on my efforts." He nodded toward the pile.

"As I fully expected," Doc said. "I think we should get one thing clear between us, Mr. Hollow. I am the investor; you the employee. It's my money invested, not yours; I take the risk, so I take the profit. When you decide to invest your own money we can talk profit sharing. Until such time, remember your status, and be grateful."

"I am, deeply," Jake said. Another silence hung uncom-fortably in the room until Jake asked, "So when do I fly again?"

"First thing in the morning."

"Same place?"

"Just up and back," Doc said, and Jake was relieved that there was no mention of a return to Adair's. "And from now on, I want you to time your arrivals back here so the whiskey is unloaded and away from the airport well before daylight. We can't have a repeat of that fiasco this morning."

"No problem," Jake said.

Doc had turned and was pulling a blackboard across the floor from the connecting room. "Tomorrow you fly with a new fellow, named Starns."

"Sterns," Art corrected.

"Thank-you. Anyhow, this Sterns flew for the US Postal Service until a couple of years ago, then several odd flying jobs around the area, most recently with some contract mail outfit in Omaha."

"What do you mean I fly with him?" Jake asked curtly. "I fly alone. There's no second seat in the Standard, you know that."

Doc turned the blackboard to face Jake. "Not in the same plane with him, but alongside him. He has his own aircraft, a De—something..."

"De Haviland D.H.4," Henry piped in.

"Thank-you. You've heard of it, Jake?" Doc asked.

"Yes. It's old, slow, doesn't carry as much as the Standard, and it's probably a worn out piece of junk ready to break apart in the sky."

"You're saying you don't like it?"

"Hey, this Strains fellow..."

"Starns," Doc corrected.

"Sterns!" Art said.

"Yea, well whatever," Jake said. "He can fly whatever he wants, but I'll tell you right now; it won't carry nearly the load of the Standard."

"It has one overriding advantage," Doc said. "It's paid for. All we have to do is put gas in and go."

Art spoke rapidly from his chair. "By figuring in the depreciation of the initial investment in the Standard LS-5 and comparing it to the cheaper investment in the D.H.4, we can actually make money with smaller loads. Of course, the newer Standard should last longer, and we are after volume here, so it does serve a purpose. However, by contracting with numerous individual operators to haul for us we can ensure both volume—through the Standard—and lower cost with the freelancers."

"You mean gypsies," Jake said.

"Exactly," Doc said. "Just like you were."

Jake glared at him.

"You seem jealous," Doc said. "There's no need to worry—you're chief pilot."

"Ha! Big deal."

Doc tapped the chalkboard where Jake's name was carefully printed at the top of a column. Below his name was Sterns', and a string of empty slots beneath. To the right of the names were columns for aircraft type, destination and loads. Jake remembered a similar board in the Army and how it seemed to reduce flying to a clerk's task, stripping all the fun out of it. Jake never even kept a logbook for himself; hated to document anything.

"You don't seem to understand the magnitude of this enterprise," Doc said. "This board will soon be overflowing with pilots and airplanes. This is big! Don't scoff at opportunity, Mr. Hollow, however it might come packaged."

"Is all that really necessary?" Jake asked waving at the board.

"We have to keep track of the gypsies, somehow."

"All two of us, huh?"

"Two now, but three by week's end, then four next week and eight the following." Each time Doc spoke he tapped another empty slot below Jake's name. "This industry will expand rapidly; I know. We must be ready to expand with it and being prepared organizationally is the key. Nothing happens without planning." He tapped the board.

"Someone may get wind of our operation—make that will get wind, and the only way we can stay competitive is by growing bigger and getting better."

He leaned closer to Jake, until Jake was staring at his yellowed teeth. "We must grow bigger; we must get faster, and we must carry more and more!"

"Canada and back's a long trip for some of the planes and pilots you'll be getting. It's no quick hop in the Standard," Jake said.

"Then you'll fly in relays," Doc said. "Like the pony express. The pilots can rest, but the product will always be moving. No time to stop. We will do whatever is necessary—whatever—because time is profit, Mr. Hollow, and that is the whole reason for any of this!"

"It's interesting," Jake said. "But you can't schedule away weather, engine failures or just plain getting lost." He stared at the board hating it for changing him from a pilot to a hauler, a driver.

"We can induce them away," Doc said. "You, for instance, get 200 dollars for a trip. That's the flat rate for up and back, one load. Now, to help you decide whether to circumnavigate a thunderstorm or bore through and save twenty minutes there might be a fifty dollar bonus waiting if the product is brought in on time, maybe sixty if ahead of schedule.

"Then there might be bonuses for hauling bigger loads or flying at night, or during colder weather. Mr. Hollow, money can fetch some amazing results."

"That kind of flying will burn a man up," Jake said.

Doc's face was expressionless. "There are other pilots. They're a dime a dozen. Our planes will fly every day. The most successful pilots get first crack at the trips. I don't believe in seniority systems, they don't produce results. If a new pilot hauls more, then he gets the trips. Finding pilots to fly seven days a week is no trick."

Jake slid into an overstuffed chair at the edge of the room facing the blackboard. Henry had found his shoes and stood near the window watching the sky. Art nodded with a bottle in his hands.

"So what about Sterns?"

"Sterns..." Art slurred.

"I said, Sterns!" Jake snapped.

"You show him the route; that's all," Doc said.

"Does he get 200 bucks?"

"Plus a fee for his machine."

Jake studied the board, the empty slots, the headings, the two names with his on top. He tried to calculate how many flights he could complete in one week and how much he could save. I've got to get my own airplane again, he thought. My own.

"I have work to do," Doc announced. "Henry, get the car. Art, I'll take Jake to the airport and check up on our deliveries. They should have been made by now."

153

Henry was out the door, and Art nodded absently to Doc.

"Time's a'wasting, Mr. Hollow," Doc said guiding him out the door. "We can drop you at the field. Mr. Sterns..."

"Sterns..."

"...is working on his machine; maybe you can give him a hand."

"I have my own work to do on the Standard," Jake muttered.

"Sterns...Sterns," Art slobbered, and the door closed.

"She says the stuff was delivered before noon," Henry said closing the car door behind him. Clara waved to Doc from the doorway of her cafe. She was all smiles and giggles.

Doc waved in return. "Get back inside and peddle the crap," he muttered through a charming smile. The Packard moved away. After visiting Sam Nyberg at Randall's Garage they headed for the airport.

"I have business with Mr. Howard," Doc said as Henry opened the door for him. "I may be a while; perhaps there's something you can help Jake with."

Jake shrugged and lit a cigarette.

"Yes, sir," Henry said, and Doc disappeared into the hangar. Jake was already walking toward the flight line.

"Can I help you, Jake?" Henry asked catching up with him.

"You're going to get your nice suit all greasy." He blew smoke in easy rings.

"I'll find some overalls."

"You do that." Jake was staring at the D.H.4 on the far end of the flight line, apart from the other aircraft. Someone stood on a step ladder and poked under the cowling, occasionally swiping at something with a rag, then hammering at the motor with the butt end of a large wrench.

Sterns, Jake thought.

"Jake?" Henry asked.

"Hmm?"

"Jake, do you think I could fly?"

The question was so simple, Henry's voice rose slightly in asking. Instantly, Jake reevaluated the oaf who drove for Doc. Maybe there was something to like in there, he thought.

"How do you mean?"

"An airplane; do you think I could fly one?"

"I don't see why not." They had stopped. Jake looked up at Henry. "Are you planning to?"

"Already am!" Henry beamed.

"Well, that's great, Henry! Are you taking lessons?"

Henry pulled him by the arm. "See that one over there?"

"The J-1?"

"That's what I'm learning in!" Again he beamed with raw pride.

"I have one of those," Jake said as they walked toward the biplane. "It's kinda buggered-up right now...you saw it when I met you in Texas."

"They've got a school right here on the field. Apparently, it used to be pretty busy, but from what I gather they're going broke, so I get lessons pretty cheap. In fact, I haven't paid a thing yet."

In the background, the Liberty motor on the D.H.4 roared, belched and was silent. Jake looked quickly at Sterns kicking the plane's landing gear and swearing.

"Have you soloed?" he asked.

"No," Henry said and plugged his hands into his pockets. "That instructor keeps harping on me to read some dumb book..."

"Did you read it?"

"Mostly...Some," Henry mumbled. "But, hell, I know I can fly the damn thing. Book's not going to help; I just have to remember to keep my airspeed up, and I think I'm getting that. Hell, any idiot can fly!"

Jake raised his eyebrows, but the reaction was lost on Henry. "Want to fly it?"

"What? Now?"

"Got something better to do?"

"No, but it belongs to the school," Henry protested, but was already taking off his suit jacket and heading for the airplane.

"You're a paying customer—almost—go ahead, get in! I'll check the water, and oil."

Henry pulled the canvas tarpaulins from the cockpits and climbed inside. Jake made a cursory inspection of engine and airframe. "Everything seems to be nailed on, all right," be said. "Switch off?" he called.

"Switch is OFF!" Henry replied. "And the fuel's on. Thing hasn't got any brakes."

"Well, keep the throttle closed while I pull it through."

Henry nodded, and Jake whipped the propeller around.

Within minutes the OX-5 was clacking away pleasantly, and Jake climbed onto the wing and into the forward cockpit. Grass clippings tumbled past the tail in the prop blast. Henry occasionally goosed the throttle when the motor seemed ready to falter. He knew machines, how they ran, what they liked; it was an innate gift. He only wished he could learn how nature worked, wind and lift, and all that book stuff that made airplanes fly.

Jake waved him toward the runway while he pulled a helmet over his head. Instructing was not his game, but he had seen enough of bad instructors in the Army, badgering students until all confidence was lost, that he suspected a more casual approach was in order.

"Go ahead, Take Off!" he yelled waving toward the sky.

With both feet hovering above the rudder bar, and his hands lightly clasping the throttle and stick, Jake followed Henry through on the take-off.

He knows what he's doing, he thought feeling the biplane float off the grass and climb. They banked gently away from the field and continued the climb over the young corn below.

Henry was crude, often pressing too much rudder without enough aileron, but close enough, Jake thought and relaxed letting Henry fly. To show his confidence he rested both hands on the cockpit's rim. The effect was not lost on Henry, and he settled into a newfound confidence and flew.

"SHOW ME WHAT YOU KNOW!" Jake yelled waving his hands in pantomimed turns and stalls. Henry nodded understanding and took him through the paces.

In a competent series of turns, climbs and a stall to the left with a smooth recovery, Henry displayed his skill. After each maneuver he looked around at Jake only to find him gazing lazily at the sky or yawning.

Henry pulled the nose high; Jake turned his attention back to the airplane, his feet once again close to the rudder bar, but his hands still clasped to the rim. Henry closed the throttle and held the nose high until it shuddered in the stall and dropped into a spin to the right.

Oh Christ, Jake thought. I hope he's learned spin recovery.

The earth spun once, then twice while Henry held the stick hard back and rudder pressed firmly. Jake was about to slide his hands off the rim when the stick came full forward and the rudder reversed.

Nothing happened quickly in the long winged Standard, but it eventually recovered from the spin, and they leveled.

Jake yawned pointing back toward the field. "SHOW ME HOW TO LAND THIS THING!" He yawned again theatrically.

Henry dove for the runway setting up a beautiful approach from downwind, to base leg, to final. Jake's feet were once again touching lightly on the rudder bar, and as they crossed the fence he felt the biplane rise momentarily. Before he could remove his hands from the rim, Henry recovered, and they squatted onto the grass in a rumble of tires and tail skid over the uneven turf.

Inside the factory, Henry's regular instructor strolled through the door, poured himself a cup of coffee in the front office—he always waited until the office was empty to avoid paying—then walked through the hangar shop and stood sipping from the mug when Henry turned onto final.

"Who's up in the airplane?" he asked one of the shop mechanics.

"Don't know," he answered, although he had seen Henry taxi the plane out earlier. Few at the factory liked the instructor.

"Nobody's allowed to fly that without my permission. I'm chief pilot here!"

"Only pilot here," the mechanic noted and walked away.

Henry was taxiing off the runway, and the instructor ran toward him at a trot. The biplane stopped, the engine at idle, and Jake climbed down and walked clear waving at Henry now clearly visible alone in the rear cockpit taxiing back to the runway.

"You there!" the instructor called. Jake walked toward him then turned to watch Henry depart.

"Who the hell do you think you are?" The instructor spun Jake around until he was face to face, noses inches apart. Jake's face was tired, and scarred from his accident, the half moon gash had dried to a crusty scab on his forehead. He needed a shave and smelled of tobacco.

"Jake Hollow. Who the hell are you?" The instructor smelled of lilac cologne and toothpaste.

"I am George Cranston. I am the only personnel authorized to fly that machine!"

"Well, when Henry gets down you might want to tell him." Jake poked his finger into Cranston's chest, while his other hand balled into a tight fist. Jake hoped the instructor would react, and he could let loose.

"I thought that was...what's he doing?"

"Looks like he's going flying," Jake said and pointed.

The biplane bounced along the grass and lifted into the air.

"That boob!" the instructor shouted. "He should never have been allowed near an airplane! He can't fly!"

"You'll have to tell him that, too—when he gets down."

The biplane crawled through the pattern and turned onto final. With a gentle popping of the motor, Henry glided in for a respectable landing and turned off the runway.

Cranston started to run toward him, but Jake stuck a foot in his path, and he dropped to the ground.

"You!" was all he could muster wiping mud from his hands. He ran straight toward the biplane waving both hands over his head.

"STOP!" he commanded in a high voice.

Henry panicked. He pushed the throttle open and kicked the rudder to avoid hitting the instructor but managed, instead,

158

to smack him solidly with the left wingtip. Again, George Cranston was sent sprawling to the ground. By now, a small cluster of factory employees had gathered at the hangar door to watch.

"STOP THAT AIRPLANE!" Cranston screeched, and Henry spun the biplane in a tight circle with the throttle wide open headed for the runway.

"STOP!" Cranston shouted and chased the airplane, Henry turned again and reattacked the now frenzied George Cranston.

Jake sat on the grass well clear of the Standard chasing its prey. He took a cigarette from his pocket, lit it and watched the biplane roar past and slam Cranston into the ground.

George Cranston lifted himself painfully, punched the ground and rubbed his neck. Slowly, his eyes met Jake's.

"Did you get a chance to explain to Henry who gets to fly the airplane?" Jake asked.

The Standard, meanwhile, came to a stop near the hangar door where the small throng of employees cheered as Henry stepped down. Doc emerged from the building beside a confused Mr. Howard. Henry waved at his employer and called, "I soloed!" The crowd let out another cheer.

Jake kept his seat on the grass, and George Cranston stormed off without another word, passed the crowd and left.

Jake heard footsteps approach from behind, but he only turned his head enough to see it was the man from the de Haviland D.H.4.

"Hello," he said quietly without making eye contact.

"Uhh," Jake mumbled, and took the opportunity to study him. He wore white mechanic's overalls, stained with oil and torn at both knees. Beneath these, he saw an orange tie— bright orange—the knot like an overripe tangerine at his throat.

"Hard to find ties that color?" Jake asked.

"I have them special made by a color blind Chinaman in San Francisco." When he spoke, he showed an even row of white teeth beneath his thin mustache.

"You the chief pilot for ALMINCO?" he asked. His black hair was plastered back with Brilliantine, but a strand drooped across his face. He smoothed it back then held out his hand.

Jake took it quickly, formally. "This ALMINCO, it's a front, and if anyone ever finds out, we all go to jail, or get killed by someone who doesn't appreciate what we're up to. Are you ready to fly for someone who gets you killed or thrown in jail?" He leaned back on both elbows, his cigarette dangling from the corner of his mouth. The smoke curled into his eyes, but he made no effort to move.

Sterns spoke. "The last place I worked was for a flying circus up in Wyoming and Montana. They paid five bucks a day. State fairs, Fourth of July picnics, Memorial Day, Labor Day. Hell, we'd fly on Guy Fawkes Day if somebody'd pay the price of admission."

He sat on his heels beside Jake and pulled a blade of grass from the ground putting it between his teeth. A large ash fell from Jake's cigarette.

"My specialty was wing walking," Sterns continued. "Not because I liked it, but it paid two bucks a day more. Mostly, I'd just walk out the wing, shimmy up the strut, grab the cabanes and stand there. But that got old; crowds wanted more, so I'd climb under the wing and hang on the bow below, and pretty soon that was passe. Crowds were getting spoiled.

"Then I tried tying a rope to my ankle—people on the ground couldn't see it, of course—and I'd walk out the wing, shimmy up the strut, and at 800 feet above the ground I'd stand up. 'Course, I'd already tied the other end of the rope to the strut where they couldn't see." He spat chewed grass and plucked another blade. Jake brushed the ash away.

"Well, we'd have this second airplane, a Jenny, fly alongside, and he'd bring the wing real close to me as I stood in the wind, one hand on the cabane, the other reaching for the bow under the second airplane's wing.

"We had to do this pretty quick, because we'd be past the crowd, and no one would see it. Anyhow, I'd reach for this wing right as we pulled in front of the crowd, and whoops!" He clapped his hands; Jake started. "Off I'd go—right off the wing; crowd thought I'd had it for sure, and they'd all be on their feet.

"'Hey, look at that, willya', they'd shout as I flew off the wing—dead for sure, they'd think. Then the rope would catch and snap! I'd be safe, hanging there in the breeze upside down, by my ankle." He rolled his pant leg up displaying a calloused scar around his ankle. "Seven bucks a day! You fellas pay 200. I don't care who tries to shoot, stab, beat, bust, jail or crap on me—not for 200 bucks!"

"Welcome aboard," Jake said.

Just before the sun appeared in the cool still dawn the following day, Jake's Standard lifted from the airstrip and climbed slowly through a patch of morning ground fog. Behind him, Sterns' D.H.4 lumbered into the sky, its Liberty engine rattling the hangar windows.

Henry stood alone beside the Packard and watched the D.H.4 join up in loose formation slightly behind and to the left of Jake. As the groan of the two engines faded to the northeast, Henry walked to the Standard J-1 and popped the cowling open.

At five minutes past eight he met Mr. Howard inside his office and handed him four one hundred dollar bills.

"Doc doesn't know I got this money," Henry said. "It's mine, though."

Howard smiled, pushed the money into a drawer and signed the bill of sale. "Congratulations. Your first airplane!"

As Henry left the office, the phone rang, and Mr. Howard answered it. The factory no longer employed a secretary. "Hello, Standard Aircraft," he said. What's left of it, he thought.

"Who? Jake Hollow? Ah, no..I don't know him. Well, Miss there are quite a few pilots around here. This is a busy place!" A bee flew through the window, turned and left.

"Well, I am sure I don't know him...Yes, I'll take a message...Ah-huh...Susan? Ah-huh...yes, he'll get it. Thank-you. Good-bye." He hung the receiver back with an annoyed shake of his head.

"Let him take his own damn calls!"

Chapter
XIII

By late July, Lincoln had become accustomed to the increased air activity around the airport. While the Standard factory seemed to stagger on its financial legs, more and more airplanes appeared, no longer confining their activities to the predawn hours. James and his cousin Herbert purchased a second, then a third truck and drove boldly through town making deliveries.

Arthur's knee healed, leaving him with only a slight limp and occasional pain easily dulled by the whiskey. His mind, however, never slowed as he shifted large piles of cash through the local banks, from one account to another with hardly a day's rest in one place. Then the funds, their trail thoroughly muddied, would move out of state, into Mexico and Canada for brief visits, then back to safe banks in Texas and Florida. Finding a clear link from any account to Doc and the liquor would prove monumental, although Arthur, himself, could envision where every dollar was at any particular time.

His deft maneuvering elevated Doc into a position of envied business personality, his fortunes rising when other businesses were feeling the initial twinges of a weakening economy.

Gossip circuits whispered of their new resident's dealings in bootleg hooch, beer, illegal gambling, even narcotics, prostitution. There was even a short-lived rumor about the airport being used as a transit point for smuggling illegal Chinese workers from Canada. Doc heard them all and savored the notoriety.

By now, the entire top floor of the Hotel Lindell was leased by him with several rooms on a lower floor, at the rear, reserved for his increased stable of pilots. Jake still kept a small room at the Cornhusker, although he was rarely in town long enough to use it.

Several of Lincoln's own citizens seemed to rise along with Doc's fortune. Clara's restaurant was transformed over a six week period from the hash house sneaking the occasional snort to favored customers into a high-priced haunt for the fraternity crowd. The city guide made no mention of its favorite speakeasy, but anyone could tell a stranger where to find it.

Clara, herself, while shunned from more polite society, took naturally to the company of her supplier, Doc, and was often seen wheeling around town with the white-haired entrepreneur in her almost new Chrysler or Doc's newly leased Lincoln Brougham.

"Never tie up your own money," Doc had told Henry when they picked up the new car. "Owning too many things merely ties one down when it's time to move on." The Lincoln dealership had not sold a car in months and was grateful when Doc strolled in, chose one in less than five minutes and drove away. His promise to pay was sufficient.

Not everyone was enthralled with Doc's presence or his boost to the economy. The Lincoln chapter of the Anti-Saloon League harassed political and law enforcement figures statewide demanding something be done about, "...blatant and often flagrant violations of the Volstead Act!" Petitions were politely ignored while the fortunes among petitionees took a noticeable turn for the better. Doc had weathered the same storms in Lake Placid and found Lincoln amusing.

"Cast the petty despot another handful of baubles!" he said to Arthur when told the mayor was expressing concern over increased daytime deliveries in the downtown area.

"Did I tell you Sam Nyberg's bought the garage—from Randall," Art said.

Doc's eyebrows shot up quickly, and he turned from the blackboard. "Enterprising young man isn't he? I've noticed he's taken to wearing three piece suits..."

"And spats."

"Shines his shoes now, I suppose."

"Hell, he even bought new ones—black to match his new car."

"Still looks like a pimply high schooler, but he does the job. That's what counts!" Doc glanced around the room. "Where's Henry?"

"Out at the airport."

"Where else?" He turned back to the board. A second one had been added with new columns depicting a more detailed account of loads in transit and estimated arrival. Each aircraft had a dollar figure chalked in the final box indicating exactly how much money could be expected. With a glance at the arrival times, Doc could estimate profits days in advance.

"These new routes out of Moyer's don't seem to be catching on," he said. "Doesn't anybody drink in North Dakota?"

"They'll improve; always do. Give it time." Art spoke evenly without looking up from his books. The questions were routine, his answers unimportant.

"What do you think of my idea of expanding into Kansas City? It's not far from here."

"Like I've said, that territory's all sown up right now, someone else is running that show. We don't need the problems."

"Ha! The problems!" Doc moved from the boards. Where's your spirit, Arthur? We didn't hesitate in New York!"

"We had Rothstein in New York."

"We don't need him."

"We'd need muscle for Kansas City."

"We fake it," Doc said and laughed. "Just like we do here."

"There's no competition here."

"Oh? Oh? Then who was running this before we came? Huh?"

"Couple of niggers and a fat broad named Clara." Art's voice never wavered from its calm assessing tone.

"Merely the point men; I mean who ran the big show? I'll tell you who. Capone, that's who! This was his territory, and I bet he's getting plenty sore about losing it to a couple of New Yorkers!"

"I don't think Capone even knows where Nebraska is," Art said.

"He doesn't have to! He just knows where his money is, and he knew the name Rothstein, so I bet he's heard of me once or twice. I'm stepping on a few toes around here." Doc fitted a cigarette into the holder he had taken to using.

"So why doesn't he just move on us? He's not the type to be interfered with, the way I read the papers."

"He's waiting," Doc said, the cigarette holder clenched between his teeth, his mouth in a tight smile. "He's just waiting; probably wants to learn how I do it, then make his move!"

Art shook his head, but Doc missed it. "I don't think he could even spell Nebraska," Art said.

Doc pulled his blue silk dressing gown tight around his body and lit the cigarette. With the smoke rising in lazy curves toward the ceiling, he reviewed the names and aircraft numbers. Mostly he read the dollar amounts. The names he could never remember; the airplanes all looked alike to him.

"Why is Sterns' name circled?" he asked. He glanced at the last box beside the name: $3555.

"We had a telegram from Vincent..." Art started.

"But I thought they were happy with Sterns; he's been running that route almost exclusively, hasn't he?"

"Yes, but that's has nothing to do with it," Art searched for the telegram in the waste basket. "Vince says he's crashed."

Doc shot a glance at the board: STERNS, D.H.4,....$3555

"Is everything all right?"

"He's suffered a few broken bones apparently..."

"The whiskey!"

"Oh, already off-loaded; it happened on take-off, no details..."

"There's 3,500 dollars up there," Doc said, his voice agitated. He ran his finger down the list of pilots stopping on Jake's name, tapping it twice. "Jake's scheduled to turn around at Moyer's tonight." He turned to Art with a knowing smirk.

"Mr. Hollow hasn't made a run to our friend, Adair, in quite some time, hmmm?"

"Not since the first," Art said, his tone amused. "Sterns has been servicing her almost exclusively."

Doc drew a thick line through Jake's name on the return flight list and chalked it in below Sterns', then with his fingertips obliterated STERNS and placed it in the column labeled, RESERVE.

"Everybody deserves an overnight visit to that ugly old woman now and then," he said.

Art shrugged. "I never thought she was ugly...or old. Anyhow, I think Jake's got some dolly over in Iowa he's been slipping over to poke now and then."

"Not in a company airplane, I hope!" Doc thundered.

"No, I don't think there's enough room in one actually."

The warm summer air had dried the two runways at Moyer's airfield into wide strips of hard packed dirt and short tough weeds. Jake was loading whiskey into the hopper tank with the help of one of the Indians when Moyer came running across the tracks from the cabin calling his name.

"Jake! Jake!"

A Travel Air biplane ran its engine in a cloud of brown dust at the end of the runway as if trying to develop the courage to take off. Jake had seen the pilot taxi to the very end of the field and then climb from the plane and push the tail into the weeds hoping to gain a few more inches of runway. Flying grossly overloaded with gas and whiskey, many of the pilots had adopted the same method.

"Telegram from Doc, Jake!" Moyer called. "For you."

The Travel Air rolled slowly over the hard dirt, bouncing several times then, in a labored heave, took to the air and passed low over their heads. Jake and the Indian ducked as the shadow flashed over. Moyer leaned heavily on the Standard trying to breathe while groping through his shirt pocket for a smoke.

"You're—gasp, gasp—not supposed to return to Lincoln—wheeze, cough, spit—you're to take a load to Adair's and pick up the money Sterns had." He drew deeply on the cigarette and coughed violently spewing gray smoke and spittle. "God, I gotta try those new menthol cigarettes—puff, puff—they say those are good for you."

"I don't want anything to do with that!" Jake snapped. "Let someone else take it."

"No, the telegram says you specifically are to take it. Anyhow, Billings there is the last one expected in for the next day and a half." He pointed at the Travel Air turning south, weaving between treetops.

Jake slid off the Standard leaving the Indian to finish loading. "I can't go back there! She's...she's...well, that Vincent's just plain nuts, too! Tell Doc I ain't going."

"It doesn't work like that, and you know it. We all do what we're told and when we're told. That's how an organization runs."

"Oh, blow the organization crap out your..."

"You want out? Well, you just take this load back to Lincoln, and you tell Doc, 'No, I don't feel like working for you anymore', and you know what happens?"

Jake shrugged.

"I'll tell you. At best. At best, you get to walk away without any airplane—without squat—and stick your thumb out along the highway and that's it."

Again, Jake shrugged.

"And where you going to go after that, eh? No airplane, and how much money have you saved up? Enough to live on?"

"I'll get another plane. In fact, I already have one."

"Yea, I've heard about that score. Some busted old clunker in some dame's barn in Iowa, for Chrissake!"

Jake stared at Moyer.

"Word gets around, Jake. If I've heard about her, so has Doc, and if he can't repay you for disloyalty, he'll take it out on someone you know."

Jake thought of Susan, and a cold feeling ran along his spine wrapping into his guts. "You don't know nothing," he said weakly.

"Look, you don't have to go near Adair. Just deliver the load and pick up the cash. It's a quick turn around, no extras.

Who knows? You might be able to help Sterns fix his plane and get him back in the air. She made no mention of his condition in the telegram, but I'm sure he's spending his time well groping around her knickers."

"Oh, shutup! Willya? Just shutup." Jake was annoyed with the flash of jealousy that seemed to come out of nowhere, although he had felt it ever since Sterns had proclaimed an exclusive right to the Adair route. He had learned how to suppress desire and the memory of desire, but when it came back, it did so with a vengeance. His thoughts tumbled out of control for a moment. He thrilled at the thought of Adair and longed for Susan. "I really don't want this shit," he said to himself.

The farm looked familiar from the air. He circled once over the silo with the white cross painted on top. Two trucks were parked near the house, and a couple of men carried boxes, ignoring his plane.

Odd, Jake thought. Anytime he ever heard an airplane motor overhead, he automatically turned to look. Maybe the world's getting too used to the sound, he mused.

Someone stepped off the back porch, looked skyward and disappeared again inside. Jake banked over the adjoining hay field and pulled the throttle back. The grass was worn in a tapering smudge where Sterns had been operating, but there was no sign of the D.H.4.

"Must have crashed somewhere else," Jake said to himself, his head over the cockpit's rim looking past the nose. "Christ, we'll probably have to go some damn place and disassemble the damn thing!" He turned onto final, the roar of the motor replaced by that familiar whistle of air rushing past wires and struts.

"Well, if he thinks I'm wasting any goddamn time working on his damn airplane..."

The wind buffeted the Standard from the right and Jake corrected glancing quickly in that direction. There it was. Ahead, through the wires and struts, the D.H.4.

About a half mile out from the end of the field, and slightly to the north, stood a grove of trees, mostly oak. Jake saw the de Haviland biplane wedged into the branches, its left wings snapped back in an evil twist, the others pointed skyward like a hand pleading for help. He turned his attention back to the Standard and made what was possibly the worst landing he had made in years.

He smacked the ground and bounced high back into the air, stalled and settled with an awkward thud, the bow under the right wing striking the ground. Jake shook his head and rolled to a stop in the middle of the field.

The bow was probably broken, he thought. Not the first time. A little glue and wire would fix it. He forced his mind to think of the broken bow, but the image of the smashed de Haviland pushed into his vision. He knew it was an awful wreck; the kind no one climbs out of.

He had no idea how long he sat in the middle of the field when a truck pulled along side. Vince stood on the running board, his pistol worn high on his hip. He waved at Jake to stop the motor.

"Well, well, never thought I'd see you around here again," he said climbing onto the wing and offering his hand to Jake. The horribly scarred face held its fixed sardonic grin.

Jake shook the proffered hand slowly. "I've been on other runs; Sterns, I believe, had the exclusive on this one." His voice came out strained.

"Sterns held the exclusive on a number of things around here," Vincent answered while still holding Jake's hand. It was then Jake decided Vince was responsible for the crash, somehow. He pulled his hand away.

"People should be careful when they think they can't be replaced. Don't you think?" The face revealed nothing to add to his vague words. Jake sat. He wanted to stand, or at least be away from this man, but Vince hovered over the cockpit, his ruined face watching.

"Sterns wasn't a careful man, Jake. You've been careful, though, haven't you?"

"I think so," Jake said, but his mouth had gone dry, so the words were choked.

"I just don't know what people are thinking about some-times when they do some of the dumb things they do." Vince seemed to lean closer, one hand resting on the grip of the fat Webley revolver on his hip.

The truck door slammed, and Jake turned quickly seeing Marvin step out and approach. He held the sawed-off shotgun in his huge hands. Jake faced Vince again still hovering over him, the face still plastered with the mangled grin.

"Where is he?" Jake spit the words. "Where's Sterns?"

"I think his problem was he thought himself capable of many things that were just beyond him. Do you know what I mean, Jake?"

"Where is he?"

"He paid little attention to his airplane, you know. His mind was always on other things where maybe it had no business. Do you follow, Jake?"

Jake snapped his head from Vincent to Marvin who now stood behind him, the shotgun out of view.

"He rarely checked his airplane for defects, things that might break in flight. Seemed a careless man that one did. Do you know what I mean, Jake?"

"No."

"The airplane was old, worn out; something could break at any time, but he never seemed concerned. You'd think when someone depends on something for their very life they could treat it with a little more pride, or at least respect. It's amazing how a lack of respect can get someone hurt."

"Where is he?" Jake demanded.

"Oh, he's inside."

Jake undid his belt and pushed roughly past him and began toward the house, expecting at any moment to hear the bark of the Webley and feel his shoulder blades separated by the .45 caliber slug.

"But he's dead!" Vincent called.

Jake froze, his back still to Vincent, a faint ringing began in his ears, and he felt his legs shake slightly. He tried to push the fear away, but the dead pilot's image combined with what he had seen of the smashed biplane blocked his way. He had to get it away, remove the image.

"You pilots seem to think you're invulnerable, completely immune from the rules the rest of us live by. Why is that, Jake? I hope you have the good sense to take care of your machine."

Jake spun around. Marvin was on the wing undoing the tarpaulin over the whiskey hopper. Jake sprinted back grabbing Marvin's belt and pulled him to the ground.

"Get away!" he screamed and thrust his arm into the cockpit and found the rag with the pistol Moyer had given him.

"This is my goddamn airplane!" he screeched waving the pistol still half-wrapped in the rag at Vincent and Marvin. He desperately tried to remember whether the safety lever was supposed to be forward or aft, and whether ON meant ON ready to shoot or ON the gun is safe and won't shoot.

"I know this airplane! Every inch of it, and you stay the hell off!" He felt his hand begin to shake uncontrollably. Vincent leaned on a strut without getting off the wing; the fixed grin seemed to become amused. Marvin slowly took to his feet, the shotgun had fallen several feet from him.

"Now, I know just what the hell is mine," Jake said in a high voice. "And if you so much as look at it sideways, I'll blow your son-of-a-bitchin' head off!" He steadied the gun with two hands. Vince slowly lifted his hands away from his belt, putting Jake slightly more at ease.

What now, Jake thought. I can't stand here forever. How the hell do I start the plane; get out of this. Maybe Adair can get out with me; sure, we'll tie these two up and off we go!

Behind him, a truck bounced its way across the field toward them. He could see Adair at the wheel, and he stepped carefully off the wing keeping the pistol before him pointed at Vincent on the other side of the fuselage. Without realizing it, Marvin was inching his way toward Jake's flank requiring him to swing the pistol in an ever widening arc to keep the two covered.

172

"I think you've misunderstood something I've said," Vincent spoke. "Is that it?"

"No misunderstanding," Jake answered. "I intend to make my delivery, get paid, and that includes the dough Sterns had on him, then I'm outa here." He heard the truck brakes squeal and the door open. "I hold all the cards."

"Agreed," Vincent said. "May I?" He pointed at his shirt pocket. "I have your money right here." Jake caught a glimpse of Adair, tall and handsome coming toward him, her rescuer. He waved quickly for her to join him and twitched the pistol at Vincent indicating it was permitted to reach for the money.

"It's all here," Vincent said and tossed a fat roll of bills with an easy underhand pitch.

Jake lowered the gun, his eyes on the money, and caught the wad. When he looked up again, Marvin held a small caliber automatic trained on his temple, his hand unwavering.

"What should I do?" Marvin asked.

"Shoot him," Vincent said coldly.

The ringing intensified in Jake's ears, and he heard Adair's voice break through.

"Now wait one damn minute!" She strode toward them.

"Yes, please," Jake squeaked.

"Oh, shut up!" she snapped at him and turned to Vincent. "What do you think you're doing?"

"Is there a problem?" Vincent asked. "I thought you were through with this one?"

Jake raised his eyebrows, insulted.

Adair spoke. "Forget him! He doesn't amount to a hill of beans."

Jake turned to her, his feelings crushed. He felt his entire rescue scenario quickly unravel. He still clasped the pistol and the wad of bills, but Marvin kept the automatic sighted on his head.

"Well, then let's shoot him," Vincent said.

"Wait!" Jake shrieked.

"What?" Adair and Vincent answered in unison.

173

"Ah, I think Adair was about to say something important...about me," Jake said.

"True," she said.

Good, Jake thought.

"Vince, we need him."

Yes, listen to her.

"What would Doc say if he lost two pilots on one run. He might cut us off!"

"You're right; bad business," Vincent said.

"Very bad business," Jake was quick to add.

"You keep out of this," Adair hissed and walked around the fuselage to continue her discussion with her husband.

"Besides," she said. "Who would fly the airplane out? They'd have to send up another pilot, and that costs dough. He isn't worth it." She glared at Jake who was having trouble remembering why it was he was so annoyed with Vincent's treatment of her. He looked her over, and instead of seeing the shapely female beneath a man's clothing, he saw a tough woman in man's pants with hair tucked roughly under a filthy fedora and a grotesque scar across her face.

While the husband and wife continued their discussion in hushed snarls, and Marvin kept a bullet poised at his brain, Jake took the time to analyze just how fickle an emotion love was. He had landed with an aching desire for this unfortunate woman held captive in a far off land, and within minutes he saw her as a repulsive scrap from a nightmare.

Funny how the mind works, he thought and glanced at Marvin behind the pistol. Marvin shrugged his shoulders and gave a quick smile, embarrassed by the whole incident. Sorry, it's just my job.

Jake stood limp with the roll of money and his own pistol dangling at the ground. Adair and Vincent snapped and hissed at each other, occasionally pointing at Jake, the plane, the house, each other. Adair planted her hands firmly on her hips and turned away once, while Vincent rolled his eyes and threw his hands at the sky.

"Women!" Jake heard him say, and Adair turned on him again wagging a finger, her jaw tight, the words squeezing out in inaudible spurts. "So!...You!...goddamn Minnesota!" He only caught snippets of their discussion.

Marvin had lowered his pistol and took a pack of cigarettes from his pocket and offered one to Jake.

"Thanks," Jake said.

"Don't mention it," Marvin answered. He struck a match and held it for Jake who leaned over the flame, both pistols pointed harmlessly away. He nodded his thanks and stood aside while Marvin once again trained the pistol at his head.

Vincent and Adair moved further away from the plane almost to the point where Marvin thought they might leave. Becoming bored, he took a cigarette for himself, lit it off Jake's and tucked the pistol back in his belt, and the two men stood leaning against the lower wing awaiting a decision.

They interrupted their argument long enough to glance over at Marvin and Jake smoking cigarettes, Jake still holding the money and his own pistol, and Marvin empty handed.

"Marvin!" Vincent shouted.

"Yes?"

"Get the whiskey unloaded and inside."

"You," Adair called pointing at Jake. "Come with me." She strode past him toward her truck leaving Vincent to help Marvin. Jake watched her backside swing as she hauled herself into the cab.

Once inside the house, she led him into the front room where arranged on a table were, a watch, a crumpled pack of Lucky Strikes, a pocket knife, Sterns' orange tie, helmet and goggles.

"Not much for a man to leave behind," she said absently running her fingers across the items.

"Where's the body?"

She pointed through the window at a truck stacked waist high with cases of whiskey and a sheet covered form stretched in between the boxes. Black flies swarmed thick over the sheet.

"We'll dump him," she said in a suddenly cold voice.

"He had no right to kill him," Jake said.

"What are you talking about?" She turned looking up into his eyes.

"Vince," Jake said. "He had no right to kill him, even if you two were..."

"Whoa! Whoa, back up! First of all, it's none of your business if I was sleeping with your dead friend there..."

"Not my friend," he said. "I don't have friends."

"And Vince did not kill him; he's never killed anyone. Maybe that's one of his problems." She looked past the window, toward the body.

Jake again saw the lonely woman with the unfortunate husband, the woman he had fallen for on a night so far removed from reality that he marveled at the situation.

"Well, how did he die?"

"Didn't you see his plane?" she asked. "He drove it right into those trees out there." She pointed vaguely at a wall. "Did that all by himself."

"That doesn't make sense," Jake said. "He was no amateur. In fact, he was a damn good pilot. He would never just fly into a bunch of trees."

"Hell, he was so drunk that night, he didn't even know he was in an airplane! We were all drunk. Yes, as usual, and what's it to you?"

Jake shrugged. None of my business.

"So anyhow, he and Vince disappear out back; he say's he's going to show him some fancy flying. I didn't pay much attention; figured the two of them would go up. I sort of fell asleep, then I hear this roar as he goes sailing over the house. Woke me up; it was getting dark then, and I hear him turn and come back around the barn and back toward the house, so I start to go outside to wave at them, and—crunch!

"It sounded like someone had just smashed a truck into the side of the barn. I ran outside, and there was Vince already running across the field. I had no shoes on, but I took off after him. When we got there it was almost dark, and Marvin showed up with a lantern. Vince was already tearing through the wreckage, just a pile of twisted wires and sticks. I cut my

176

feet. Gasoline was dripping all over everything, but nothing caught fire."

She hugged her arms tightly to herself and stared out the window at Sterns' body on the truck amid the whiskey. The flies were thickest near the head.

"He wasn't even in the wreck," she said and laughed softly. "It took us twenty minutes to find him. He was still alive, but unconscious, which I guess was good. His arms were turned back behind his head, bones stuck out here and there; one leg was ripped so badly it just kept bleeding. Vince wrapped a tourniquet around it, that stopped the bleeding, but it was obvious he would lose the leg.

"Of course, none of that really mattered as he died the next morning. Most of his head was bashed in on the left side." Her hand ran absently along the left side of her skull. She turned away from the window to face Jake.

"You know," she said. "Vince really liked him. Oh sure, he knew we slept together, but that's my business. Vince understands."

"I don't," Jake said.

"You don't have to; it's none of your concern. Oh, don't look hurt, loverboy. There's a great deal in this world that's none of your business." She patted him lightly on the cheek, her fingertips lingering along the the base of his thin jaw. She kissed him, and he slipped a hand around her waist.

Funny place, he thought, and outside, the truck with Sterns' body started and drove away.

Jake declined Vincent's offer to spend the night, even though it meant finding Moyer's airfield after dark. Both runways were lined with smudge pots that could be lighted when he flew over. The fact that there was almost a complete lack of lighted ground objects the last fifty miles of the journey paled in comparison to the possibilities that loomed if he stayed another night with the demented trio. He could always spend the night under the wing somewhere.

With several thousand dollars of Doc's money in his pocket, Jake lifted the Standard's nose and climbed past the silo with the cross and over the house then turned northwest. He took a long look at the de Haviland biplane squashed in the woods as he flew over. Its wings pointed in helpless pleading not to be left behind, or at least that was how Jake saw it.

The D.H.4 was never removed from the crash site in the grove. Winter storms shook it from its perch in the branches, and time pressed it into the weeds and earth beneath. Decades later the rotted traces of the biplane were bulldozed into a pile of stumps and set alight as the grove was cleared to make way for a highway.

Jake climbed the Standard through a broken layer of stratus clouds about twenty miles northwest of where Sterns had died. When he centered himself over a long stretch of cloud he reached into a bag between his feet and took out Sterns' helmet, goggles, watch, cigarettes and even the bright orange tie.

He took one cigarette, for later, slipped it into a pocket and threw the other items—the last of Sterns—overboard. They dropped quickly and became swallowed in the cloud layer below. He then reached into his shirt pocket where the money was and took out two hundred dollar bills. He held them over the rim flapping in the slipstream and thought for a moment.

Then with a snicker he stuffed the money back into his pocket. "Hell, Sterns, my very dead friend, you don't get a day's wages for crashing your airplane!"

Moyer heard the engine roar over the cabin and turn back and circle. He had been half asleep, but pulled on his pants and boots and ran to the runway with a handful of wooden matches to light the round smudge pots. There was only a sliver of moon, but Jake knew he was over Moyer's by its location from the lake and the rail line visible in the pale night light.

Circling overhead, he saw the flicker of the first pot being lit, then thirty seconds later another sparked, then another, and finally the entire runway was outlined as Moyer ran from lamp to lamp lighting the oil.

Jake had no lights on the biplane, but he lined himself up between the flames, stayed a little high as insurance against the unseen trees and telegraph wire, and glided to a hard landing in the darkness. Again, he smacked the bow beneath the right wing knowing this time it had to have snapped.

Moyer waved a railroad lantern in the darkness guiding him to a spot beside another biplane. Jake noticed it was another Standard LS-5, the company's third. There was no ALMIN-CO painted on its side.

Things must be going well for Doc, he thought, or poorly for Standard.

The motor shuddered and stopped, clicking in the warm night air. Moyer held the lantern in Jake's face.

"Honestly didn't think I'd be seeing you for a while," he said, his breath stale from cigarettes and sleep.

"Or at all, eh?"

"Thought had crossed my mind, eh!" Moyer grinned in the reflected glow of the lantern. "How's Sterns?"

"Dead," Jake said. His tone made it clear he wanted to leave it at that. "Whose plane is that?" He pointed with his thumb. Moyer turned the light on the other LS-5 as if he could only answer by looking at it.

"Doc's latest—brand new from the factory. I hear they're practically giving the things away. Maiden Voyage. Some kid from New Jersey flew it up—I wasn't expecting him. Big kid. Made a terrible landing, but I suspect he'll learn soon enough." He turned the beam on Jake who blocked it with his hands.

"You must be beat," Moyer said. "We finished the guest house while you were gone. The light from the cabin door across the tracks illuminated a crude wooden structure beside the water tank. Moyer and the Indians had been nailing it together over the previous week.

"There's blankets and beds inside, you know where the latrine is. Be quiet when you go in; that kid's in there, asleep."

"Thanks, Moyer," Jake said and lowered himself from the cockpit stretching his legs after he dropped to the ground.

"Don't have to thank me! Just give me three bucks a night for the bed; I told you this was a business enterprise."

"Fair enough," Jake said and took his bag and headed for the cabin while Moyer walked along the runway snuffing out the lamps.

"There's a lantern by the door, on the outside," Moyer called after him.

Jake found the lantern, lit it and went inside. The room was large, beds arranged in a horseshoe along the walls. In the center of the room stood a table with a lumpy pile of clothes on top. The air smelled of fresh cut pine. Jake moved the lantern around until he spotted the other pilot half covered with blankets, back toward him and snoring.

Jake chose a bunk, and after dropping all his belongings on the floor in a heap, set his head on the thin mattress and vanished into sleep.

In the middle of a confusing dream about airplanes with wings that melted like wax whenever he pushed the joystick, something shook Jake's unconscious world, and he awoke to the cabin bright with morning sunlight. He raised himself on elbows and looked around. The clothes from the table were gone, but a few items were scattered across the other pilot's empty bed.

"Aaiiee! Bluubalbla! Blah!" Animal cries of anguish rose from outside the window. Jake leapt from bed, stuck his head out the window and saw a large figure with a hairy back standing naked under a shower faucet and yelling at the top of his voice as cold water from the tank sloshed over him.

Jake sat back on the edge of his bed and rubbed his neck. He found an empty pack of cigarettes in his jacket, crumpled it and threw it into the corner. The smell of frying bacon drifted through the open window and twisted at his empty stomach. He wanted a shower, but the smell of breakfast told him it could wait.

He reached for his trousers off the floor when the door opened and a half-dressed man filled the frame. His wet hair hung straight across his forehead.

"Henry!" Jake exclaimed.

"Good mornin', Jake. You slept late. Breakfast is cookin' down at Moyer's. One of the Indian women cooks it. You pay her a dollar, and she gives you all the eggs, flapjacks, bacon and biscuits you can eat. Interested?" He rubbed his head with an old shirt, a puddle of water collected under his feet.

"Henry, I say again! What are you doing here?" Jake thought a second and added. "Did you fly that other Standard up?"

Henry grinned and nodded while he dried his hair.

"When did you get a license?" Jake asked.

"Never did; don't need one. Hell, what we're doing ain't exactly legal, so I'm not worried about offending anybody by not having a piece of paper in my wallet that says I can fly the plane I just landed in." He sat on his bed and dried his feet while Jake considered what he had just heard and came to the conclusion that it made sense.

"Does Doc know you're here?"

"Of course, I talked him into it. I said I'd check the operation for him first hand. Just an excuse to get away and go flyin'"

"Somehow, I never thought he'd let you out of his sight. Who'd he find to replace you?"

"Kid named Sam Nyberg, runs one of the outlets for Doc. Works out of a filling station. Smart kid, not real big or anything, but for what Doc needs he don't really have to be. Anyhow, this is just sort of a vacation for me. I've been buggin' him for weeks to let me fly, so when we got the new plane, I made the run.

"To tell the truth, I think I was such a pain in the ass he wanted me out of his hair."

"Have any trouble?" Jake asked.

"I got lost real bad right off the bat and ended up flying in a couple of circles until I figured out what to do. I landed on a road and asked a guy where I was, and he told me some place in Iowa, then showed me on the map. So I took off again and found my way pretty good."

Jake pulled on his boots, laced them, then slipped a dirty shirt over his head and headed for the door.

"You hungry?" he asked.

181

"You bet," Henry answered and finished dressing. "You flyin' back to Lincoln today?"

"Yea, I don't plan to hang around here all day."

"Mind if I fly along with you?"

Jake thought briefly of Sterns and the first time they flew the route. He wanted to forget him. The beginning of the summer seemed a long time ago. "I don't mind," he said finally. "Might be fun."

"Thanks, Jake. Thanks." Henry stood. "Let's get breakfast." They started to leave when Jake saw Henry reach under the blanket on his bunk and take a pistol in its holster and clip it to his belt. He moved casually as if he were donning a hat, then followed Jake outside.

"Hey, just one thing," Jake said stopping Henry before they reached the kitchen. "Moyer said you told him you were from New Jersey. Are you? You sure don't sound it."

"No, I'm from Arkansas, but I tell everybody I'm from New Jersey, 'cause if you say, Arkansas, first of all half the folks don't know where it is—they think it's somewhere next to Alabama—then they just think you're some kind of backwoods boob, so I tell 'em I'm from New Jersey. Sounds more sophisticated."

"People buy it?"

"Sure, mostly. Hell, most people that know where Arkansas is don't know beans about New Jersey."

"Sounds reasonable," Jake said.

Breakfast was everything Henry had promised. The table was covered with potatoes, eggs, biscuits, and meat still bubbling from the pan. A pot of coffee perked on the stove, giving off the aroma of morning. A young, plump Indian woman near the stove grinned as they entered.

While they ate, the morning train rolled into the stop, and Moyer marched along the cars waving and calling while the train took on water and disgorged liquor and gasoline.

"Fuel for the modern world!" he would bellow now and then as Indians hustled the barrels away.

Henry easily ate twice as much as Jake. As the train pulled out amid hisses and whistles from the steam, Henry mopped the last bit of yolk from his plate and signaled to the cook that he was full.

She smiled looking confused, uncertain what to do with the leftovers. Henry slapped a dollar bill on the table, nodded to her and left. Jake followed. Immediately, two small children shot through the door and devoured the remnants.

"You already fueled?" Jake asked.

"Did it last night," Henry said. They walked to the biplanes, Henry carefully mimicking Jake's preflight inspection routine.

Jake ran his hand over the many fabric patches and shook the cracked and patched bow beneath the right wing. It rattled loosely in his hand, so he wound an extra strand of wire around it and slopped glue over the whole thing.

"Good enough," he muttered wiping his hands in a rag.

Henry walked around his plane, but instead of patches and mud found taut new fabric and shiny paint. Jake's plane was streaked with mud, oil, exhaust and grass stains. Henry's showed a blush of stain behind the exhaust stacks and a thin wisp of mud along each side of the fuselage. He envied Jake the ragged look of the seasoned pilot—the true professional.

The caps were replaced on the whiskey and fuel tanks, and Jake and Henry buttoned their jackets and slid their helmets on. Henry kept glancing at Jake to make sure he did everything right. Moyer appeared with receipts for fuel, booze, bed and breakfast. He thoroughly enjoyed his role in the air transport business.

The pilots slid into their seats and strapped on their belts. Henry dropped his once and had to fish around for it under the seat while the lineboy waited with obvious impatience.

"Go ahead," Henry called. "Pull it through."

The two propellers snapped through one blade at a time. Jake called something and looked over at Henry. With a big grin, Henry pointed, thumbs up, and the two Hispano Suizas fired, first Jake's, then a moment later, Henry's.

Jake led the way from the parking area toward the runway's end. A cloud of brown dust rolled across the open ground and disappeared into the trees as the two biplanes taxied, slowly S-turning their way across the field.

They waited several minutes at the end of the runway for water and oil to warm, then Jake took the runway followed by Henry. The gentle idle of the motors changed to a clacking roar, and the two machines lifted into the air in a loose formation.

They floated over the cabins, their engines vibrating the Canadian woods as they slowly cleared the trees and turned south. The air was smooth a thousand feet above the ground, and they leveled, Jake settling into a comfortable position for the hours ahead, Henry tense behind the controls intent upon keeping a good formation. Jake signaled him to back off. He did, embarrassed.

In the distance, a third plane appeared headed in the opposite direction at the same altitude returning from the U.S. Drawing closer, Jake could see it was Billings in his Travel Air. Slowly, he rocked his wings, and the Travel Air rocked back, passing off to his left. Henry joined in the courtesy with a timid rock of his wings and returned to straight and level.

Arthur sat at a small desk in the hotel room with the flight status boards behind him. Jake lounged in an over-stuffed chair in the corner smoking the stub of a cigarette. A stack of money was beside the ledger on the desk. Art counted, made a notation, counted some more, made another note, then he removed several bills from the pile and made still another entry. This apparently satisfactory, he recounted the entire mass and tapped the figures into an adding machine at his side. He gave the handle a sharp pull and tore off a strip of paper and sighed.

"Jake, Jake, Jake." He shook his head. "You are just going to have to start getting receipts for your fuel. I can't go on these vague estimates of yours."

"Oh, forget about the receipts; you got most of them, at least the ones from Moyer. We're only talking about plus or minus ten bucks," Jake said with listless indifference. He crushed his cigarette in a glass ashtray on an end table.

"That's not how it works, Jake."

"I know. I know; ten dollars here, ten there and pretty soon your books are off twenty bucks, and the whole empire collapses."

"Listen, I'm responsible for this!" Art thumped the ledger. "It may not mean much to you, but it's my job, and you aren't about to screw it all up." Immediately, he felt sorry and took a softer tone. "Just get the receipts for me, huh? Please?" His sad pale face gazed at Jake.

"Sure, Art. I'm sorry. Now, can I have my dough?"

Art counted out 200 dollars and pushed it across the desk. Jake stood and took it. "That's for the run to Moyer's." He then counted out another hundred and slid it across. Jake took it. "Doc says to give you that for Adair's." He glanced up at Jake. "I'm sorry about Sterns."

"Why tell me? I barely knew the guy. Anyhow, I can't say much about someone who gets himself killed. Pretty stupid thing to do." The words rang false, but it was Jake's way of handling Sterns' death. He pushed it away, deep inside where it would stay out of his way. At least that was what he hoped.

The door swung open, and Doc strode in wrapped in a swirl of scarlet silk bathrobe and blue smoke from the cigarette at the end of its holder. He slapped Jake on the back with a politician's enthusiasm and moved straight to the boards.

"Hello, Jake! How was your trip? Arthur pay you all right? Arthur, has Mr. Hollow been paid?" He picked a felt eraser off the board's ledge.

"I've been paid," Jake said.

"Excellent!" Doc replied, and he wiped the name, STERNS, from the board in the reserve column, and chalked a new name on the active list: HENRY.

Chapter
XIV

Henry was on a run to Canada and would be gone for at least two days. He was scheduled to turn around and fly straight back, then make two side trips to Adair's where demand was beginning to compete with Lincoln. His schedule would tie him up for at least a week.

The autumn weather had been pleasant so far—cold but clear and dry. Jake was crossed off the active list for a few days, at his own request.

"You can't make money sitting on the ground," Doc had chided when he asked.

"It's personal," Jake had said.

"Well, give her a squeeze for me."

Henry's plane, the J-1 he bought from the school, would be free, so Jake called Susan, or actually just left a message for her at the train depot in Luther.

"Just tell her I'll be in tomorrow," he yelled into the mouthpiece. "No, she doesn't need to pick me up—I'm flying in...yes, flying...uh huh, this is Jake...Fine, thank-you! Will you give her the message? I know, but everything looks good this time; I should be there tomorrow...Huh? Jake, Jake Hollow!"

The room was crowded and noisy when he hung up the phone and stepped out of the phone booth at Clara's. One of the newer pilots, still on reserve status, sat at a table waving his arms and slobbering an explanation of some impossible aerial feat to an overweight waitress looking bored.

"Really, that was me! I ain't kiddin' ya," he slurred. "Jake, Jake! You tell her! You saw my landing, didn' you?"

"No," Jake said and walked past him into the evening and down the street toward his room at the Cornhusker. He ached to see Susan, and toyed with the idea of setting out right then in the dark, but without a car it was a long walk to the airfield. He would wait for morning and catch a lift with the pilots scheduled for dawn. Besides, he realized, the whiskey from Clara's was slowly pressing his eyelids into his cheeks.

The night clerk barely stirred when he asked for his key and disappeared upstairs. Once in the room he flicked the switch on the wall illuminating the tiny room in pale yellow from a single bare bulb in the ceiling.

"I wonder what idiot they got flying my plane tomorrow?" he asked the empty room and turned out the light. The window over his bed was cracked open letting in a cool breath, and he stared at the ceiling and lit a cigarette. Holding it over his head he stared at the orange glow.

"I hope whoever it is doesn't go and break something on it; or worse, go and fix something." Then he had to laugh at himself. One moment he longed for Susan, and the next he pined over being unable to fly a run. "Not my damn airplane," he said to the room. "Not much that is mine, anymore," he added and fell asleep. The cigarette dropped from his fingers and died a cold stump on the floor beside his clothes.

Jake poured hot oil from the can into the OX-5 motor in Henry's plane. He could only hold the can for a few seconds in each hand before he had to to set it on the cowling. The mornings now were almost freezing, and one by one the pilots left the warmth of the stove inside the hangar to carry the heated engine oil to their planes.

A Jenny started, and a Travel Air 2000 taxied behind him toward the runway. The factory had all but ceased building airplanes, so Doc had virtual control of the field. Jake listened to the Hisso motor on his regular plane, the ALMINCO LS-5, warm up somewhere off in the darkness. There was no moon, and the sky to the east seemed reluctant to give up night and progress beyond its anemic shade of pinkish gray.

FENTON, that was the name Jake had seen on the status board scratched in where his normally was. He felt left out, the kid who couldn't play with the others. The Hisso motor gave a rasping blare, and he saw the biplane's outline move down the runway past the lights and into the sky. He caught a brief glimpse of the red flash from the exhaust stack. "Lousy take-off," Jake mumbled. "Looks like they put some rank amateur in my seat."

Doc's Lincoln pulled under the yellow cone of light outside the hangar door. Sam Nyberg, his new driver, hopped out and ran around to open the door. Mr. Howard appeared, wringing his hands and bowing, to lead Doc inside. When he thought no one was watching, Sam rehearsed grabbing for a pistol he had taken to carrying beneath his long overcoat.

Jake was slow about readying the airplane, and the sun was peeking above the horizon when he lifted the needle valve in the carburetor to prime the engine. 'Tickling the carb', they had called it in the Army, and Jake repeated the phrase aloud as he did so.

Gasoline dripped onto the yellow grass beneath the engine, and he pulled the propeller through feeling for the best stroke. Gas still dripped in a steady flow onto the dry grass. His thoughts drifted 200 miles away to Susan, and how she would react when at last he fulfilled a promise and actually arrived.

He walked back to the cockpit, flipped the mag switch on and turned. "Now don't forget to untie the damn thing, Jake, or you're going to make a silly ass fool of yourself." He had seen so many pilots start their engines, hop in the cockpit, then despite their best efforts to blast their way from the ropes with power they had to swallow pride and climb out again. Not him!

The last rope was in the grass, and he glanced into the cockpit again. "Yep, switch is still on." He walked around to the prop.

Taking it in both hands, he lifted a leg and pulled. It snapped through and quit. He took it again, repeated the swing and nothing. A third time, then a fourth, a fifth, and when he had pulled it through a dozen ties without the slightest hint of a spark, Jake, heated with anger, swiped off his helmet and tossed it on the wing. "Son-of-a-whore."

He walked around to the cockpit. "Yea, switch is on, gas on." He checked the valve. "Throttle..." He tapped the throttle forward slightly. "That's a bit wide, but it should fire." He started for the prop, when he stopped, reached for the rope near the tail and looped it around the tail skid and tied it. "Not biting me in the ass this morning," he said.

Jake took the propeller, flexed his fingers over the blade, adjusted his weight, raised the leg and...it fired.

He knew heads turned when the little engine barked and raced away. Jake was around the wing tip and leaning into the cockpit to pull back the throttle. The airplane tugged at the tail rope. He hoped whatever heads had turned his way had seen how casually he had moved, how completely in control.

With the power back, and the OX-5 idling gently, Jake leaned against the tail, taking the pressure off the rope to untie it. Then over the cockpit rim and finally seated, he buckled his belt and reached to adjust his helmet...Oh!

His eyes focused on the helmet on the wing, so far away, so obviously wrong. Jake reached for his seat belt, hoping to get out of the plane, retrieve the helmet and get back in before anyone... The voice pierced the idle of the motor, and Jake saw the figure slide past him and take the helmet from the wing.

"I swear," the lineboy snapped, "you new pilots are the biggest collection of weak sticks I've ever seen." He handed the helmet to Jake.

"Thank-you," Jake said and tried to smile. There was no need to defend himself.

"I swear you'd all forget your heads if they...hell, maybe they ain't attached!" He walked away leaving a sheepish pilot to pull his goggles over his face.

"That woman's on your brain, Jake Hollow," Jake muttered. "She's messing with your thoughts. Watch yourself."

A cold flight, it gave Jake time to let his mind run dry. This was an easy flight, a short hop across safe terrain without the fear of breaking the law, getting caught and losing what little he did have, his freedom. If the engine did quit, he'd pick a field, land, hitch a ride into town, and after whatever had failed was put to rights there'd be rides to give and a meal to share. "Jake, they'll never take the gypsy out of you, will they?"

The OX-5 never missed a beat, and he arrived over Susan's and dropped low across the house to announce himself. She ran through the back door, eyes skyward and both hands waving. If you can't make a flashy entrance, he thought, there's no sense even going. He dropped a wing and dove again at the house, his wheels almost brushing the shingles, but he knew just how much room he had.

Gusts near the ground were strong and unpredictable, and he had to work for the landing. Over the fence, and he chopped the power. Still too fast, he touched, bounced and floated into the sky. Nothing she would notice. "Damn," he muttered and, head over the side, he tucked the stick back into his lap, bleeding speed, feeling for the ground. The biplane settled like an exhausted goose.

He pushed the thought away, before it went too far, "That was an easy one, Jake. You blew one of the easy ones. Can't afford to do that in your business."

Three men working on the roof of the new barn stopped to watch, expecting to see their first airplane wreck. Disappointed, they returned to work, each one competing with the other to describe their version of the landing automatically adopting the pilot shorthand of hand waving when words failed.

Susan ran to Jake before he could stand, hooked her toe in the step below the cockpit, reached over the rim and kissed him hard and wet on the mouth. Jake slipped a gloved hand around her neck to keep her from falling back. It also steadied his trembling fingers.

"Mmmm," she moaned grinding her face on his and laughed when their lips separated.

"Well!" he said. "I should have welcomes like this at all the places I land."

"What? They don't know how to treat pilots in Canada?"

"Not in Canada."

"Elsewhere, then?"

He laughed rolling his eyes.

She kissed him again, quicker with less mush and dropped off the step while he undid his seat belt and pushed his goggles onto his forehead. Red dents rimmed his eyes. Someone catcalled from the barn roof teasing them, but the comment was lost in the wind.

"They're making progress on the barn, I see," Jake said.

"Yes. They've finished the other side of the roof, and should finish the whole thing by next week. Daddy keeps saying what a wonderful man you are for lending him the money."

"Merely a business loan," he said, but she knew it was more than that. She had seen Jake give her father money on other occasions. Whether as loans, or start-up capital for ideas they had discussed about getting him into the whiskey trade she never knew, nor would she ask. Other than the barn, the money seemed to vanish.

"We put your airplane inside yesterday. When can we start working on it?"

"Maybe this spring," he said climbing down. "Depends on the weather."

"What's wrong with during the winter? You won't be flying then, will you?"

"No, I doubt it, but winter's too cold to work on airplanes. You have to heat everything. The dope's about as useful as cold tar, and your hands get so numb you can't hold onto the tools. No, warm weather's a must."

"We could put a stove in there."

"There's no rush. Spring will be best. Besides, I don't like to hang around up north during the winter, I can't stand the cold."

"I gathered," she said feeling uneasy. A silence dropped between them.

"Did you miss me?" she finally asked knowing he would be forced to answer in only one way.

"Did I?" He took her into his arms pressing her face into his leather coat and resting his chin on her head while he caressed her back. They both felt the emptiness of his response.

"What will you do then?" she asked without looking up.

"Do?"

"This winter—when it gets cold." She pushed back, her brown eyes sad and anxious.

"Oh, I usually make my way down to Texas, then along the Gulf coast across Mississippi and Alabama, then eventually down into Florida, wherever the weather's best. Then, as winter unwinds, I retrace my route north. I just fly in one direction as long as I feel warm air; when it turns cold, I turn."

"Jake, you know I don't really like what you're doing, I mean with this..."

"Bootlegging?"

She nodded. "I don't understand why you're all not in prison. It's dangerous..."

"Ohhh..."

"And will they simply let you pack up and take off like that any time you want to? Will they let you go?"

Jake was still caressing her back, but it was purely mechanical. "They've got nothing to say about it. I go wherever I want!" He thought that sounded corny, then realized he had read the line in a cheap western novel once— something about cattle rustlers and a rancher's daughter.

"Besides," he added. "I doubt if they plan to run operations right through the winter. Hell, they'd lose more pilots than the Post Office has, and they've lost a bunch!" He suddenly remembered Doc wiping Sterns' name from the status boards. "Winter flying would never be profitable."

A cold wind blew off the recently harvested field behind them. Traces of corn stalks scratched against the plane's skin, and Susan drew her sweater tight around her shoulders.

"I worry about you," she said. He started to protest. "Not about your flying; I assume you're left to fate on that one, but I worry about you, you, Jake."

He stopped caressing her, fighting a growing urge to say, "excuse me," and climb back into the biplane and leave.

"You're so much like my father," she said.

"What's wrong with that? I like him."

"Nothing, and he likes you, very much, but you two are..." She pushed free of his grasp and walked along the the fuselage with her back to him looking up at the barn. "You're both

193

lovable, unrealistic men bumbling through the real world without grasping the truths of the demands life can place on you!" She turned.

"Did you rehearse that?"

"I'm serious!"

"I'm sorry."

"Daddy works very hard, but at what? Beer-making!"

"It's in demand."

"But he virtually ignores the business end of things. Okay, he wants to be a bootlegger...that's fine, but he acts as though the customers are an imposition rather than the reason for making the stuff in the first place!"

"He's a craftsman, not a salesman," Jake said lamely.

"If it wasn't for me, the stuff would simply stockpile in the cellar, or he'd drink it all. I run the business end."

"Bootlegger's daughter, eh?"

Susan frowned turning away again. Jake reached for her.

"And you talk about casually giving notice to those smugglers in Lincoln as if you were working for General Motors! 'Oh, I've taken a better position smuggling for Ford!'" she said sarcastically. "'But first, I shall winter in Florida!' Jake you're not being realistic."

"So?"

"So, how do you expect to make this winter sojourn?"

"Where do you learn these words?"

"You have no airplane, and you've been spending money like mad on this...damn barn." He saw tears well in her eyes.

"Oh, Sue..."

She wiped quickly at her nose desperately fighting back the tears. Damn, she thought. Why do I always have to cry when I get upset. Why don't men cry?

"We can get another airplane," he said softly. "This one, for instance."

"We?" She turned.

"Of course, I want you to come with me," he said.

She was thunderstruck. We! It rang like a proposal, but, and she missed this, without any of the commitment. It was a ridiculous comment, part of a scheme completely without reason, but she felt the pragmatic woman inside give way to the farm girl stuck on this man from the skies, seven years her senior.

She only smiled without saying anything, knowing if she did it would be irrational, something stupid like, "Yes, of course, darling, dear, sweetness, love, honey," or any of the countless romantic bromides that squashed common sense and led to disaster.

"Until then," Jake continued, "I'll be flying every day; sock away as much money as I can before winter. Then we can go. You'll love it.

"Think of it, Sue." He gently clasped her face in his hands tilting her face toward his. "The two of us, and no one to bother us as we fly wherever we want, with a pocket full of nickels. If we want to go somewhere, we will. Dallas, Houston, and how 'bout Miami; pal around with the millionaires on the beach, and if we feel like it, we pop over to Havana to play the roulette wheel.

"No snow! No stoves to fire, no wind howling at five a.m. telling you to go milk the cow in a cold barn."

She unbuttoned his long leather flying coat and wiggled her hands slowly around his waist pulling him toward her. With her face buried in his chest, she cried, the tears soaking through his shirt.

What a confusing lot, women are, Jake thought.

"Teach me to fly!" she said suddenly pushing herself away from him and wiping at the tears with her sleeve. "I've been studying. I did a good job last time you were here, didn't I? You said so." She spoke with a bubbly exuberance recovering her composure.

"Yes...Oh, yes, you did," Jake lied and recalled the previous lessons when they had bounced across the pasture, wingtips slapping the tall grass, and Susan working the controls like she was competing in a butter churning race.

"Well then, let's go! Is there enough gas?" she asked.

"Not much, but enough."

The biplane ran along the pasture into the wind, Susan at the controls in the rear cockpit, Jake poised ready to correct in the front.

"Rudder, Susan. Rudder! Rudder!" Jake yelled as the airplane drove toward a stand of poplars, then veered 45 degrees in the other direction directly at the barn. Luckily, the Standard required little runway to become airborne, and it sailed over the building, the wheels merely nicking the new roof. Not nearly hard enough, Jake thought, to cause the carpenters to vault, as they had, straight off the edge.

"It's going to be a chore picking all those nails out of the grass," he said to himself before Susan yanked the stick violently to one side without benefit of coordinating rudder and he slammed his shoulder against the rim.

"Doing fine," he called with a smile and a wave. Her face was a serious mask behind the rear windscreen. "Just relax..."

Whamm! He hit the other side of the cockpit, and the nose pitched up. She rammed the throttle forward.

"A little smoother with the..." The stick came back with an awful jerk, the nose shuddered, and the Standard stalled and spun to the left. "Well, that's one way to lose altitude."

The earth spun once, twice and started into its third revolution. "Okay, Susan. Stick forward. Stick forward!" He pressed the stick, but her grip was impressive holding it back.

"Push the stick forward! Forward!" The ground drew closer and whipped under the nose. "Gimmee the God Damn..." Suddenly, she released the back pressure and corrected the revolution with rudder, and slowly the Standard regained airspeed and swooped low across the brown fields and trees.

Jake turned, and she still grimaced with that intense glare. "Doing fine!"

The carpenters saw her coming and scrambled down the ladder, again, taking up seats safely on the ground where they could watch her return. The biplane skidded across the sky and dropped evenly toward the field. All looked well, when about 20 feet above the ground the nose suddenly rose unusually high, then the whole plane seemed to shake, and the nose dropped. The carpenters were on their feet. The Standard was headed far to the right side of the strip.

Susan's father had taken a position on the back porch upon hearing her go over. "The trees, Susan! The trees!" he shouted and ran down the steps.

"The trees! Watch the trees!" the carpenters joined in without moving.

"The stick, Susan! Gimmee the stick!" Jake shouted marveling at her strength.

Susan recognized her error and mashed rudder and stick realigning herself with the field. The biplane slipped over the grass and rolled gently to a stop beside the barn opposite to where the carpenters now hid, their faces peeking cautiously around a corner.

"How was that?" she asked.

"You're improving, Susan. You're improving."

Jake and Susan walked toward the house, hands laced together at first then with arms wrapped around each other's waists, when she stopped and pulled away to face him.

"You haven't even looked inside the hangar yet!"

"So it's officially a hangar now is it?" he asked.

She pulled him toward it. The carpenters had disappeared leaving wood scraps and a few sawhorses scattered around. Jake gazed up at the plain wall and the two small windows above where the huge door would be.

"It has an airplane in it, so it might as well be called a hangar. You act as though you have no interest in it at all. I declare," she said trying a phrase she had read in a magazine.

"You declare?"

"You haven't so much as peeked at your airplane since you crashed it here four months ago."

"Has it been that long?" he asked.

"You've gone on about fixing it, and you've poured a ton of money into building this hangar, but you never come near it. Don't you want to fix it?"

"I'd rather fly it. There's not much thrill in working on the damn things."

She led him inside, around a pile of lumber and a keg of nails. His wrecked Standard was parked square in the middle, surrounded by debris and coated with dust. A slanting shaft of light shone through the incomplete roof. Two sawhorses and a pyramid of cement blocks held the biplane off the dirt floor.

"I thought you'd have it all patched up by now," he said letting go of her hand. "Isn't there a chapter in one of those books you're reading about repairs?"

He walked around the wingtip bending to check the gear. "Not much left of them is there?" Jagged stubs of gear legs hung beneath the fuselage. The propeller was turned parallel to the ground, and a dusty puddle of oil had formed under the OX-5.

"I've covered a bunch of miles in this thing," he said and dislodged a hard chunk of mud from the wing, crumbled it, letting the dirt fall between his fingers. "Been all over the midwest."

He no longer spoke to Susan but more toward the plane itself, his voice soft as if a harsher tone would disturb it somehow.

"I'd charge five bucks a ride in most places; ten at the big fairs. I'd talk to some other pilots that'd been flying since the war, and they'd tell me they used to charge 15, even 20 bucks a ride! That was before everybody and his cousin had an airplane. I missed out on that.

"But even so, you should have seen the people line up to fly with me. They'd wait all afternoon for the chance to go up with the great air ace from the war, the one who shot down twelve Germans! Did I ever tell you I shot down twelve Germans?"

"You did?" she asked. "I didn't know you flew in the war."

"I didn't. Never shot down anything. It was just a ploy, a gimmick to get the customers away from their money."

He laughed. "I could never understand why people felt safer flying with someone who claimed to have murdered Germans, than with someone claiming merely to be a good pilot.

"'Ride with the pretty good pilot!'" he called like a carnival hawker. "Wouldn't sell a single ticket. But! 'Fly with the great Hun Killer', and they'd be climbing over each other to get in. I guess people are still afraid there's Germans hiding behind the clouds waiting to pounce. Funny how most people have no idea what flying's about. I wonder if they ever will?"

"Did you like barnstorming?"

"Didn't like calling it that."

"In the book I got at the library, the author says all barnstormers are nothing but a collection of renegades, not a real part of the air transport world. He says the future's in highly regulated airlines with great eight-engine transports run by the government, and everybody on a flight plan and on regular routes, like railroads in the sky."

Jake scowled at her.

"The author says," she continued, "the days of the outlaw gypsy pilot are numbered. Soon, the government will send out inspectors, like policemen, to sweep them out."

There was a long pause while Jake ignored her and quietly ran his hand over his old plane.

Susan asked, "what's a flight plan?"

"I haven't the faintest notion," he replied.

"Don't you have one?"

"Shocking isn't it? I've been flying around all these years from Canada to Mexico; from the Rockies to Key West, and never once did I have a flight plan. It's a wonder I wasn't killed long ago!"

"But you did crash here," she said pointing at the Standard as proof. "Maybe a flight plan would have prevented it."

Jake rolled his eyes and shook his head. "There's no arguing with logic is there? Yes, I flew without a flight plan, and yes, I crashed. Therefore, the two must be related. You win! The government wins! Everyone wins! From now on we'll all add a couple of pounds of flight plan to our airplanes, and no one will ever run out of gas or fly into thunderstorms or crash in people's back yards, so they can fall in love with their daughters and have conversations in barns about government interference in things about which they know nothing!"

"Who knows nothing?" Susan asked, confused.

"You know nothing," Jake snapped. "The government knows nothing. Hell, I know nothing! Otherwise, I wouldn't be a pilot for crying out loud."

"What's bothering you?"

"Government," he muttered and ducked under the tail. "Never mind. Forget what I said. Forget flight plans." He popped his face over the elevator and smiled. "Friends?"

She shook her head. "Silly!"

"You know," he said. "I haven't had a chance to really go over this thing, but from what I can see, I think I can rebuild it. The wings are a little scraped up, but I think the spars are intact and most of the ribs." He took a knife out and cut a small hole through the fabric and pulled it back like a surgeon exploring a patient on the table.

"Can you do that?" Susan asked.

"Do what?"

"Cut away at the airplane—just like that? I mean, don't you have to be a mechanic, certified or something?"

"Maybe I should inspect that book you're reading." He continued to probe the wing. "Airplanes fly because nature lets them, not because a bunch of chuckleheads in Washington, D.C. stamp a pile of forms in triplicate authorizing it." He tore another strip of fabric, less to see what lay beneath, than to emphasize his right to do it.

"This is my airplane," he said, his jaw tight. "I paid for it; I fly it. I pay for the gas that goes in that tank, and I will damn well do any damn thing I want to it!" He tore ever enlarging strips until his hand snagged a splintered edge tearing a shallow cut across his thumb.

"Damn Government!" he swore and pressed the thumb into his shirt to check the bleeding.

"Are you all right?"

"Fine, thank-you," he said.

She crawled under the wing to him and took his hand. He yielded sheepishly like a child. "Let's see." She pressed the wound with a handkerchief.

"Hey, be careful!"

"Oh, hush!" she said curtly. "It's just a scratch; you'll survive. Is that the way you fix everything? Just rip into it until all that's left is a pile of junk?" She looked at the shredded wing. Her face pressed into a tight frown.

"Look, it's my airplane, and I..."

"And you can go to hell," she snapped. "I get so blinded by you sometimes, Jacob Hollow!"

He recoiled.

"You just pop in and out of here whenever you feel like it, sprinkle a few gifts around, kiss the ladies, and without so much as a, 'thank-you, miss', you fly off expecting everyone to hang motionless until you return!"

"Huh?"

"Don't play innocent. You're a selfish child sometimes. You're so...so...ohhh!" She turned her head, saw him cringe then felt like giggling.

Jake looked around slowly mouthing something but remained silent.

"Keep that handkerchief on until we get inside," she said and turned completely away. Her shoulders quaked with stifled laughter, seeing how uncomfortable he now was.

Jake gazed at the torn fabric and felt immensely foolish. "You're right," he said. "I'm sorry."

"It's your airplane," she said, back still to him, her voice tight, hiding more giggles.

"The fabric was no good under the wings, anyhow," he added. She remained silent. "It won't take long to repair. Maybe we could start on it tomorrow. We could get the fabric in town, and some glue and egg whites mixed together makes a passable dope."

No response.

"Do they sell muslin in town? How 'bout a stove? I think we could hook one up in the corner; maybe run the chimney out the wall, there."

She moved away, ducking under the wing toward the door. He chased after her on all fours bumping his head into the wing's trailing edge.

"Ooo, son-of-a..." He crawled out and stood behind her. "Maybe you could help me remove the old fabric, then we could find a lumber yard. Is there one in town?" No answer. "We'll need a couple of pieces of ash for the gear legs."

Her shoulders began to quiver noticeably more. "Look, Susan, I'm sorry...it doesn't have to be ash, you know. Spruce might do, although I don't know if it'll hold up." He spoke faster; her shoulders shook more, and he heard distinct sobbing.

"Or...even mahogany, or hedge, or iron wood, Christ, even a tall cottonwood would be fine. That way I could throw a rope over a limb, tie a noose and hang myself! I said I was sorry!"

"Oh, Jake!" She spun around spitting laughter at his discomfort and threw her arms around his neck kissing him hard on the mouth, intimately not passionately.

"This isn't fair," he said. "I've been had! You let me babble on..." She kissed him again.

"Oh, hush up!" The kiss lingered a while longer. The beam of sunlight crept along the fuselage and moved slowly up her legs, over her arms and onto her face. She felt the warmth rub against her cheek. When she finally opened her eyes, she and Jake were framed in the beam, wrapped in each other's arms.

"Teach me to fly?" she asked softly, he face inches from his.

They walked entwined out of the barn into the cool autumn afternoon. The last of the carpenters stood on the back porch drinking beer with her father, their voices only a faint murmur on the wind. Jake began to explain a few points about airplanes and broke free of her grasp to animate his lecture.

"Don't tense up on the controls. You're fighting the airplane. There's no need for that. Remember, it wants to fly; you're just along to tell it where to go."

They reached the plane, and he plucked the helmet off the joystick. "Here," he said giving it to her. "Now, remember to look off to the sides, not forward, when you're ready to flare." His voice droned on and his hands danced before her while she buckled the helmet and listened intently, or at least appeared to listen.

God, he's cute, she thought watching him wave his hands and point at various parts of the machine. Miami, I'd like that...

"Any questions?" he asked. She shook her head. "Remember to watch that airspeed! Got that?"

Again she shook her head, but she only thought about the blue in his eyes.

"Oh, we'll need gas," he said.

"I'll get it!" She ran off with the helmet still on and returned soon with a five gallon can.

"It's all yours," he called over the idling motor and buckled himself in. She opened the throttle a crack and taxied out. Jake kept his hands in plain sight on the cockpit's rim, but his feet lightly on the rudder bar.

The sun was about a hand's width above the horizon when the Standard lifted from the pasture and climbed over the house, missing both the poplars and the barn.

Better, Jake thought. Susan held a straight track toward the sun and climbed. Jake occasionally tapped the rudder, and soon Susan was doing well on her own. He kept his hands clasped to the rim and yawned. He signaled her to level, and she did. Then with a couple of waves he indicated she should try some turns.

She banked and turned and generally flew the biplane displaying less of the ham-fisted performance she had earlier. Not completely smooth yet, but improving.

By the time they returned to land, the sun was a fat orange ball behind the house. The carpenters had all left, and Susan's father sat alone on the porch watching.

She turned onto final approach, and the engine popped at idle. Ahead the sky was awash with bright orange from the sunset blocking her view. Jake held his hand against it and waited to grab the stick. He glanced back at Susan, her face still intent as she swung her head from side to side feeling her way down to the ground. A strand of blonde hair whipped across her face, and she brushed at it quickly, finally stuffing it back under the helmet.

Jake slid his hands into the cockpit and lightly clasped the stick while the plane glided toward the field. Once, he started to correct her, but she caught the mistake herself, and the biplane bounced onto the grass, lifted as she relaxed back pressure too much and settled again rolling to a stop.

The sun had vanished by the time they stepped onto the back porch. Smoke rose lazily from the chimney. They had said nothing on their walk across the field after tying down the airplane. Susan held his hand and reviewed the landing over and over in her mind.

"Not a bad landing," Jake said on the porch. "Something smells good."

"Chicken," she said. "And it was a very good landing!" She squeezed his hand, again surprising him with her strength. "Damn good!"

"Oww! Okay, Owww, a good landing. Oww!"

"A very good landing?" She squeezed with both hands.

"Best damn landing I ever saw! Now, lego, you damn farmer!" She released him. "A damn lucky landing," he said grabbing for the back door.

"What?" She was right behind him as they crashed through the door into the kitchen where her father stood scraping chicken from a skillet.

"Close the door! You'll let all the flies out." Jake obliged. "What's all this?"

"I'm a better pilot than Jake, that's all!" Susan skipped to her father and pecked at his cheek. "We're hungry. Burn the chicken?"

"Only a little," he answered. "Get the plates, and Susie, you might want to fetch us some beers."

"Jake, open that cabinet there, yea, that one," he said. Jake took down an unopened whiskey bottle he had presented several weeks earlier. "There's some glasses up there; get three out and pour us a round, willya." Jake poured, and Susan disappeared into the cellar.

*****

"Had enough?" Susan's father asked.

"Yes, thank-you," Jake answered, but Mr. Bowers poured another glassful of whiskey for him. "Whoa, that's plenty!" he said, the whiskey overflowing onto the table. "I haven't finished the beer you gave me, yet."

"Oh, there's plenty where that came from. Drink up." The father was quite drunk, but then so was Jake. Susan had two whiskies and was into her second beer by the time the chicken was gone. She rarely drank, but the giddiness she had carried with her since the flight continued over supper, and she bowed to her father's influence.

"Daddy, you're drunk," she said with a giggle and rubbed her leg against Jake's under the table.

"As are you, my dear," he answered, his voice thick with liquor. "Jake, have I ever thanked you for helping with the barn?"

"Many times," Jake said. He propped his chin on his hand over the beer glass.

"No, I mean it, Jake," the father said. He put his hand on Jake's arm causing his head to bob. "You've been a hu..huge help, but remember! It's only a business loan. I fully intend to repay."

Jake waved away the suggestion like a foul odor. "That's unnec...unnecec...not required; after all, I have my airplane in there."

"And God only knows how long that will stay," Susan piped.

"You keep out of this," her father chided. "You stay as long as you like, Jake."

"Thank-you." Jake was back to resting his head on his chin and gazed foggily at Susan who returned the same stupid look. He half listened to her father ramble into a dissertation on friendship and common bonds. Then somehow, this led to an appraisal of the rising value of farmland, or was it falling value, and the prospect of the farming community falling into ruin when world corn prices collapsed along with pork and beef prices. Jake lifted his heavy eyebrows and muttered a monosyllabic grunt and continued to gaze at Susan and rub her foot under the table.

"Then there's the stock market, Jake; big money to be had there. That is if one has money to invest." Jake listened in confusion as Mr. Bowers drew graphs and scribbled figures on the back of an envelope, now and then tapping a figure with the point of the pencil to emphasize some theory completely lost on him.

Susan gave up and cleared the dishes leaving Jake to bear the full brunt of the lecture.

"I get the newspapers from Des Moines; they come on the train. Susan picks them up for me. Of course, they're a couple of days old sometimes by the time I get them, but I still read them, every page. I think I've a pretty healthy grasp of what's happening in this economy.

"The future's bright, Jake. Indicators are up. 1930 will be a banner year, and it's time to get on the bandwagon!"

"Bandwagon," Jake said weakly deciding how to politely announce his pressing need to urinate.

"And a person doesn't have to be rich to get on, either," Mr. Bowers continued. "Why, the brokers will even lend you the money to buy. Margin buying, they call it."

"More debt, I call it," Susan said from the sink.

"No! No, it's good sense. The market keeps going up, so you take someone else's money, buy stock, and when it grows you pay off the margin and keep the profit. Everyone does it."

"What if it goes down?" she asked.

"Can't! All indicators are up. Up!"

Jake smiled politely. "Well, if you will excuse me..."

"And I watch that aviation business, too," Mr. Bowers said. "Did you know, Jake, that there are something like a hundred and eighty airplane manufacturers in this country? What do you think of that, Jake?"

"Well," he said. "People say the future's in airlines, big eight-engine jobs run by the government, flying government routes, government pilots. Oh, big business, I hear." He glanced at Susan.

"Is that a fact?" the father asked. Susan looked back at Jake.

"That's what some people seem to think, anyhow," Jake said. "I figure it's more in small planes, the personal types."

"What for?"

"What for? You mean, what will people do with these planes?" Jake rubbed his face working blood back into the skin. "Well, they'll be used like automobiles. You know, to go places."

"Hmm," the father answered. "I somehow doubt it; sounds like a rich man's hobby. Not everyone has a landing field in their back yard, but everyone does have a road leading to their house, or will, anyhow. No, I think those folks that say the future is in airlines are right. Get people from city to city, then let them fend for themselves in autos."

"Oh," Jake muttered hearing his opinion squashed.

"Now, of course, what you do interests me," the father said.

"You mean falling asleep?" Jake said, the alcohol tugging his eyelids toward the floor. The only thing keeping him from falling face onto the table was the dull ache in his bladder.

"Bootlegging," Mr. Bowers said. Susan spun around. "Now, that's where the money is, good times or bad."

"It's dangerous," Susan said.

"Not really," Jake said. "All the flying's quite routine. Schedules aren't nearly as tight as say, flying the mail, and the pay's a hell of a lot better."

"Exactly!"

"Why? Are you thinking of taking up flying?"

"Jake!" Susan called. "Don't encourage him."

"What?"

"Susie, leave him alone. He makes good money; he's a smart man."

"Oh, you two, I give up. I'm going to bed. Good night." She leaned over and kissed her father. "Don't bore people with your schemes, and you should go to bed, too."

"Bed sounds good," Jake said and stood. The ache in his bladder turned to pressing anguish as his organs shifted.

"Excuse me!" He waddled through the back door and across the porch to a lilac bush in the darkness.

"Walks kinda funny. I've never noticed that before."

"Go to bed," Susan said and left.

Jake stood facing the bush, its branches bare for the season, when he heard Susan's father take a similar stance opposite.

"Jake, I'll only ask once; feel free to say no, but I think it's a good idea." There was a pause. "What are the chances of running one of those planes right through here? I know I can peddle a fair amount of booze out of here. We could keep it small to avoid detection, like my brewing, but still make some money. We could be partners, you and me."

"I really don't have anything to do with the business end of the game. That's not what I do. I'm just a flyer."

"You don't have to answer now, but think about it. It could be big."

Jake stood alone at the bush and watched Mr. Bowers climb the steps and go inside. A light glowed briefly in an upstairs window, then went out. Jake returned to the house and found blankets neatly folded on the worn couch in the front room. He made himself a bed, took off his clothes and slid under the covers in the darkness. He knew where Susan's room was at the head of the stairs.

When the house was silent, Jake heard the creak on the stairs without seeing her descend. A robe fell away, before she slipped under the covers with him and pressed her lips to his mouth.

She was warm. She was soft, and she stayed until he slept.

Chapter
XV

When Jake awoke the following morning the sun was already bright filling the room with dusty light. His eyes stung as they opened wide, his brain sluggishly regaining consciousness through a haze of metabolized liquor. The house was empty. Susan, he reasoned, must still be on her morning delivery. Who knew where her father was.

Jake picked his shirt off the floor, and sniffed at it before slipping it on. His whole body smelled of stale beer.

"I'll change tomorrow," he said.

Walking through the kitchen he tore a chunk of bread from a loaf on the counter, rinsed his mouth with cold water and stepped outside heading for the barn.

The morning sky was partially overcast with a slight breeze attempting to move out of the northwest. Jake buttoned his coat to the collar just as the sun disappeared behind a cloud. He felt the damp cold creep under his shirt and up his spine. With a glance at the sky, he wished the sun would stay out, but the clouds were extensive and flat; slow moving clouds, the kind that brought trouble in winter time.

"But it's not winter yet," he called to the clouds. The cold needled into his sleeve, and he shivered. To the east a V formation of Canadian geese moved south. "Call yourselves Canadians! They ought to call you Panamanian geese. Cowards!" he hollered. "Got more sense than me," he muttered under his breath.

He looked around at the rolling Iowa landscape, all browns and gray, trees bare, fields empty except for the occasional cow or the numerous sparrows that collected on the wire fence, now and then darting into the barn to crap on his airplane.

Only one workman was on the barn tacking shakes in place on the roof. The flat cracking of his hammer sounding like distant gunfire. It reminded Jake of the rifle ranges at Fort Sam Houston. Texas could be cold, he remembered, but nothing like the craziness of an Iowa winter, or worse, Canada.

"December first," he said tucking his chin into his collar. "That's time enough; after that, I'm gone." He thrust a finger out as if Doc stood before him cringing. "I go where I want; when I want!" It was another lie he told himself.

Turning the corner of the barn, he waved to the carpenter on a ladder against the south wall. He worked in shirt sleeves, protected from the wind and seemed to enjoy himself.

"Good morning, Mr. Hollow! Getting up kinda late?" he called with a hissing snicker. "We'll have the roof done by tonight, and that back wall, on the west side there by tomorrow; God willing and the creek don't rise." He hissed again. "Plan to fly that thing?" He indicated the Standard inside.

"Might as well, can't dance; too wet to plow; feed a cold and starve a fever; think the rain'll hurt the rhubarb; vaya con Dios!" Jake flung back, irritated. The carpenter nodded without answering and pounded another nail.

Inside the barn the air was colder, and dampness rose from the dirt floor adding to his mounting depression. With the roof nearing completion, the sunlight inside was slowly constricted and would vanish by tomorrow if the carpenter's account held true. He knew what it was like to work in poorly lit unheated shops, and he hated the prospect. The exuberance he had felt the day before with Susan now faded when he stared at the tall broken biplane perched on the sawhorses, its fuselage covered with brown dust, its wings ripped and splintered with shreds of fabric dangling like soiled tassels.

The sun poked through the clouds sending a bright shaft through the roof. The carpenter slid a shake over the hole cutting it off, but before it vanished Jake saw a distinct bend in the fuselage. The hammer tapped, and Jake ran his hand along the fuselage.

"Oh, Craminie," he said. "The trusses are busted." He pressed and felt the structure crunch under the pressure. "That adds another month to the rebuild."

A sparrow flew into the barn, landed on a cabane strut and dug at its feathers with short, choppy movements. Jake reached down and chose a large stone. He hurled it at the bird but missed entirely striking the wall with a sharp crack.

"Drop something?" a voice squeaked through the roof.

"Just testing the structure," Jake answered. The sparrow flew to the tail, and Jake left.

He heard the brakes squeal. Susan pulled to a stop beside the house. She wore a red plaid coat, a size too large with the sleeves rolled up, and a wool cap perched at an angle. The door slammed behind her with a hollow crack, and she looked quickly around heading for the porch. Jake stood without calling making no effort to meet her.

She caught sight of him, and with a quick wave ran toward him. Feeling trapped by the weight of promises he had made the day before, he wished he had taken off before she returned. She waved again. He waved back.

He glanced past her at the landscape seeming to get colder before his eyes. Frustration, laced with vague fear, welled inside him growing as Susan drew closer, as though she carried anxiety with her. He wanted to be away. Control was slipping from his life with too many others making too many demands. Without his own airplane, he was compelled to fly where others chose carrying what they commanded and when. He thought of the money Doc could pay; he thought of December first. Susan skipped the last few steps to him, and kicking one leg behind her, she pecked at his cheek, her lips cool, her own cheeks flushed.

"Good morning, sleepyhead," she said and hugged him. Reflex demanded he put his arms around her, but he wanted to push her away, to fight off the crush of her presence, of his own helplessness. She felt the stiffness, the formality. "What's wrong?" she asked.

"Nothing," he said, the only thing he could say. Then, "I have to leave."

"When?"

"Today—right now." He gently removed her arms from around his neck and walked toward Henry's plane.

"Why? Why today? You said you'd be around for a few days." She held a smile and tried to avoid a nagging tone in her voice, but the result was a pained expression with a tinge of panic. It hurt Jake to look at her.

"Weather's moving in," he said and waved vaguely at the high stratus layer. They both looked up, but neither seemed convinced. "It'll probably start to sleet or even snow; if not today, then by tomorrow. No sense getting caught in it, and this isn't my plane, you know."

"I know," she said. "It's Henry's, but he doesn't seem like the type to complain if you're late. He was nice, I liked him." She had met Henry in September when he and Jake were returning together from a run. A line of thunderstorms had crept in from the west forcing their flight paths east toward Susan's. Rather than spend a night under the wing, Jake suggested the reroute, and they dropped in. They left early the next day, and she never told Jake he returned the following week claiming weather had detoured him again.

"I just have to get back, that's all," Jake said. "That's all."

She stared at him, her mouth drooping into a frown. "Why are you being this way?"

"What way? I'm not being any way; I just have to get back, that's all. I can't make any money sitting around here!"

"Is that all it is, money?"

"Yes, what else is there?"

Her face started to crumble, and he could see her holding back the tears. "I'm sorry," he said and reached for her without touching. "I didn't mean to bite your head off." She shook her head and forced a smile, her eyes bloated, ready to flow.

Oh Christ, he thought. Why, the hell, am I always the one apologizing?

"My head's still attached, see?" she said bravely. "When will you be back?"

"Ah, I don't know. Probably not until December. You know, when we head for Florida. I'll be flying every flight I can get until then...to build up some cash. Look, I don't know exactly when, but..."

212

"December?"

"That's only, what, six weeks away. We can spend Christmas together."

"And head for Key West?"

"Yes."

"And Havana to play the roulette wheel?"

"That's it. I'll have plenty of money by then; a whole pocketful of nickels!" He smiled at her seeing the tears dry in her eyes. The door was opening again for him to escape.

She looked straight at him, a thin trace of a smile tight on her lips. "I can't go with you," she said.

"What?" he asked. "Why...Why not for Chrissake?"

"Jake, you are far too innocent. You let whatever happened yesterday cloud your thinking. Oh sure, it was fun talk all this about flying airplanes to Miami to mingle with the suit and tie crowd, but honestly! Do you think I could simply pack up and leave? Leave my father all alone in that house through the winter? Why, I'd come back, and there'd be this frozen corpse propped up in the kitchen chair surrounded by empty beer bottles.

"No, Jake, I can't go." She pulled the plaid jacket tight around her, fastening the top button. "Besides, you might be in the money come Christmas, but I still have a business to run here. That beer doesn't deliver itself."

She reached out and patted his cheek, her hand rough and cold. "Sometimes, Jake Hollow, you amaze me with your unrealistic view of the world." She shook her head with a deprecating laugh putting him at a conversational disadvantage.

He stammered a few, "ah's" and a throat-clearing, "well," but only became further confused and uncomfortable. Nothing in his life seemed to be under control, even the things he took for granted. Getting close to another person only made his own life muddier, not his own anymore. He made no move to leave, that urge having been squashed by this young woman planted firmly before him, arms folded across her breasts and a fixed smile on her lips.

213

"Maybe, I could come back sometime before December," he said. Christ, Jake! he thought. Leave, just leave. She's giving you an out, take it.

"You take care of business first, my friend," she said. "There's plenty of time to visit come next spring."

"Spring? Sue, you know I love you!" The words flapped out of his mouth and fell flat in the cold air.

She laughed, ha, ha, and drew her coat even tighter around herself. Jake remembered June when she wore the thin summer dress, and he could gaze at the curves now hidden beneath layers of wool. He started to reach for her when she pressed a hand against his chest, smiled quickly and kissed him dryly on the cheek. She was cheerful but detached. Jake knew he had blown it, instantly changing from dashing cavalier, the light in the farm girl's life, to an awkward lover cutoff by his own hand.

"You get your plane ready," she said. "And I'll run get your fuel." She took off at a trot toward the house leaving him to drag around wondering how he had managed to so completely bungle this affair. When she returned, they filled the tank in silence except for the incidental, "thank-you" or, "excuse me". It was an empty moment.

Jake pulled the propeller through, it taking more effort than he was used to. Finally, the engine was ready to fire.

"I wasn't kidding when I said I loved you," he said.

"I know," she answered, and he thought he noticed her voice soften. "And I love you, Jake Hollow." She squeezed his hand. "I'll see you in December; now go away!" She moved aside before he could touch her.

Jake stood for a moment until he realized nothing more would happen, then he climbed into the cockpit.

"Switch on?" she asked. Her voice cracked, and she blinked.

He nodded. "On."

She took the prop, the way Jake had taught her. Before her leg came up, her eyes met Jake's, and she pulled the prop through.

When the motor coughed, and the propeller sent the cold blast across the windscreen and around his head, he knew he was back in his own environment where he could think. Rimmed by the cockpit and alone with the instruments, throttle, rudder and stick, there was no need to speak, no chance for mistakes. It was his world, a perfect world, detached and lonely.

He looked over at Susan, isolated from him by the wings and spinning propeller. She waved and said something, but of course he never heard it. He nodded and pulled his goggles over his eyes; the man from the sky returned to where he belonged leaving a girl, in the pose of a woman, to deny that her heart was broken.

That night the temperature dropped below freezing, Susan drew an extra quilt over herself and slept without dreaming. Outside, a barn owl screeched as it plucked a field mouse from the floor of the new barn. After swallowing it whole, it sat perched on a cross beam, eyes fixed as dozens of other mice discovered that airplane fabric and rib stitching made excellent nesting material as well as an almost endless food source. Winter would be easy for everyone inside.

By the end of the week, the carpenters finished the barn, and Susan's father slid the heavy doors along the greased roller tracks and closed the latch. That night it snowed.

In New York, Nick Barchek returned from a temporary assignment at the Treasury Department's Washington headquarters. It had been a dull six weeks of wading through tax records and field reports from Chicago and Miami on a tax evasion case the department was building against several Chicago area gangsters with Alfonso "Al" Capone heading the list. Most of the field work was headed by an agent named Ness, a nobody from the Department Nick had met several years earlier. He considered him a dull man with a plodding mentality, a stickler for paperwork and minor details, in short, exactly the type personality needed to take on Capone.

Nick carried his bag through the Penn-Central railroad station and stepped out on Seventh Avenue to the familiar din

215

of New York City traffic. He looked up at Gimbels and smiled.

"Ahh!" He inhaled the cold noisy air, glad to be home. He was tired of hotel living and was anxious to get to his own apartment off Gramercy Park and see if his neighbor had watered the small Austrian pine he kept on the window sill in the bedroom. Mrs. Levy was an old woman, a widow with a thick Eastern European accent in spite of having been in the U.S. for almost 30 years. She often stood at her apartment door across from Nick's to ask how his work was going "...at the Treasure Office...all that money!" she would say and shake her head. Then she would laugh and ask about, "...that tree growing in your room."

She had a son, who lived in New Jersey, but seldom visited her, and although Nick had been raised in a Greek Orthodox home in the Bronx, she treated him like a nephew to the point of inviting him over for shabbot. Twice he had gone, and each time she exclaimed, "to bring goyim inside the house for shabbot! What would my husband say?" Then she would shake her head and fill him with stuffed fish and chicken.

He glanced at his watch, 3 p.m.; enough time to take a cab downtown to the office and check in before he went home. He waved at several passing cabs before he managed to flag one down.

"Federal Building," he said and settled back in the seat. The cabbie grunted something and pulled into traffic after thumbing his nose at someone in the crosswalk.

I will miss the expense account, Nick thought. It would only be a nickel on the subway from the office back home. Meanwhile, he enjoyed the ride. The day was bright with a blustery wind tossing litter along the streets and through the crowds. Everywhere masses of people rushed in a continuous swirl, mostly along the edges beneath the tall buildings, but now and then one would break away and dart through the automobiles and vanish amid claxons and horns. Nick was at home.

The cab dropped him off in front of the Federal Building, and Nick handed the fare to the driver ignoring the scarce concealed comments about tipping and headed for the self-service elevator to his office.

"Hello, Nick!" the short Arnold Gruyer called from the desk near the front door. He always smoked a pipe, although it was seldom actually lit. Mostly, he played with it, tamping the tobacco, running pipe cleaners through the stem, and always rapping it on things, desks, his shoe bottom or wherever.

"Hello, Arnie. Everything still here?"

"Hasn't moved; we hardly noticed you were gone. Have you been in to see Fitz yet?" Fitz was Patrick Fitzpatrick, the boss, and Nick was uneasy with the way Arnie asked. There was a note of, "I know something and I won't tell," to it.

"No, just got in."

"Oh well, I guess he'll find you all right," Arnie said and rapped his pipe against a file cabinet, the ashes dropping to the bare floor where he brushed at them with his toe. Nick shrugged and headed across the roomful of government desks to his own by the far window. The day was bright and cold, the steam radiators hissed intermittently in the corners. A typewriter clacked a slow cadence from a far off desk as if someone was tacking nails into the environment to keep it from moving. He dropped his suitcase beside his desk and hung his hat on a hook behind.

"Hello, Nick," a large woman at the nearest desk called. "How was your trip? Catch Capone?" She wore thick red lipstick and routinely stopped whatever task she was at to check herself in a pocket mirror and spread another layer of red grease across her mouth.

"Hello, Elly. The trip was fine; Washington's cold, boring, and I think Mr. Capone had better find himself a stable of fancy lawyers."

"Jews," she said.

"Huh?"

"He should get Jews—Jewish lawyers. They make the best lawyers, you know. My George says, when you want a smart lawyer you get a Jewboy." Her voice was thick New York, all the r's being squashed into ah's.

217

"Uh huh," Nick muttered and turned his back. A note was taped to his lamp: BARCHEK, SEE ME. FITZPATRICK. Nick plucked it quickly from the lamp shade crumbling it into a wad. He knew Elly was watching in her mirror; he suspected that was why she really carried it.

"Fitz says you're to see him as soon as you get in," she said, her voice just loud enough to carry to every desk in the room, but politely hushed to carry the pretext of being discreet. She knew her politics.

"Thank-you," Nick said distantly.

"He's in his office, now." She plugged a short cigarette into her mouth and sucked until the ember glowed bright. Then with a flick of her head, a quick smile and a puff of smoke, she set it back in the ashtray. A red greasy ring stuck to the paper.

Nick looked back at Arnie dismantling his pipe on his desk and slowly walked to Fitzpatrick's office and knocked.

"Come in!" a voice growled.

"Hello, Fitz!" Nick said cheerfully, the only one in the office bold enough to use the nickname to his face.

"What do you call this?" Fitz snapped and thrust a hand printed letter at him. Nick looked at it and saw immediately it was the anonymous letter he had received during the summer concerning the whereabouts of someone named, The Doctor. He had stuffed it into his desk on that Friday and forgotten it entirely. Now it dangled accusingly in front of his nose held by the gruff Patrick Fitzpatrick.

Nick took the note cautiously trying to look as though he now saw it for the first time. "Mmmm," he hummed.

"It was in your desk," Fitz said. "According to the postmark you received it over a month ago, before you went to Washington."

"What were you doing in my desk," Nick asked hoping to turn the tables.

"Looking for bootleggers! It's what I get paid to do; what you get paid to do!"

"I..."

"And why I was going through your desk is no concern of yours. That is not your own private desk," he snarled leaning over his own desk, the top deep in papers. "It belongs to the U.S. Government, you are merely assigned to it." He sat. "Now, tell me about this letter." He lit a short cigar and waited for Nick to speak.

"Actually," Nick said and giggled. "I'd forgotten completely about this." He was about to toss it back on the desk, but a pair of tiny cold eyes stared through a cloud of cigar smoke without blinking.

"I don't know anything about any doctors, and I've never even been to Lincoln—that is in Nebraska, isn't it?"

Fitz nodded slowly twirling the cigar between his lips. The room was hot, the radiator hissing steam in short even bursts behind him. He leaned slowly forward seeming to enjoy Nick's discomfort.

"Well, I know who The Doctor is, and you are going to Lincoln—tomorrow."

"I just got in!" Nick protested. "Hell, I've got a suitcase full of dirty laundry sitting out in the next room right now. I can't go to Lincoln, for Chrissake." He sat.

Fitz nodded, a smile played at the corners of his mouth.

"Whatever it is, and whoever The Doctor is, can't someone in—what the hell is Lincoln even near—can't someone in Kansas City or Chicago handle it."

Fitz shook his head slowly. "No. I want you to go. Kansas City has its own problems, and Chicago wouldn't be interested."

"Omaha," Nick said hopefully. "I know we have a federal office in Omaha, I hope. We can't go snooping in someone else's jurisdiction."

Again Fitz shook his head, slowly rolling the cigar between his lips. He reached into the mound of papers at his elbow and fished out a thick folder tied with a dirty string. He dropped it before Nick.

"Look."

Nick slowly unraveled the string as Fitz stood and walked to the window. A handful of photographs and newspaper clippings fell onto his lap. The clippings were from a newspaper in Lake Placid, New York, the photos were mostly of a gray haired gentleman posing with a variety of successful looking types.

"Hey, that's John Barrymore, isn't it? And that's Fanny Brice!" Nick said. "Isn't this Mae Murray?" He held a glossy photograph of Mae and Doc arm in arm on a golf course somewhere in the mountains.

"That's Mae Murray, and Barrymore, and Brice," Fitz said.

"Who's the gray-haired geezer?"

Fitz turned from the window and snuffed his cigar in a crowded ashtray. "That's The Doctor—Doc."

"Now I remember," Nick said.

"Uh huh, worked for Rothstein, ran the tightest bootleg enterprise on the East Coast."

"I thought he was dead."

"No. Just vanished. And now for some reason he's in Lincoln, Nebraska, for crying out loud."

"You were after him when Rothstein was killed, weren't you?" Nick asked. He stared at the photo of Mae and Doc. She had her head back, nostrils pointed at the camera and her hip pressed firmly against Doc. Nick flipped the picture over. 1926 was printed on the back.

"I put a lot into getting him," Fitz said. "You were in the other office then, I believe. At first I was just nosing around Rothstein's affairs, but Christ, there wasn't a trace of him connecting to the booze. That's when I found Doc, and he reeked of it. He made no pretense of hiding it, either. But I couldn't get at him. Always seemed to have himself well padded from the actual operation. How? I don't know. Except he had that son of a bitchin' accountant Art something..."

"Reynolds?" Nick asked reading from a data sheet.

"That's the one. He could find more ways to juggle money..." Fitz gazed out the window, his hands clenched behind his back.

220

Nick had worked for Fitz less than a year and was perceived as a constant source of aggravation, but he actually respected him knowing that for some reason authorities higher up had stymied Fitz's career. Fitz had an impressive record of tracing tax cheats, especially in the underworld. Nick displayed an unusual ability to unravel schemes and was being watched closely and groomed for better things. Fitz had recommended him for the temporary assignment in Washington and kept close tabs on his performance. He liked what he saw.

Fitz reached over the desk and took the file from Nick. With restrained energy he flipped through the pages dropping paper clips on the floor as he walked around the room and read.

"Doc, The Doctor, a.k.a., Warren Phillips; born 1872, Greenfield, South Carolina; entered medical school at Duke University in 1893; was withdrawn from Duke in 1897." He raised an eyebrow without looking up. Nick lit a cigarette and listened.

"This was three months prior to graduation. Withdrew at his own request. No one knows why. Found his way to Central America where he went to work for the Canadian Fruit Company until the war when he drifted back to the New York area.

"Was believed to have been involved in narcotics smuggling during this period, no substantial evidence, however. Took up with Rothstein right after the war and ran a successful bootleg operation from the Lake Placid area." He slapped the file on the desk, and picked out a thinner one tapping it with his finger.

"That was where he hired an accountant named Reynolds, small time hood from the Chicago area. How the hell he got to Lake Placid of all places, I don't know or care." He poked Art's file back at Nick to read.

"This Arthur Reynolds," he continued, "is somewhat of a genius with money; kept us from finding anything substantial on Doc. Hell, we could see the liquor trucks on the roads in New York state, but when it came to tracing any of the money

221

for taxes..." He waved his hands through the blue smoke of a freshly lit cigar. "The trail would simply evaporate."

"Were the local police on the take up there?" Nick asked without looking up from the file.

"Local cops are always in on the skim, you know that. Just part of the cost of doing business."

"Is this Reynolds with Doc in Lincoln now?"

"Yes, and reports say he's running the same operation out of there but not quite as big."

"Why Lincoln?"

"Who knows. He probably just wandered in there keeping out of harm's way, found a void and filled it. I hear Lincoln's a college town, lots of boozers." Fitz produced a handful of letters printed in the same style as the one Nick had discarded in his desk. "These came while you were in Washington; nobody knows about any of this but you and me, and I want to keep it that way!"

Nick looked up taking the letters. "I detect a personal interest in this Doc."

"I just hate to have one slip through my fingers; ruins my batting average."

Nick glanced through the letters.

"My guess is someone in his enterprise, maybe a follower from the old days, has an axe to grind with him," Fitz said.

"Man or woman?" Nick asked.

"Can't tell." Fitz was pleased with Nick's quick grasp of detail. "Whoever it is just sends us little clues, not the whole picture. Look at the postmarks."

"I have, they're all different. Sioux City; Sioux Falls; Rapid City; St. Paul. Somebody who gets around."

"Or somebody who knows somebody who gets around," Fitz said.

"What do you have so far?"

"I know he's using airplanes to run the stuff in from Canada..."

"Airplanes?" Nick's eyebrows rose. "Interesting angle."

"Oh, it's been done before around Detroit, Erie, up in Vermont, Maine, just about everywhere, but always small time. No one's ever organized it the way Doc is, or at least as I suspect he is."

"But airplanes can't haul much," Nick said.

"Ah, but think about it. No competition. No hijacks. It must be making money or he wouldn't be involved." Fitz opened a lower drawer in the desk and produced two short glasses and an almost full bottle of Canadian whiskey.

"Whiskey," he said pulling the cork and filling each glass, "weighs about eight pounds per gallon." He slid one through the papers toward Nick who took it. "I've made a few phone calls to the Commerce Department, Aeronautics Branch; talked to some inspector over there, and he told me all about what airplanes can carry as far as weight goes. Turns out a lot of these crates flying around can haul quite a load."

Fitz twisted the glass before his eyes peering through the amber liquid at Nick. "At eight pounds a gallon, some old war surplus airplane can easily carry around over a hundred gallons of hooch. And I'm told, the midwest is loaded with deadbeat pilots willing to fly anything for a buck. Cheap labor, cheap transportation, no competition. It all spells big profits for Doc. Cheers." Fitz drained his glass in one swallow. Nick took two.

"Good whiskey," Nick said.

"Got it in a raid in New Jersey. We lost the case in court, so I kept the evidence. That was after we had a big photo session showing us pouring the stuff into the Passaic River; had to sacrifice a little for the newsreels."

Nick thumbed through the photographs again lingering on the one of Mae and Doc. He noticed Doc had his hand resting on her ass. "Nice work if you can get it..."

"Nick, you're familiar with Capone's operation."

"Vaguely."

"Don't give me that modesty crap! You know it better that anyone in New York. Now you tell me, do you think Capone's wise to Doc's operation in Lincoln. He runs everything in that area doesn't he?"

"Not really," Nick said setting down the empty glass. "He stays mostly in Chicago; he's got that pretty well to himself.

Kansas City is run by a couple of brothers, can't think of their name, but as for Nebraska...Well, he might have some say as to who runs things, but I doubt he pays it much mind. Sounds like small potatoes. Who knows." He blew a smoke ring and poked his finger through the core.

Well, you're going to get out there and see what you can find out. I want Doc!"

"But Christ!" Nick whined. "Lincoln, Nebraska. Why don't we just let the Omaha office handle it?"

"No. You!" Fitz pointed. "The Omaha office is expecting you, but they don't know why you're there. Don't tell them. Just check out a car and get down to Lincoln."

Nick crushed his cigarette out in a glass ashtray on the desk. "This Doc sounds like he's past his prime, a nobody. We're up to our elbows working on guys like Luciano and Lansky. Why screw around with some old has-been a thousand miles away? A little out of our territory, I'd say."

"Doc was a pain in the ass when he was out here before Rothstein died. This office spent a lot of time and money on him and got nowhere, a big zero. Someday, somebody's going to put a bullet through the good Doctor's brain, but first I want him. Call it personal, but I want him and that Arthur Reynolds, too."

Suddenly, Fitz stood, gathering Doc's profile into the folder with Art's and thrust them at Nick. "Read these and get your butt down to Penn Station by 5 p.m. You're on the Broadway Limited to Chicago. There you transfer to Omaha, pick up the car and go to work. Elly has your expense money. And get a goddamn receipt this time for everything you do! I'm tired of trying to straighten out your travel vouchers. Now get lost!"

Nick left the office with the two folders under his arms.

"Sign for your money, Nick," Elly called. She counted out three stacks of small bills, pausing between each stack to lick her fingers. "Maybe now you'll have enough to get your laundry done." She wrinkled her nose toward the suitcase at Nick's desk. A mildew odor had crept through the room.

"Get a Chinaman. My George says, always get a Chinaman to do laundry."

"And Jewboys for lawyers, thank-you, Elly." Nick gathered the money into a wad and stuffed it into his pocket. He picked up his suitcase, hat and coat and left. Elly took out her compact, and Arnie blew a sharp toot through his pipe stem.

The ceiling was low and ragged when Jake approached the airport from the east. He could barely see downtown Lincoln until he dropped within 300 feet of the ground. He knew where the airport was and followed the railroad until he picked up O street and turned for the runway passing through a low hanging cloud on final approach. A trace of ice formed on the struts.

Factory looks kinda quiet, he thought flying low over the hangar. None of Doc's four LS-5's were to be seen. Normally, at least one would be outside the maintenance shop awaiting repairs. The only airplane he saw, as he turned onto final approach, was a Travel Air sitting in a twisted heap by a hangar.

"Scrap for the junkman," he said to himself giving the Travel Air little thought as he corrected for a blustery northwest wind kicking him around on short final.

Once down, he taxied toward the hangar where he was able to study the Travel Air more closely.

"Christ," he muttered and pushed his goggles up. He recognized it as belonging to Billings. He also realized now it was more than a simple groundloop that trashed that airplane. It looked more like Sterns' D.H.4, the last time he saw it in the trees at Adair's.

The wings were all torn into ragged pieces twisted together in a mat of tangled wires and cracked struts. The fuselage was bent at least 30 degrees, and the bend centered at the rear cockpit where the pilot would have been. The windshield was shattered and the top of the rudder was crushed. It's been on its back, he thought. The landing gear was tucked under the

225

belly like the front paws of a sleeping cat. A swirl of barbed wire was coiled around the broken wheels.

"Somebody really screwed up," he said and taxied well past the wreckage to the hangar. Swinging the airplane's nose into the wind, he shut down the engine. The wings rocked in the uneven gusts, occasionally feeling as though he would lift off. He hurried from the cockpit and was tying it down when someone approached from behind.

"You can't park here!" Jake started. It was Mr. Howard, the sales manager, holding his brown suit jacket tight round his waist by the pockets. Loose strands of hair from his temples whipped across his bald head in the wind. He looked nervous, tired, scared of something, but Jake thought he always looked that way.

"What?" Jake asked holding a rope tight against a strut. The wind wanted to pick up a wing and drop the plane over. "What are you talking about?"

"None of you pilots can stay here anymore, you must leave. You have to go over to your new field." He waved in the general direction of west, away from town. "All of Doc's planes are over there now."

"What new field?" Jake asked. "I've been gone a couple of days, and I have no idea what you're talking about."

"Doc leased a farm, house and everything, out west of town, that's where the new airfield is. He's moved the whole operation out there. You can't miss it. Just go about 15 miles or so straight west, and you'll see it. Just look for the others. Now go, please!"

"Hold on there," Jake said and grabbed Howard by the sleeve as he tried to leave. A cold mist dripped out of the clouds obscuring the treetops.

"Things aren't too good here right now," Mr. Howard blurted. "Your boss has caused a bit of a stink around here, and he had to move. And on top of that," his voice rose sharply, "things aren't good with Standard Aircraft, either. We may be moving ourselves. Business is just drying up!"

"How can that be? We've bought, what, four planes from you just since the summer."

"Four planes does not a factory make, and Doc bought them at quite a discount." His shoulders slumped, and he dropped his hands from his pockets looking around at the empty field and quiet hangars. We've got just enough staffing to finish the few orders we have left—two more to you folks and one to a company in Mexico. After that, there's nothing. We'll have to let everyone go." Mr. Howard seemed to be talking past Jake, airing his problems to the cold mist.

"I thought things would pick up, but they've only gotten worse. Money's tight and competition's serious. These old biplanes just can't compete. Just look at Ford. Those damn bastards are building that Tri-Motor big enough to carry God knows how many passengers. How can we compete?"

"Those big planes aren't economical," Jake said with authority, although he had absolutely no idea what they cost to operate. "Things will pick up again. Is that company in Alaska, ALMINCO, still buying? They were a pretty big customer, weren't they?"

"Went belly up weeks ago. Company president, I hear, committed suicide, too many discrepancies in the books," Mr. Howard said and looked sideways at Jake as if he might know something about the mining industry. "We had a pretty big order crated up and ready to ship off to them, all on faith, no cash. We sold most of it to one of your pilots for pennies on the dollar. That big fellow, used to play baseball? In fact, he owns this plane here." Howard pointed at Henry's J-1.

"Henry?"

"Yea, that's the one."

"What would he want with a bunch of airplane parts?"

"I don't know. Maybe he wants to start his own airline. Ha! That'd be a laugh. Anyhow, he bought a whole load of Hisso engines, couple of OX-5's, wires, spare wings. Jeez, a whole slug of junk. He and a couple of colored folks hauled it out of here over the last couple of weeks. He didn't mention it?"

"No. I don't see him too much," Jake said.

"Looks like you saw him long enough to borrow his plane."

"Actually, I just left him a note saying I was taking it. I knew he wouldn't be using it for a while."

"Whatever. That's your affair, but like I said, you're going to have to move out of here, over to the new strip." Mr. Howard started to leave, but again Jake stopped him.

"Hey, what happened to that Travel Air? A kid named Billings flew it."

"Crashed it obviously."

"Yea, but how?"

"We had some low fog here the other night. Your friend tried to get in under the stuff across town. Well, it must have been lower than he anticipated, because folks all over town heard the motor barking across the rooftops as he circled once or twice downtown. Must have been trying to get his bearings, I guess. Pretty soon a crowd gathers, all pointing up and speculating what the pilot would do. Must have been around seven o'clock.

"Anyhow, a whole bunch of them starts heading for the airport here to see what they can do to help, or at least to see a good crash. Well, your friend..."

"He's not my friend; don't call him that," Jake said.

"Sorry. Well, this Billings comes shooting out of the fog along the road leading out here, right over a string of cars. He must've seen their lights. Mostly, everybody just heard him, it was so foggy and dark, then there was this crunch! Right alongside the road he sets it down, but he doesn't see the wire fence, and his gear catches on that and over he goes."

"What happened to him?" Jake asked.

"Got busted up awful bad. He's over at the County Hospital. The police are acting real funny about him and have a guard on him in his room. He's going to have some explaining to do when he wakes up."

"Why?" Jake asked. "What'd they find on him?"

"What do you think? Whiskey, and it was running all over the place and soaking into the ground. Everyone saw it before the police could get there. What I hear, they only salvaged about a pint for evidence, the rest just vanished."

"So Doc's just packed up and moved everything out of town?"

"Oh, he still lives at the Lindell. Life goes on as normal there, but the police are in a tough spot. They can't just ignore it, too many people were there. Too many questions being asked. Some reporter started asking questions in front of everybody about Doc's connection to the whole thing since he runs such a big flying business out here."

"But Doc doesn't own the Travel Air; that belonged to Billings," Jake said.

"That's what he said to the police. He said he knew nothing of the whiskey, and the pilot was merely an employee obviously acting on his own and would be dismissed immediately. That kept the police happy, for now, anyhow."

"Think he'll wake up so he can be dismissed?"

"It'd be better for everyone if he didn't," Howard said. "So your boss might have weaseled his way out of this one, but as soon as we towed the wreck out here I told him to pack up the whole lot and stay away!" He struck his hand with his fist. "I guess he got the message, because he leased that farmland right that day, paid cash, and everybody moved out." Howard turned to leave and called back over his shoulder, "Now, I suggest you get out, too. I don't need any trouble from any of you." He walked across the damp field and into the hangar. Jake restarted the Standard, untied it and left.

Finding the new airfield was easy. Landing was easy. Finding someone in the commotion inside the farmhouse who might know what was happening was difficult.

"You just landed out of that crap?" someone Jake had never met called to him after he shut off the motor. The ceiling melted in thick wet drips through the barren trees bordering the landing strip. Airplanes were parked haphazardly in the muddy yard between the house and barn.

"Where's Doc?" Jake asked, but the other man shook his head and disappeared inside the house.

Jake pulled his helmet off and felt the damp mist collect on his face. Once inside the house, he wandered from room to room stepping over parachutes, propellers, spare engine parts, a seat from some airplane—he thought maybe a Jenny—beer bottles, a gas tank reeking of fuel and a snoring pilot face down on the hardwood floor.

The large house was filled with the din of men waiting for something.

"Christ, did you see where we're supposed to sleep, for Chrissake?" a short thin pilot with red eyes and pitted skin called from the top of the stairs. Jake recognized him from one of the Travel Airs. "They got about six mattresses scattered on the floor like a flophouse, for Chrissake."

"Don't worry, Mel, the way you fart you'll have the whole room to yourself!"

"Oh for Chrissake already!"

"The hell he will! I'll kick his ass out with the chickens, he farts around me!" a voice shot back from somewhere.

"You ain't about to kick nobody's ass!" the short pilot at the top of stairs called as he bounded over the railing and crashed into a cluster of pilots, sending furniture and beer bottles crashing through the already over crowded room.

Jake picked his way through the melee past two men waving hands and ignoring each other's accounts.

"Anyone seen Doc?" he called.

Crash! A bottle hit the wall, tearing the wallpaper and leaving a foaming streak. Someone walked in the front door carrying a toilet seat and a hammer and proceeded to nail it to the living room wall without any further explanation. No one seemed to question it. One pilot, dead drunk, sat alone in a corner with a fat dog combing its coat looking for fleas. Occasionally, he would find one, pinch it between two fingers and drop it into the flame of an oil lamp at his side.

Crash! A window broke, and Jake turned in time to see the short pilot accused of farting sail through the glass shards into the front yard. Someone adjusted the toilet seat on the wall, stood back briefly and readjusted it.

A few of the faces were familiar to Jake. The drunk with the dog he recognized from an overnight stay at Moyer's but had never spoken to him. Jake rarely bothered with more than a nodding acquaintance with any of the pilots. Sterns, Billings, and Henry being the only real exceptions, and Billings was gone, Sterns dead, and Henry always seemed to be heading north when he was heading south.

He noticed the older pilots seemed to keep to themselves, always flying alone, never swapping tales or bragging like the younger ones did. They did, however, drink every bit as much as the younger ones but privately, rarely in the squadron style the young ones were so careful to emulate. But then, Jake noticed, few of the pilots stuck around for more than a couple of weeks at a time. The ones with their own aircraft took fewer chances, and once they had a fresh roll of money in their pockets they simply vanished, and other gypsies would appear to take their place. There was no shortage of pilots or planes. Word just spread.

"Is Doc around anywhere?" Jake stopped a pilot walking past carrying a case of whiskey. Someone snatched a bottle and ran.

"You son-of-a-bitch, that's two bucks you owe me!" he shouted, but with little real concern. "Who?"

"Doc. Is he around? Or Art?"

"Don't know any doctors around here. Maybe you want the hospital! Ha! Ha!"

"Doc's your boss. The guy who runs all this!" Jake snarled, but the younger pilot looked nonplussed, anxious to hurry off with his whiskey. Another fight broke out as someone crawled through the broken window, and a roar of swearing and laughter rose from the next room followed by a dull thump against the wall.

"I have no idea who's in charge here, bud. I just heard someone was looking for pilots—good pay and a place to sleep. Well, it ain't the Ritz, but it ain't bad. There's some guy drove up in a Lincoln gimping around with a clipboard telling people where to put things. He's got some kid with him, looks real snotty, but I wouldn't mess with him. Are you a pilot?"

Jake ignored the question and turned away to leave.

"Well, you're welcome!" the pilot shouted after him. "And don't expect to jump over me for a flying slot. I was here first!"

Jake stepped out the back door onto the flight line. The ceiling had dropped no further, but a gray cold rain fell in a slant over the ragged collection of biplanes and one monoplane, a Curtiss Robin painted a bright red with LUCKY LUNDY-PILOT/OWNER stenciled to the side. The plane appeared to have been recently painted with a poor quality brush and cheap paint. The wings were mismatched, one red, one green. The landing gear had obviously been buckled once and rebuilt as the plane sat with a distinct list to port.

All the airplanes sat like ghosts in the cold gloom unable to move. Doc's four LS-5s were parked in a straight line apart from the others. These were the pride of the fleet, the flag-ships, but with closer inspection they too bore the scars of heavy use—patched skin, cracked glass, oil-stained bellies and gear. Two mechanics poked through the cowling of one of the machines, the flat tapping of their tools making the day seem gloomier.

"Art, hey, Art!" Jake called and ran toward the Lincoln parked around the corner of the house under a bare chestnut tree. Art was seated alone on the back seat, papers spread across his lap, the thick fur collar of his coat turned up against the damp cold. Sam Nyberg sat at the wheel and saw Jake running toward the car. With a practiced threatening air he stepped from the car and planted two feet wide apart blocking him.

"Art!" Jake called a few steps away, ignoring Sam. Sam pulled the wide brim of his hat low over his eyes, and as best he could swelled his narrow chest inside the camel's hair coat. He exhaled a roll of blue smoke through his nostrils and flicked the unfinished cigarette at Jake saying nothing.

Lately, his strutting with the ubiquitous display of the .32 caliber automatic he carried was enough to get respect, keep pilots in line. Jake, however, remembered Sam from his days not long ago in the filling station drinking pop and reading comic books.

"Your gun's going to get wet, Sammy," Jake said brushing past him.

Sam gripped Jake in a rough hold from behind squeaking into his ear, "That's far enough!"

Art looked up through the wet glass, embarrassed, and about to roll down the window when Jake spun around and sliding a foot behind Sam tipped him over into the mud in a blur of flailing arms and overcoat. His brown hat bounced into a puddle.

"Don't get stupid with me, Sammy boy!" Jake said and Sam fumbled into his coat for the pistol spitting rage. Jake dropped a foot hard on the teenager's arm pressing it to his chest. Sam groped for Jake's leg with his free hand, his face contorted with anger. He tried to get up but became ensnarled in the massive overcoat with Jake's foot still pushing him into the mud.

"You son-of-a-bitch!" he spat.

"Sam!" Art hollered, a smirk barely concealed. "Just calm down and get your goddamn hand away from that gun!" Sam settled back on the ground, his eyes closed tight with rage and embarrassment. Luckily, no one else had seen.

"Let him up," Art said, and Jake removed his foot leaving a gray heel print in the center of the coat. He stood back and offered a hand to Sam, who ignored it getting to his feet on his own, brushing mud from his coat and, retrieving his hat.

Jake smiled and started to speak when Sam thrust his scarlet face, the pimples appearing ready to explode, within an inch of Jake's and said, "Stay out of my way. You got that? Just stay clear!" He turned and left.

"Now, Sam!" Art called. "Sam!" He ignored them. "Moody little turd."

"He's what replaced Henry?" Jake asked.

"Oh, he actually does a good job; you might not want to screw with him too much. Doc certainly likes him, plus he runs a good business of his own. He has some cousin of his running the store while he's out." Art looked into Jake's face and waved at the farm. "So? What do you think of the new place?"

233

"What's going on, Art?" Jake asked and leaned on the door blocking the drizzle from Art's face.

"We had to move out of town; it got a little too complicated. You heard about Billings?"

"Yes. Is he still in the hospital?"

"Still unconscious." Art looked away. "With luck, he'll stay that way."

"Seems to be the general opinion," Jake said with a shrug. "Sounds like he just plain screwed up. Don't need that crap." His tone was flat, unconcerned. "When do I fly? I see four idle LS-5s on the line. At least one of them must be flyable."

"You don't fly; nobody flies, at least not for a couple of days. There's been too many people poking around the hotel asking questions. We might start attracting too much outside attention. The locals we can handle, but I don't particularly want the Feds getting too curious. They can be a pain in the ass."

"Why doesn't Doc move out of town, avoid the mess?"

"Too many connections in town, that's where the business is. A visible profile actually helps in this business...to an extent."

"So when do I fly?"

"Soon. Meanwhile, if you need a place to stay, find a spot inside." A sharp crash and shower of glass came from an upper window in the house, and a body bounced off the porch roof hitting the soft earth with a squishy thud.

"Bear with us, Jake. It's only temporary."

A truck pulled beside the Lincoln, steam creeping out the radiator cap. The driver leaned out the window calling to Art.

"Where do you want these?" He pointed to three fuel storage tanks strapped to the bed.

"Anywhere over by the barn, west side." Art checked something off a list on the seat beside him.

"Where's Henry?" Jake asked.

"In town with Doc at the hotel. I have work to do, Jake. I'll see you later." He waved for Sam to return. "Just stick around; something will come up." The window rolled up again, and Jake left passing Sam with a friendly grin. Sam returned a cold, and he hoped, withering stare and mentally fixed his pistol sights between Jake's retreating shoulder blades.

The Lincoln left, and an hour later Jake caught a ride into town with the driver who brought the fuel tanks.

Chapter
XVI

Nick Barchek never slept well on trains, and the ride from New York to Chicago was no exception. Whether it was due to the clicking of the wheels beneath his bunk or the stress of traveling or just being in unfamiliar surroundings, something prevented him from snatching more than the briefest nod when away from home. He did manage to drop off as the first glint of dawn crept over the gray Indiana landscape, but it was fleeting and he snapped awake staring out the window at the red slag heaps from the steel mills in Gary.

After a greasy breakfast in Union Station, he caught the train to Omaha sitting all day in the coach, occasionally reviewing the files he carried on Doc, Art, Moyer, Adair and Vincent. There was also a hastily compiled dossier on a man named Henry, Henry Townsend, an ex-major league baseball player, age, 29.

His record showed a few minor encounters with the law in Arkansas as a juvenile, then a short statement: "...while Townsend was little more than a long shot hopeful for the Chicago White Sox, he is suspected of complicity in the 1919 World Series Scandals (someone scribbled underneath this: CHECK CONNECTION BETWEEN ROTHSTEIN, WHO FIXED SERIES, AND DOC AND TOWNSEND)"

Nick stared blankly at the moving Iowa landscape. "Townsend? Mmm, Townsend. Never heard of him," he muttered. A well-dressed woman seated across the aisle thought he might be addressing her.

"Beg your pardon?" she said.

"Mmm?"

"Did you say something?"

"Umm, yes, Townsend. I said the name Townsend, a baseball player. Ever heard of him?" Nick asked sarcastically.

"Hank Townsend?" she asked. "Haven't heard that name in a long while...relief pitcher for the White Sox. Not worth a plug. Why's he making a comeback or something?"

Nick only stared and shook his head.

"Funny you should pull a name like that out of the air," she said, talking more to herself as Nick raised the folder over his chin. "A third rate nobody like Hank Townsend..." She shook her head and snapped a newspaper open to the sports section.

Nick inched closer to the window and continued to read the file concentrating on a more recent entry on Henry: "Arrested and convicted...second degree assault and battery, Cook County, September, 1924. Served six months Cook County Jail; released March 3, 1925." A handwritten note attached described the battery as a minor quarrel between Henry and a small time gambler, named "Muffles", on Chicago's southside. Police reported Henry was seen pounding the victim's face repeatedly into an iron railing outside a known whorehouse when they arrived. The note added it took four policemen to subdue him and, therefore, he should be considered extremely dangerous.

"Muffles," Nick said a little too loudly and glanced at the woman across the aisle.

"Eh?" she asked.

"Muffles, Muffles. You ever heard of him?"

She thought for a minute and shook her head returning to the racing section of her paper. "I knew a bookie named, Mumbles, once. Could that be who you were thinking of?"

Nick stood and gathering his files moved to another car after first dropping his briefcase in the aisle spilling papers in a cluttered lump.

"If it's Mumbles you're trying to find, you might as well give up, because he's dead. Got a bullet in the ear sitting in a barber's chair. Whoever shot him gave the barber four bits before he left. I think it was Jack McGurn did it, of course he always leaves a nickel in their palms when he..." Her voice was abruptly cut off as Nick left the car.

Nick also carried transcripts of the anonymous letters his office had received on Doc over the last weeks. Together they provided a teasing, but annoyingly incomplete picture of the enterprise, mentioning only the key players, and the fact that Doc was bootlegging again and using airplanes.

Nick leaned back in his seat. The aroma from the dining car drifted back toward him wrenching at his empty stomach and taking his mind off the files.

"Lunchtime," he said and started to gather the files dropping each one into his briefcase after first reading the names slowly, "Moyer...Henry...Art...Vincent...Adair...and the weasel himself, Doc. One of you is sending us letters!" He snapped the case shut.

By the time Nick arrived in Omaha it was dark, unfriendly dark, the kind of dark that says, "Go away, you're not wanted." Nick found the local office where he was supposed to pick up a car, but everyone was gone. After shuffling around town until he found a phone booth in a dank cafe he managed to contact someone, who knew someone, who might know what-the-hell he was talking about.

"Meet me at the office," the suspicious voice said at the other end of the line. "Yes, we have the car for you! What, for Chrissake, is someone from the New York office doing out here anyhow? This isn't normal procedure, you know!"

"It should have all been cleared already," Nick whined.

"Oh, it is! It is; it's just...not standard operating procedure, you know. Besides, I'm having my supper. Just meet me out there!" Click.

Nick waited an hour at the Federal office and another half hour for a disgruntled clerk with gravy on his chin to rummage through the office for the keys, check Nick's identification, fill in the travel voucher forms, and call someone higher up to take the blame.

"It's the Ford out back. Watch the water level. Don't gun the engine. And write down the mileage." He handed Nick the keys attached to a small block of wood. "You sure you know what you're doing out here? All the way from New

York, for Chrissake." He ushered Nick out the door and locked it without waiting for an answer.

"Thank-you," Nick said. The clerk nodded and started to leave. "Oh, is there gas in it?"

"Might be; who knows." He was gone.

Nick found the car, got it started, found a filling station (the tank was almost empty), and after finding the station owner and convincing him to sell him gas before morning, he drove twice through Omaha looking for the highway, found it and an hour later was in Lincoln, cold, mad, hungry and wrung-out tired.

BWANG! He thumped the bell on the counter at the Hotel Lindell. The lobby was dimly lit by the yellow glow escaping from a back room. A bent figure with thin gray hair sat in a stuffed chair staring out the plate glass window at the passing traffic. He looked momentarily at Nick, blinked, and returned to staring.

BWANG! BWANG! Nick rang.

"All right! All right already!" A round man, haloed by the light from the back room, stepped behind the counter. Nick stood limp and red-eyed, his face in need of a shave and a damp stale odor clinging to his suit from two days of travel. The night clerk hesitated, then shoved the registration book across the counter, switching on a small brass lamp as he did.

"I have a reservation," Nick said taking the pen.

"I have several," the clerk said and laughed, but Nick only stared. "It's a joke! Oh, don't be such a Gloomy Gus."

"Barchek," Nick only said. "Nicholas Barchek. Do you have a room for me?"

The clerk wiggled his fingers through a card file and pulled one out with a flourish holding it under the brass lamp to read. The glare shown into his face from below, making his eyes into deep sockets. He was a tall man with pink hands and a bald head. His face was moon round with a tiny mouth and thin mustache between smooth fleshy cheeks.

"Here it is!" he announced. "You're Barchek, Nicholas from New York, oh my! That is a long way!"

"Yes, and I'm..."

240

"United States Treasury Department! That is something, now isn't it."

"How did you know?" Nick growled snatching the reservation card and reading it.

"I suppose whoever made the reservation for you told us. Probably asked for government rates; we take off ten per cent for those on government business..."

"Elly," Nick muttered. "The idiot, I'll kill her."

"Who will you be killing, sir?"

"Nevermind, just give me the room."

"Yes, sir, but promise you won't be murdering anyone in the bath..." Nick stopped him with a cold frown. "Just kidding! Oh, you're so gloomy, you are." The clerk rapped Nick's knuckles with the card and spun around reaching for a key. For a large man he moved with amazing grace, Nick noticed, possibly too gracefully.

"Your room is on the ground floor in the back. They've only just been vacated, and frankly between you and me..." He leaned into the light bringing his face close to Nick's. "The previous tenants, a collection of airplane pilots of some kind, left it just a mess!" He rolled his eyes and brought a pink hand to his cheek. "We haven't had a chance to clean it too well, and there are a few holes in the plaster. God knows what they were doing!" Again, his eyes rolled. "But if you can live with that, I'm instructed to take off another five per cent. Otherwise it will be morning before we can ready another room for you."

A car horn blared beyond the plate window, and Nick turned to see a Buick round the corner, the driver waving to someone on the sidewalk. The old man in the chair was staring at Nick. The Buick drove off, and Nick turned and took the key.

"Straight through there, past the elevator and beyond the closet. I'll have the bellboy show you to your room," the clerk said and rapped the bell waking another figure from the shadows.

"Why? You just told me where it was."

"Ha, very funny," he said. "Room One Eleven," he said to a skinny boy in a baggy Phillip Morris costume who swiped Nick's bag off the floor and vanished back into the shadows.

"Sleep tight," the clerk called and himself disappeared. Nick started to say something, then saw his bag being whisked across the lobby, turned to follow and caught his sleeve on the brass lamp knocking it over.

"Please don't take that," the clerk's voice called from the dark. "Each room has its own lamps." He hissed to himself pleased with his humor.

Jake set the lamp upright again and, after banging his knee painfully into the corner of the desk, followed the bellhop to his room.

"Room One Eleven, sir," he said in a flat midwestern voice unlocking the door. Nick followed him inside taking in the room in a swirling glance.

"I hope everything's to your satisfaction," the bellhop said dropping his bag on the bed where it bounced several times on the overly soft mattress. Nick counted the pock marks in the plaster, some the size of fists others head size, and one near the window the size of a healthy rump. The mirror over the dresser was cracked and the rug was speckled with black cigarette burns. The whole room smelled of beer, sweat, cold burnt tobacco, soap and urine.

"You didn't have to give me the honeymoon suite," Nick said, but the bellboy only stared at him with confused sleepy eyes holding out his hand, palm up. Nick took it and, shaking it vigorously, said, "Everything's just dandy! Thank-you ever so much!" He ushered the boy out the door and pressed it closed behind him.

"Christ, what a dump," he mumbled and dropped onto the bed sending the bag floating on the waves of bouncing mattress. He fell into a deep sleep waking hours later long enough to undress, drop his clothes on the floor and crawl under the covers.

When he awoke the following morning he pulled on his wrinkled trousers, slipped his shoes over bare feet without tying the laces and stumbled from the room hoping to spot a bellboy. He poked his head around the corner to the front lobby, but the only one present was the day clerk busily sorting papers behind the desk and brushing dandruff from his collar.

"Hey," Nick called quietly. "Psst..." The elevator suddenly dinged beside him and he ducked behind the corner as four men dressed in suits stepped off.

"Good morning, sir," the desk clerk chirped.

"Good morning," the oldest of the four, dressed in light gray with matching spats over shiny black shoes answered with an imperious nod of his white head. His voice was tinged with a refined southern accent.

Doc, Nick thought.

One of the four, a skinny kid much like the bellhop, rushed ahead of the others and opened the door pushing aside someone trying to enter. A large square-shouldered man helped the fourth one who limped slowly behind Doc.

"Art," Nick said to himself. "And the big mug is Henry—has to be." But he had wondered who the skinny kid was now arguing outside at the curb over who would drive the car. Henry apparently prevailed as the skinny kid seemed to bounce momentarily off the fender from a shove and, after retrieving his hat, ran around to the passenger's side and got in. They drove off.

Nick was about to return to his room when he noticed a second car, plainer and older, start up and pull away from the curb a short distance behind Doc's. There was nothing unusual about the car, except Nick was curious why there was one driver in the front seat and two men in the rear, hats low across their faces.

"Did you see them, Art?" Doc asked as they pulled away.

"Yes, looks like a Chevrolet, about a '27 Capital Double A, Nebraska plates, couldn't read the number."

"Are you sure they weren't Illinois plates?" Doc asked.

"No, Nebraska," Art answered. "Who knows where it came from."

243

"Stolen, most likely," Doc said and turned briefly in his seat looking at the car following. "Someone from Chicago is on to us."

"Maybe, but I don't think so," Art said.

"I'm not surprised though," Doc said ignoring Art's comment. "It's about time someone took notice of us, we've been squeezing out someone's business around here maybe a little too long." He smiled as he spoke.

Art raised his eyebrows briefly without contradicting.

"Capone, maybe," Doc said. "He's heard about me."

"Think that's him back there?" Art asked nodding toward car behind. A smirk played at his normally enigmatic face.

"Now, you know damn well that's not him...in person. Capone sends out others to do his bidding."

Henry drove in silence. Sam practically twisted backwards in his seat listening to the conversation.

"What car are you talking about?" he asked.

"Turn around!" Doc snapped and rapped him on the hand with the silver tipped cane he now carried. "You should notice these things, Samuel. We are being followed. Henry, where are they now?"

"Still about five lengths back; three men, one up front, two in the back. Could be a shooter in the front seat below the window, but I doubt it. I think they're just watching us."

"Shooter?" Sam shouted, his voice breaking into a squeak. His hand slid inside his coat groping for his pistol.

"Calm yourself!" Doc said and caned him again. Henry smiled without turning.

"Maybe it's that Treasury guy the clerk told us about," Sam said. "I think we should get rid of him."

"No it's not him," Art said. "First of all he came alone; second, the bellboy checked his room before we left, and he was still asleep. And whatever you've got in mind about the Treasury man, you just put out of your head right now. He is off limits! A single one of those Feds can bring more crap down on our heads than you can imagine."

Sam turned away with a pout but not without catching the amused grin Henry did little to hide.

"No, whoever is following us is somehow connected with Chicago," Doc said. "We've made too large a dent in someone's business for too long now. It's obvious. Otherwise, why would the Treasury Department be sending out a task force to investigate us? Art, you remember all the commotion we stirred in New York."

"Yes, but that was bigger, much bigger, and we were there longer."

"But everyone's been waiting for me to make a comeback. They're not wasting any time coming after me this time." Doc's tone let the others know there was no reason to argue. He leaned over the seat putting his face close to Henry's ear. "Henry, take us around town a bit then back to the hotel. Park the car out front, and after we all go in you stay with me, and Sam will take Arthur out to the airfield. Sam, you take good care of Arthur. The heat's starting to rise, and it's up to you to keep him safe."

"Let's not go overboard here," Art said.

"Oh, I will!" Sam responded fixing a serious look on Art. "Don't you worry."

"Oh Christ," Art said. "You watch yourself, sonny and just do what you're told, and goddamnit, don't go pointing that gun around where anyone can see it! He doesn't really have bullets in that thing, does he?"

"Leave the boy alone, Arthur. He does good work."

"He's going to shoot himself in the foot one of these days."

"Try not to shoot your foot off, Samuel," Doc said. "Now, Art check on the airfield and make certain everyone's ready to fly."

"We taking off again?" Henry asked hopefully. He missed flying more than he thought he would. Even driving the new Lincoln Brougham was dull compared with the satisfaction of flying 700 miles in the LS-5 across prairies and rivers, one day in one country, the next in another.

"Soon, Henry. Although, I can't understand why you like it when you could be making just as much right here on the ground with me."

"You have Sam," Henry said.

"Yes, we do."

"It isn't the money," Henry said.

"Everything's the money. Don't kid me!" Doc interrupted. "I haven't seen you turn any down yet, Henry."

"No, I still like it, but..."

"And you took my advice and invested in the Stock Market the way I told you, too. You like seeing that money almost double in less than a month's time don't you?"

"Well..." Henry stammered looking at Art.

"Well, what?"

"Actually," Art jumped in. "I advised Henry to sell, to get out of the market."

"Oh?"

"Yes, I suggested to him that maybe things were getting a little confusing. Maybe not the best time to be fully invested."

"That's not the way I see things," Doc said. "Sam, are you playing the market?"

"Oh, one hundred percent, sir."

"Now there! You see. Things aren't so gloomy."

"Still," Art added. "I view the indicators as turning downward, and I myself have withdrawn, and Henry is wise to follow."

"That's surprising, gentlemen, but you certainly do as you please with your own money. The market's not unlike our own line of work—no place for weak knees!"

Sam laughed showing his appreciation for the attention Doc was paying him at, he sensed, Henry's expense.

"Samuel," Doc continued. "You and I will be cutting up a fat hog after the first of the year while these others are cursing their shortsightedness." He smiled pleasantly tapping the silver tip of his cane. "No sir, gentlemen, I always say, never get off a train when it's moving full speed."

Art and Henry sat quietly while Sam twisted in his seat beaming with pride and fingering the pistol's grip under his coat. He still watched the Chevrolet following far behind.

Hours later, Nick Barchek ate meatloaf in a small diner across from the hotel, when Sam and Art returned from the farm. The street was damp, and a cold drizzle blew between the buildings clinging to the diner's window distorting Nick's view. He moved from the counter and rubbed a small hole in the fog created by the warm moist air inside. Through the smear he could make out Art hobbling away from the car while the driver brushed angrily at what appeared to be mud on an otherwise expensive looking overcoat.

He sat back at the counter and opened a notebook he had purchased earlier at Woolworth. He noted the time and wrote, LINCOLN BROUGHAM RETURNS; NOTHING UNUSUAL THERE, SHERLOCK! Then he closed the book and finished his meatloaf.

Art limped into the hotel room as Doc set the telephone receiver gently on the hook and stood thinking, his long silk robe wrapped loosely around him. Art went straight for a liquor bottle on a sideboard below a large window and poured himself a drink, swallowing it off quickly trying to burn the cold aches from his joints.

"Christ, how I hate the cold," he said and swallowed another glassful.

"How's everything at the farm? Can we get started tomorrow?"

"Things are a mess, but livable..."

"I'm not concerned with the the the social life. I want to know if the airplanes are ready, and if there are pilots ready and willing to fly!"

Art hesitated before he took another swallow. "You don't have to worry about the pilots; they're ready."

"And the planes?"

"They'll be ready. The field's a little muddy..."

"Cleanliness doesn't count in this," Doc said.

"A muddy field can cause accidents," Art said.

"Not my concern. I want whiskey moved, if pilots are worried about getting hurt, they can pack up and leave."

"We need those airplanes," Art said. "We can't meet all commitments with just our own."

"You worry about the books; I'll manage the personnel!"

Art sat in a stuffed chair, rolling the glass between his stubby fingers. Doc walked around the room staring at the carpet. "I'm sorry," he said. "I'm just a little tired."

Art nodded.

"I've been on the phone, and we have more orders than we can handle at the rate we've been working in the past. Everyone wants to stock up for winter. They don't think we can deliver after it snows."

"We can't," Art said leaning forward. "You know we agreed to pack up before the end of the year. The risks are too high in winter, it won't be profitable. We discussed this."

"Why give up a good thing? If it's still making money, why quit? We ran trucks all winter in New York, and winter up there was no picnic."

"These aren't trucks, they're airplanes! They can't pull over to the side of the road and wait out a storm. It's too risky."

"I know the risks, and there's no reason to even discuss it now. Winter is a long way off."

"Not so long," Art said waving at the gray sky outside the window. "Besides, weather's not the only thing closing in on us."

"Capone I can handle," Doc said.

Art laughed. "First of all, I doubt Capone could give a shit even if he knows anything about us, and even if, for the sake of argument we say it is Big Al." He laughed again. "What's our defense? Sammy boy?"

Doc glared at him but said nothing.

"Our real threat is from that Treasury man who checked in downstairs. Give him enough time and he'll be up here poking through the books, and when he figures the whole scheme we'll be dead."

"What could he know?" Doc asked.

"I don't know, and I see no reason to linger too long letting him study us. Our original plan to pack up when weather sets in is still the best plan. We can always set up again somewhere else, Cuba comes to mind," Art said and pushed out of the chair and poured another tall drink, toasting the cold gray sky and swallowing it off in one long tip.

"This is all academic, Art. Tomorrow we start up again, full steam. Everything that flies takes off tomorrow morning and heads to Canada. I've already roughed out the schedules." He pointed to the blackboards where freshly printed names and aircraft numbers appeared.

Art saw Henry's name printed at the top of the list of pilots across from the number for the newest Standard LS-5. He also noticed the payloads penciled in were larger than normal, although only by a few dozen pounds.

"Something bothers you," Doc said, but Art merely shrugged. "You've noticed the increase in loads." He tapped the WEIGHT IN TRANSIT column on the board. "We pick up an extra hundred gallons or so on each run. You tell me the advantage in terms of pure profit without any increase in cost of goods sold."

Art was silent.

"That's like free money, Arthur!"

Art moved wearily to the window and pressed his head to the glass. He had been cold, bone cold, but the whiskey was burning his system, and his face was flushed. The cool glass felt good against his forehead. He knew he was drunk, again, and unable to argue for more than a few sentences.

"Have you told the pilots?" he asked. Of course not! he thought immediately. Doc doesn't talk to pilots. He leaves that to Jake or Henry or me, he thought, still pressing his head to the glass.

"Jake or Henry can tell them, and they'll say, 'yes, sir' and hop into their airplanes like good little aviators or they can hit the road!"

Art knew the pilot's opinions were of no consequence, never entering into what Doc had been calling his, "Big Push", his autumn offensive. He enjoyed his role as commanding general, the brains behind the machinery.

Doc rarely slept for more than an hour at a time anymore. He would stay up late into the night scribbling figures on a notepad, rearrange the numbers using a different variable and calculate a new sum. Only he knew what the numbers implied, and only he knew why a pilot's name would suddenly be erased completely from the board and yet reappear hours later on a different route flying a different airplane.

Art rolled his head along the glass and leaned on the sill watching Doc pace the room, his scarlet robe swirling as he turned, the blue smoke from his cigarette at the end of its holder winding toward the ceiling.

"Where's Henry?" Doc asked. "Henry!" he shouted sticking his head through the door to the adjoining room. "Henry. Sorry to wake you. Come in here, please." He walked past Art, and a large ash fell to the carpet. He crushed it into the soft pile with his toe.

"Did you want me?" Henry shuffled into the room rubbing his eyes and pulling a suspender over his shoulder.

"Tell Arthur what you told me," Doc ordered impatiently. "About the airplanes—about what they could carry! Tell him."

"Oh, don't bother. I trust you, whatever it is," Art said but was silenced by a wave from Doc.

"Tell him."

Henry looked at the two men and adjusted his weight from foot to foot. "Well, all I said was that these planes can carry a helluva lot more than the book calls for."

"And what else?" Doc demanded. "About cold weather."

"Well," Henry said scratching his rear. "I said that now that the weather's turning colder the airplanes should fly better, more air or something...denser air, that's it. Anyhow, if they perform better, then they should carry more."

"See that?" Doc said to Art.

"Course, that doesn't mean they're safe," Henry added.

"Whatever," Doc said and turned back to Art. "You see, the pilots have no complaint. We can transport an easy 13 percent extra at no extra cost. That combined with the stepped up schedule..."

"Stepped up schedule?" Henry asked.

"Henry, you will get to do all the flying you ever dreamed of, seven days a week!" He shook Henry by the shoulders smiling brightly.

"Now, combine this with the extra airplanes, and extra pilots, and we will carry more in the next 30 to 40 days then we did all summer!"

He swirled past Art in a blur of silk and blue smoke stopping before the blackboard to stare at the columns and names. He spoke directly to the board, his back to Art and Henry.

"There isn't anyone, anywhere, who's running an organization as efficiently as this one. Everything's paid for, in cash; markets expanding and supplies unlimited."

Art pressed his face into his hands and massaged his forehead pressing a dull liquor headache down.

"Demand! Demand! Demand!" Doc said. "Always up; always they want more, and I am here to deliver." His head nodded up and down, eyes darting across the columns, his mind rearranging the names and figures exploring new strategies.

Suddenly, he rubbed Jake's name from below the column for one of the newer LS-5's and reprinted it under the original ALMINCO biplane, long overdue for an overhaul. Art and Henry exchanged wondering glances watching the change.

"No sense putting an inexperienced pilot in the older machines when there's a well-seasoned pilot available. Hollow knows how to nurse those finicky engines. He can squeeze that extra life out of it before we have to waste money on repairs." Another ash fell from his cigarette, and again he ground it into the carpet with the toe of his slipper.

"If we plan to kick off in the morning I'd better get out to the farm and give the orders," Art said.

"No, I want you here," Doc said. "Henry, you go. Take the Packard. Sam has the Lincoln."

"What's he doing with that?" Henry asked. He felt slighted being assigned the firm's second car.

"I told him to check on Billings at the hospital," Doc said.

251

Henry shrugged and picked his jacket off the back of a chair when the door opened and in walked Jake Hollow dressed in his full length leather coat, two pairs of wool pants, boots and even without his helmet and goggles there was no disputing his occupation. The white skin around his eyes was rimmed in red, and his lips, although thick with grease, were chapped from constant exposure to the wind and rain and oil. His nose ran as it always did hours after a cold flight.

"Hollow!" Doc said, all smiles.

"Thanks for the use of the airplane," Jake said to Henry.

"Anytime. How's Susan?" He tried to sound casual asking after her, but he felt his heart quicken merely speaking her name. He was ashamed of the immense jealousy he harbored toward Jake.

"She's fine," Jake answered quickly. "When do we fly?"

Doc took Jake by the arm pulling him toward the board. "Jake, you are going to like this. Take a look." He waved his hand across the board as if Jake would see an image appear revealing the inner secrets of the enterprise. Instead, Jake only stared blankly, impatient.

"Think demand," Doc said. Jake wiped his nose on his sleeve. "Demand, Jake! It's up, way up, and that's good for everyone. From now on loads are up. Everyone carries more; everyone makes more."

"How much more?" Jake asked. "Money, I mean."

"Much more," Doc said. "We fly round-the-clock starting tomorrow. Every airplane in the air every hour of the day."

"We already do that," Jake said.

"Too many gaps in the schedule, too many machines sitting idle while product waits to be moved. You fly in shifts. You fly up, pick up a load, refuel, deliver, get out for a rest period, someone else jumps in takes off, and the engine never even stops turning. It's like the Indianapolis 500, keep driving, driving, driving. Only this race runs for a month or more."

"It'll kill pilots."

"Only the poor ones," Doc said.

"True," Jake answered coldly, making Henry wince. "But the more planes crash, the more questions are going to be asked as police start picking through wrecks."

"You let me worry about that; you just keep to schedule," Doc said.

"I hear Billings is somewhat of an embarrassment to your paid police force; can't quite figure what to do with him if he comes to. I doubt you own all the reporters in town, and one of them might decide to try and make a name for himself."

"Forget Billings," Doc said. "He won't wake up." There was an icy finality in his voice lost on Jake but not on Art or Henry.

"C'mon, Jake," Henry said. "I'll drive you out to the field."

"I'm in no hurry," Jake said. "I'll stay in town tonight, that pigsty out at the field is so stuffed with bodies there's nowhere to sit."

"You'll spend the night at the farm," Doc said. "As long as you fly for me you will stay where I tell you."

"The hell I will! You can't squeeze 15, 20 pilots into that cracker box and expect them to survive. No, that's one of your problems, you've got too many bodies floating around, and that makes me nervous. Nothing but a bunch of strangers sliding through this operation, who knows who the hell anyone is..."

"I know!" Doc shouted and thumped the blackboard. "I know everything about this operation. Everything! Every detail! And you, Mr. Hollow, are becoming a pain in the ass!" His southern accent became noticeably thicker the more agitated he became.

"You work for me, Mr. Hollow, and when I say fly, you fly, and when I say sleep, you sleep, because, Mr. Pilot, I pay you very well..."

"I'm not talking about money..."

"I'm always talking about money! There is nothing else, you understand? How much have I paid you so far?"

"Hell, I don't know."

"Well, I do, and it's a sight more than you could ever have expected to pick up selling rides at carnivals, which is where

253

I found you a few short months ago. Everything you have now, Mr. Hollow, is because of me. And just as quickly as I gave, I can take away."

Jake tried to think of an argument, but he knew Doc was right. He had earned thousands since June and could easily earn that much again before December. Money was the issue, the only issue, anymore.

"Now, Mr. Hollow, do you wish to continue with this organization, or should Henry drive you down to the train station and put you on the next train to Des Moines. I believe your little trollop is somewhere in hog country."

Jake's face burned, but he said nothing. Henry turned toward Doc suppressing an urge to pound his face into the blackboard for slighting Susan, but that would betray all.

"Well, Hollow, are you with us?"

"Yes," Jake answered weakly, his eyes rolling to avoid Doc's gaze.

"Yes, what?" Doc asked sharply.

Jake studied him for a moment saying nothing. He had thought of Doc as an old hood in a silly robe playing a role unfit for him, a comic parody of a gangster. He put up with the charade, because the money was good, but, now, he saw in Doc the vicious quality of one at home with cruelty and violence. He considered the stakes.

"Yes, what?" Doc screamed, his mouth close enough to Jake's face for him to smell the tobacco breath. He was reminded of drill instructors in the Army.

"Yes," Jake said slowly. "I wish to continue flying."

"And from now on, Mr. Hollow, you will call me, sir, and you don't ever think of entering my room again without knocking, because if you do, I'll cut your face off!"

Jake believed him. The room was heavy with the embarrassed silence that follows an argument. Doc knew exactly how long to play the tension for maximum effect. After a minute, or what seemed to Jake as ten, he walked to the board and slowly erased Jake's name from the slot below the old ALMINCO biplane. Then with careful lettering chalked him in below the newest LS-5.

"We need good pilots flying good airplanes," he said, and without explanation chalked Henry's name below where Jake's had been.

His tone was friendly when he turned back from the board. "Jake, you are a vital member of this organization. I dare say, you are even more essential to it than myself or, say, Arthur. After all, without you there would be no product. You are the link, the lifeline.

"Why, without you this whole thing would collapse." He tumbled his fingers to illustrate imaginary debris dropping to the carpet.

"There are other pilots," Jake ventured casting a quick glance at Henry.

"Yes, but none with your experience, your skill, your record of achievement. Arthur, who is our biggest money maker?"

"Hollow," Art said without looking up.

"Hollow," Doc repeated the name with reverence. "Jake Hollow—the very best!" He stood close to Jake pouring on the warmth until Jake was overwhelmed and subdued.

I should just walk away from all this crap, Jake thought. Not quite the lark it started out to be, although the money keeps getting better, and what-the-hell is there, if not the money. It's only until December, he thought.

"Jake," Doc said putting a hand on his shoulder and kneading the neck muscles, something Jake hated, but did nothing to stop. "When we step off tomorrow morning every airplane out there will be heading for Canada. There is no commercial airline venture anywhere that has tried anything quite like this. We are doing, right now, what others cannot, dare not! We are running a growing, profitable air transport business, and we are doing it completely without government assistance."

"Seems to me we're doing it despite the government," Jake said.

"Ha, ha! Yes, that's true. I like that, but don't you worry; I've dealt with them before! Remember, it's all out for one month, then even if the FBI and the IRS come banging down the doors, they will only find ghosts, because we shall have packed our suitcases full of cash and vanished.

"You see, Jake, the government has to play by the rules, but we are not so restricted. They cannot do anything in a month's time, while we, unfettered by regulation, can do everything."

Doc's sudden warmth and candor made Jake feel guilty, a reaction he found confusing. He knew the speech was calculated to keep him off guard, on the defensive, but he also knew it worked. Jake said nothing and took Doc's proffered hand, shaking it as Doc guided him toward the door.

"Good boy, Jake," Doc said. "Henry, take Jake back to the farm. Maybe stop by a restaurant on the way and get something to eat." He started to reach inside his pocket and said, "Do you have enough money?"

Henry nodded. "Yes," Jake said.

"Good! Now, run along, and good flying!" The door clicked behind them, and they walked in silence to the elevator and rode it down without a word, until they almost reached the lobby, when Henry giggled, then blurted a spitting laugh, leaning against the door.

"What?" Jake asked feeling the giggle infect him. "What-the-hell?"

"Henry was in tears as the door opened, and he fell into the lobby roaring like a drunken bear.

"What's the matter?"

"You!" Henry blurted between sobs of laughter. "You looked like a little kid about to pee his pants!" He clapped Jake hard on the back.

"Oh, for..." Jake said but knew Henry was right.

"Doc can be a mean son-of-a-bitch," Henry said. "Ah, just forget him; let's go fly." They walked through the lobby, past Nick Barchek seated in one of the chairs beside an old bent man with the thin gray hair. As they disappeared down the street he made another entry into his notebook: HENRY TOWNSEND LEAVES WITH PILOT. BOTH APPEARED DRUNK; WISH I WAS. MAYBE THEY COULD SELL ME A BOTTLE.

The night nurse at the County Hospital on the ward where Billings was kept noticed he had stopped breathing at 8:14 p.m.. The resident doctor was summoned at 8:26, and at 8:32 p.m. Aldo D. Billings, pilot, address unknown, was officially pronounced dead by Willis B. Bleim, M.D., who later became a successful dermatologist in Omaha.

By the time Billings was pronounced dead, and Dr. Bleim had returned to bed in the residents' quarters, Samuel Johann Nyberg was already five miles away from the hospital drinking black coffee at a counter stool in a roadhouse outside Lincoln.

He drank his coffee and kept grinning, occasionally talking to himself, causing the few customers in the diner to view him with curiosity.

"Not bad for my first," he said. His heart still raced, and he drank four cups of coffee without tasting one.

He retraced the murder in his head from walking into the hospital with the bouquet of flowers, past the admissions desk, up the stairs and onto the ward where Billings slept. He savored the excuse he gave the nurse about being the patient's brother, only in town for the night, catching the very next train out.

"Oh, please. Mother would be so upset if he didn't get to see him! Oh, thank-you!" He remembered.

The room where Billings slept was empty, the guard away. Doc had come through as planned.

He drank coffee and stared at the polished urn across the counter from him. The short order cook leaned sleepily against the wall picking at a scab on his thumb. Suddenly, Sam wanted the chocolate layer cake under the keeper in front of him. Without asking, he lifted the lid and took the cake with his fingers. Crumbs spread across the counter top, and the cook looked up, hesitated and added 15 cents to the tab. He returned to picking the scab.

Sam bit into the cake, and inside his mind the door to the hospital room closed with a soft click. It was dark except for the yellow halo from a reading lamp over the bed. The room smelled of disinfectant. From down the hallway Sam heard the nurse's voice whisper, arguing with someone, then the two voices faded as they moved down the hall.

He stuffed large chunks of cake into his mouth washing them down with black coffee. He could not eat fast enough to satisfy his appetite, and before he finished the first piece he took a second leaving the lid off the keeper. The cook lazily scratched another 15 cents onto the bill.

Sam choked, spitting cake crumbs across the counter, and he reached for the coffee taking several gasping swallows. There was Billings, face up on the bed, terrible cuts and bruises across his cheeks, his nose broken and crooked, eyes swollen. A brace supported his chin. Sam banged the coffee cup on the counter, and the cook pushed heavily away from the wall and, without a word, poured him another cup.

Sam swallowed. The coffee was hot, but he would only feel the burn on his tongue tomorrow when his brain had cooled. Now, he remembered Billings and how easy it had been to open his mouth and stuff the washcloth past the broken teeth and deep into his throat until the automatic gag response made him retch. Then he clipped his nose tight and holding the rag in place watched him convulse and drown.

Sam finished a third piece of cake and paid the bill adding five dollars as a tip. The old man grinned showing missing teeth and told Sam to come back—anytime.

Dr. Bleim fell asleep casually wondering what it would be like to drown in one's own vomit.

Chapter
XVII

Tuesday morning, October 24, 1929 was cold, although no colder than usual for that time of year. Jake and Henry left the farmhouse, Henry in a new bearskin flying suit he had bought through the mail, Jake wore his knee-length leather coat, buttoned tight around his throat. Underneath, both men wore an array of wool sweaters, long underwear, anything they had to keep the 80 mile per hour slipstream from creeping onto their skin.

Their boots crunched on the frost covered grass; steam rose in puffs as they spoke.

"How'd you sleep last night?" Jake asked.

"Fine," Henry answered. He walked faster than Jake and had to wait every dozen or so steps for Jake to catch up. "I wasn't too crazy about breakfast, though. I hate oatmeal."

"I'll eat when we get to Moyer's."

"All day without eating?"

"I brought something along," Jake said holding out a whiskey flask and a BABY RUTH candy bar.

"I pitched against him once," Henry said pointing to the bar.

"Ruth? Really? Strike him out?"

"Nope, homered off me. We lost," Henry said. "How'd you sleep?"

"Not so good, that damn Lundy, the guy with the Curtiss Robin, his damn dog did nothing but fart all night—BRAP! PHWEW! BRAP!"

"No problem, just put a candle next to his ass."

"Naw, might burn the house down, then Doc'll make us sleep in the barn."

They walked along the ragged line of aircraft being readied for flight. One of the Travel Airs popped to life, complaining in the chill air. Elsewhere, propellers were being pulled through, and fires glowed in several places where pilots warmed themselves and the crankcase oil before pouring it into the cold engines.

CAPLUH, PLAH, BRAKaBRAK! Another motor hammered away at the night. Across the dark airfield, engine after engine struggled to life. Pilots and mechanics moved like ghosts through the shadows, the only clear voices coming from mechanics swearing at cranky motors that refused to fire.

Henry and Jake turned down the row of Standard LS-5's parked separately from the others. The immense biplanes, painted maroon and silver, were the pride of the fleet. Jake never admitted it, but he enjoyed the recognition he received for being one of the senior pilots, flying the company's own aircraft.

He waved to a mechanic standing on the landing gear pouring hot oil into the motor. All the planes had been fueled the previous night, so it was only a matter of getting them started and leaving. Jake was anxious to go, although he knew it would be a long cold ride, even after the sun came up. Passing the old ALMINCO plane he called to the pilot just strapping himself into the cockpit.

"I hope you brought along extra oil, she uses it faster than gas it seems."

The pilot, Richard something—Jake had forgotten his last name—waved back.

"What was his name?" Jake asked Henry. "Hirshman, Hirshstein, Hirsh-something."

"First name's Robert, isn't it?" Henry asked.

"No, Richard; Richard Hirst, or Hirsh..."

"You sure it isn't Robert?"

"No, Richard...I think."

"We could ask him."

"Ah, screw it; I'll ask him if he's here next month," Jake said blotting Richard-Robert Hirsh-Whatever out of his mind. "Anyhow, I thought Doc had you scheduled to fly that one."

"He did; I switched." Henry grinned. "I told the kid he'd been promoted and didn't have to fly the new planes before they were broken in yet."

"And he bought that?"

"I was persuasive, very persuasive. I told him I didn't think he could fly an airplane with broken thumbs. He became agreeable."

Jake felt he could only get so close to Henry, then there was this wall. Something about the ex-baseball player told him that Henry could be a good friend or a terrible enemy, a complex man of gentle violence.

"See you up north," Henry said stopping at his plane as Jake continued to the end of the line for his. Already, the mechanics had Henry's cowling unbuttoned and the cockpit cover removed. Doc had hired several mechanics from town, mostly friends of James, the black man who ran the delivery trucks around town. He reasoned that black mechanics would be less likely to stumble into idle conversation with white authorities.

Henry performed his walk around inspection checking oil, water and fuel, as well as tugging at flying wires and shaking wingtips to assure himself they were still attached. The Standard to his left barked to life and sat idling, the grass behind laid flat, with frost crystals whipping away in the blast.

"Is everything to your satisfaction, Henry?" a mechanic called from under the plane's nose.

"Good morning, Andrew. Too damn early for anyone to be up on this cold a morning," Henry said.

"I don't get paid, if I don't get up, Henry," Andrew said while pulling the propeller through. "Now you go jump inside, and we'll get this thing started, so I can get back home for breakfast, and you can get to work."

Henry climbed into the cockpit and settled himself in the seat that would feel like a slab of marble to his rear at the end of 700 miles. Andrew climbed onto a small ladder alongside the motor and poured gasoline into the petcock valves on the intake manifold. As gas ran out the bottom of the cowling, melting frost on the grass below, he again pulled the propeller through until he was satisfied it was in a position to fire.

Henry waited nervously in the cockpit for the command to start. Looking down the line he saw Jake's cowling still

261

undone, and Jake leaning far into the engine compartment tapping away at something. A Jenny taxied past heading for the runway.

"It's ready, Henry! Give her a whirl!" Andrew called.

Henry cranked the booster coil waiting for the familiar shudder of ignition. It came with a violent shake of the fuselage, and he adjusted the throttle keeping the motor running while water temperature slowly rose. The cold morning air rushed past his face behind the windscreen, goggles, and scarf. Little needles of ice were already sneaking under his clothes. After a few minutes he saw Jake's plane warming up, tugging at its chocks.

Henry followed two Travel Airs taxiing out to the departure point in the darkness at the end of the pasture, far away from the faint lights of the house. The runway was outlined with a mixture of oil lanterns and smudge pots placed haphazardly along the edge of the field. No lights marked the end of the strip, but any pilot would know if he had gone too far as there was a sharp drop into a rock strewn gully somewhere beyond the lights.

The racket of an OX-5 engine whirred over Henry in the darkness, and he turned with a start catching the outline of a biplane skimming over him, silhouetted against a weak half moon. It vanished again into the darkness.

"Shit!" Henry blurted. "Someone's going to get killed. Maybe if someone stood at one end with a light or something telling when it's safe to take off." He spoke easily with himself whenever he flew. He hoped no one ever caught him doing it.

Ahead, the two Travel Airs pivoted in unison. The pilots ran their engines up checking magnetos before turning into the light westerly breeze. They took the runway as a pair, one slightly ahead and to the left of the other. The front pilot turned and waved to his companion. The two biplanes shook with open throttles, and they rolled down the bumpy strip lifting into the sky as a pair and turned east over the parked planes, shadows in the pale moonlight.

It was Henry's turn, alone at the end. His hands were damp with cold sweat inside the huge gauntlets, and his heart raced. He hated night flying. With his head peeking over the cockpit's rim looking down the converging rows of lights he slowly opened the throttle feeling the cold vicious propblast against his face as the Standard lumbered over the rough surface, bounced, the tail lifted, and the biplane tried to turn to the left. Henry pressed the right rudder, the gear smacked a rut, and the whole machine rose into the sky passing low over two other LS-5's, one containing Jake, the other Richard-Robert What's-His-Name.

The lights slid quickly beneath him, and he was over the few remaining aircraft waiting to start. Most sat in the darkness, engines at idle, pilots checking instruments, testing controls or just mustering nerve to taxi down the runway hoping no one would plow into them on takeoff.

Henry turned east, then north parallel to the field. Looking over his left shoulder he watched a black shape move slowly along the ground, momentarily blocking out one edge of the runway lights one at a time as it gathered speed.

"It's either Jake or Richard-Robert," he said and loud enough to be heard over the roar of the Hisso engine. He knew Jake would take a direct heading for his first fuel stop, cutting a straight line across northwest Iowa, past Sioux City and into western Minnesota. Henry preferred to find something a little more solid to navigate by than his own dead reckoning skills.

With the airway beacon at Lincoln off to his right flashing its beam, he turned further east heading directly toward the Missouri River. Once over it, he would turn left to Omaha and start to look for Sioux City. He knew Jake would pass him somewhere in the night beating him to Canada. Idly, he wondered what the odds were of Jake actually bumping into him in the darkness.

"Ridiculous," he said and slowly rocked his wings looking for all the other planes that had taken off before him. All he saw was blackness and the beacon in Lincoln amid the shimmer of the city's lights. The half moon sank toward the western horizon, and the sky toward Iowa began to glow slightly. He knew by Sioux City the sky would be light enough to start reading water towers.

Doc had violated his own rule of staying clear of the actual operation, when he had Sam drive him and Art out to the farm to watch his squadron depart. Sam parked the Lincoln on the dirt road near the house in a position where Doc could watch and, yet, keep out of sight. The planning was done, it was time to let the machinery run. Sam kept the motor running to keep the car warm and fell asleep on the front seat.

Down the road and well clear of the house or any of the aircraft, a second car, a Ford, was parked in a small grove of leafless trees. Nick had taken up an observation position closer to the airplanes in a low brushy spot beyond the fence.

He sat for about two hours shivering in his overcoat, waiting for the first plane to depart. Occasionally, he shifted position to keep from cramping but was unable to move around too much lest someone should hear. His presence was known in town and virtually ignored. He thought it best, however, to keep out of sight so close to the actual operation.

As each plane started, he tried to catch some glimpse of whomever might be around, but the pilots were all swathed in scarfs and long coats, making identification impossible. The types of aircraft were more readily discernible in the weak moonlight and glow of small fires scattered around the field. Besides, he cared little about the pilots, it was Doc he was after.

Voices drifted toward him—snatches of technical conversation and sudden oaths condemning some part of a machine to the deepest recesses of Hades. He watched a Jenny lift off, proud that he could recognize the type through a book from Woolworth entitled, EVERY BOY'S BOOK OF AEROPLANES.

Two Travel Airs lifted off in tandem, narrowly missing a flatbed truck driving across the runway with its lights off.

"It's a wonder no one's killed," he said through chattering teeth.

The distinct roar of Henry's Hispano-Suiza motor vibrated the ground, banking over Nick. He took advantage of the racket to scurry, as best his stiff limbs would allow, back to the car. Emerging from the underbrush, he stretched his legs and reached for the door handle, when another plane opened its throttle in a pounding roar.

"Another one of those big bastards," he said to himself.

Gradually, the sound moved through the trees and lifted, held that pitch for a few seconds, then with a heart-stopping cough—instant silence, followed by the sound of crunching wood and metal across the field.

Jake was behind the ALMINCO LS-5 and saw it roll away from him down the runway. He taxied his plane into position and waited for the biplane ahead to bank past the half moon, before he would open his throttle.

The moon, however, never blinked. Jake looked off to the side for the departing Standard. With his own engine at idle, he had no trouble hearing the sudden change in pitch of the ALMINCO motor, the brief silence, then the twisting crunch of the wreck.

Richard-Robert Hirsh-Whatever knew he was too low to turn back to the field, when the engine suddenly quit. He also knew there was no more runway ahead, only darkness and vague shades of gray outlining the house, a few out-buildings and some trees beyond.

Doc and Art saw it happen. Nick heard it. Andrew was busy helping Lundy start his Robin when the LS-5 dropped from the sky into the black void behind the house.

Richard-Robert remembered a short field just beyond the house, and knew if he could keep the plane under control and aimed for the darkness without hitting the roof he could drop into the field, perhaps roughly, but at least without serious trouble.

There was no time to panic and barely enough time to realize what was happening, when his landing gear caught the tops of the tall cottonwoods. The machine was plucked out of the sky as if a giant had reached a hand and grabbed it. He plunged almost straight at the ground, hitting right wingtip first, instantly shearing it off without slowing the impact of the nose.

The propeller hit next and shattered. The impact then tore the Hispano-Suiza engine from its mounts, slamming the whole tangle of hot cylinders, ruptured water hoses and fuel lines back into the cargo compartment ahead of the pilot. The empty whiskey tank collapsed as Richard-Robert's seat broke free from the floor and carried him straight through the instrument panel and into the front compartment. It was then the aircraft pivoted on its compressed nose, and the tail came over crushing him into the hoses and hot cylinders, but by then he was already dead.

Jake opened the throttle half way and raced down the runway between the parallel rows of lanterns. The tail lifted off the ground, the main gear crowhopping toward the wreck. He skidded to a stop, fishtailing sideways into a handful of figures running past, ignoring his spinning propeller. He killed the engine and scrambled from the cockpit to join the others running into the shadows behind the house.

Doc smacked the seat in front of him awakening Sam Nyberg. "Son-of-a...there goes a hundred gallon capacity!" he shouted. Art had the door open, but stopped to stare at Doc.

"That could've been Henry," he said.

"Let's get out of here," Doc said, poking Sam in the shoulder with his cane. Sam hesitated, wanting to be a part of the crowd at the scene, less to help than to gawk.

"Shouldn't we see...?" Sam asked.

"No," Doc said calmly. "Take us home. We have planning to do; the schedules will need readjusting."

"But what if it's Henry?" Art asked, his voice strained.

"Then his name goes on the reserve list until we can evaluate his status," Doc said.

Art was already out the door limping away from the car, before Doc could finish what he was saying. He reached across the seat and closed the door behind Art and settled back in his seat.

"Let's go, Samuel."

The Lincoln Brougham backed across the road, and the gears ground before Sam found first, and they left.

Art was the last to arrive at the sight. Beside him stood Nick Barchek, staring at the twist of wires, splintered wood and torn fabric. A handful of pilots and mechanics tore through the debris probing for the pilot. Someone waved a lantern, but the sky was becoming bright enough to see without it, still he held it aloft anxious to assist.

"Here!" Andrew called, his arms deep in a tangle of ragged plywood and sheet metal. "I found him!"

"Bring a light!" Jake hollered and scrambled across the mess. Pieces of airplane snapped and creaked under his weight. "Where is he?"

Before Andrew could answer, the lantern appeared, illuminating a hole where the weak dawn had not yet reached. There was the body. Andrew reacted by doing nothing, only staring. Jake took one look at the broken form that minutes before had been a pilot whose name he could not remember. He pulled back in horror. In the sharp glow of the lantern, one dead eye gazed straight at him from behind the cracked lens of the goggles, still strapped neatly to the crushed head. The rest of the corpse was mixed in a bloody weave with the biplane's mangled structure.

Jake gasped for breath and felt the nausea rise from his stomach. He interrupted the reaction by violently kicking the inverted fuselage.

"Bastard!" he cried, and kicked again, splintering the plywood panel. "You bastard!" The words were strange to his ears, faint, distant as if from one of the other mouths gaping around the wreckage. Jake stormed off the plane, through the crowd and pushed aside two pilots still in leather togs. He broke into a run passing Nick and Art standing almost shoulder to shoulder.

"Jake!" Art called, but Jake hurried away toward his plane. "Who was it? Was it Henry?" he called.

"Who is it?" someone asked from the blur of the crowd.

"I dunno; some new guy; yea; maybe...No, I think it's the big guy, Henry. No, it's that Richard Something. You mean, Robert What's-His-Name...?" Art felt the wave of relief swell over him hearing it was other than Henry. Nick slipped back into the trees before anyone knew.

The dawn was almost upon them. They heard Jake's plane howl from beyond the house. Impatiently, he taxied onto the airstrip, opened the throttle and lifted into the sky with the light breeze at his back. They saw him rise from beyond the cottonwoods and level a few feet above the distant hedges before turning north, his throttle still wide open. The engine was soon a dull murmur, and the sun peeked over the horizon.

To an observer, high above the northern Minnesota landscape, the arrival of Doc's air fleet of smugglers would scarcely warrant notice against the seemingly endless stretch of snow mottled woods and lakes. What had departed Lincoln in a stream of roaring machines were now specks creeping independently across the border. First, the two Travel Airs in loose formation low across the trees and marshes, then Henry some 20 minutes later came angling in from Fargo, North Dakota where he had wandered after following the wrong river fork. The others straggled, one by one, into Canada at odd intervals, having lost sight of each other almost immediately after take off.

Jake flew a straight line, throttle wide open, motor howling. He tried to suppress the image of the crushed pilot from his brain by running the engine fast, watching the needles in the gauges lean toward redline. He wanted the oil hot, the water boiling. He wanted the dead staring eye burnt from memory.

"THE GOD DAMN BASTARD!" he screamed into the vicious wind, but the image would not leave. The eye only stared at him from the scattered clouds, the half-frozen ponds and marshes below. He was a prisoner, strapped to his seat, unable to escape until he landed and could pass the image to others or simply hide from it in a crowd. Alone, in the sky he was defenseless against the gaze. The sky had always been his hideout, his safe haven. Now, the world was coming aloft with him. He had to keep it away, to keep his sky free of what bound others to the ground. He had to, or he would be no better than those below.

"The stupid idiot," he said after a long quiet pause. "Where did he screw up?" By merely asking the question, he felt the gaze shift away from him and turn inward upon itself.

He ran the engine too fast and used more fuel than planned forcing him into an extra stop before Canada. Again, it was a small town where airplanes had never been seen, at least up close. A crowd of beaming faces asked the usual questions about speed and range and Charles Lindbergh. Jake blankly told the same lies convincing them of his familiarity with all three. He was still a barnstormer at heart knowing how to work the crowd.

"How 'bout a ride?" the voice that was always in the crowd asked.

"Sorry, can't. Company policy."

The nods of agreement were followed by the dozens of hands all anxious to help turn the machine back into the wind, so it could leave them to remember and retell.

Airborne, Jake waited for the dead eye to stare again. It did, from the blur of the propeller tip.

"I didn't want to know your name," Jake said to it. "I still don't want to know."

The image faded momentarily from the propeller then reappeared. "You screwed up! That has nothing to do with me," Jake said. "If you want to screw up, that's your business, not mine." The eye no longer looked at him, although the image was slow to leave. It lingered, but Jake refused to acknowledge it anymore. Somewhere, a few miles from Moyer's, it melted deep into his brain where he could leave it and go about his business.

The Indian lineboy waved both arms directing Jake to taxi behind the other biplanes waiting to be fueled. With a slashing motion across his throat, he signaled Jake to kill the engine and ran back toward the runway.

The plane ahead jerked from a dozen pairs of hands pushing it closer to the pump. Jake was still in the cockpit when the same hands grabbed his struts and yanked him forward.

Ahead, a Travel Air was pushed away and abandoned while two hoses were plugged into the next machine.

"Faster there, lads!" It was Moyer's voice calling from behind. Jake saw the red-bearded giant striding, with clipboard in hand and clods of muddy snow kicking off his boots.

"Get that beast all the way away from the pumps in one push. No reason to make two trips, eh. Charlie! Charlie, where-the-hell are you...Oh, there." A small boy in blue wool trousers, green coat and a hat a size too large stood close to his side. "Don't be hiding on me, lad. Run inside and get those lazy buggers moving, eh. I've got two planes ready to fly and no pilots. Now, run along!" The boy ran straight to the cookhouse where Moyer's dog and a gaggle of children lingered near the door awaiting hand outs.

"Put a board under the blessed thing if it won't move!" Moyer called to a crew trying to dislodge a wheel from the mud. Someone grabbed a board jabbing it into the goo beneath the tire, while Moyer waved at the rest to push. Setting his clipboard on the wing he leaned in with the crowd, and together they rocked the plane clear.

"Don't slop that gasoline the way you've been doing, eh!" he called to the fueler. "I saw you do that once while we were still pouring whiskey, and I don't think even the Americans can stomach too much of that mixture, eh!" He broke into his deep coughing laugh, bringing smiles to the faces working the line.

He lit a cigarette and wrote out a receipt for the gas and whiskey, recording the aircraft number and pilot's name on his clipboard. He also noted the time of arrival, length of refuel, and any minutes wasted waiting on pilots to wolf down a meal, take a leak and scramble back into their machines. He caught sight of Jake and gave a friendly wave and a shrug.

A pilot strolled out of the cook shack, stretching his legs and lighting a cigarette. It was one of the Travel Air pilots (there were many), a Canadian named Austin. Moyer flipped through the papers on his clipboard and saw that he was six minutes away from being behind schedule.

270

"Hurry your arse up!" he shouted in a strange mixture of British Army sergeant and North American twang. The effect, however, seemed to work. The pilot looked thoroughly confused. "You don't get paid to stand around and fart!" Moyer stood ramrod straight, shoulders squared and arms akimbo.

"In your machine, lad! You're running late. Get a move on!"

The pilot was alone, unprepared to argue. Moyer knew even the most troublesome recruit could be tamed by cutting him away from his peers. The pilot donned his helmet and scrambled into the plane while someone pulled the propeller through once.

Austin stood in the cockpit when Moyer barked the commands, "Switch on! Throttle cracked!"

"Ah, it's on and..." The pilot's complete response was lost. Moyer nodded, and the lineboy kicked his leg high and swung the propeller through. The motor caught, and the biplane lurched forward dropping the pilot into his seat. The engine roared briefly, while Austin groped for the throttle, and the tail skid enscribed a lazy S in the mud and patchy snow heading for the runway.

Another of the Travel Air pilots came stumbling out of the cook shack, having apparently heard the engine start. Moyer stood legs apart waving his clipboard at the departing aircraft.

"Well?" he called in his parade ground voice. "Your friend's up and left you. Said you'd fallen into the crapper and would never get out."

With a dozen tiny Sioux hands tugging at his sleeves and coattails, the pilot considered Moyer with that same glazed look the other pilot had. Austin's biplane lifted over the trees heading south.

"You're almost two minutes late, my friend," Moyer called. "Not good, not good." He wrote something on the clipboard as he spoke, and the pilot having no idea what it could be ran to his plane and was inside, while an expressionless crew swung his propeller through.

271

"SWITCH ON; THROTTLE CRACKED!" Moyer shouted. The pilot nodded, and the engine popped. Moyer waved him away like a foul stench with his clipboard.

With the sound of the second Travel Air fading beyond the trees, Jake's plane was shoved into the fuel pit, and the hoses inserted. A light snow blew across the railroad tracks and swirled around faces and wings clinging to cold surfaces refusing to melt.

Jake stood inside the cockpit and stiffly climbed over the side. He glanced at the sky, stomping his boots against the soft ground.

"Why aren't you inside getting something to eat?" Moyer asked without looking up from his papers. Snowflakes were catching in the red thatch around his face.

"Not hungry," Jake said. The greasy column of smoke from the cook shack twisted in a gust and pointed toward them. The smell of fried meat passed through them and vanished again.

"Well, at least it's warm in there," Moyer said.

"Not cold," Jake answered and stamped his feet again. A beam of cold sunshine broke between the fast moving clouds, making the snowflakes almost glow. "That wind's getting strong out of the north; I might make it all the way back to Lincoln tonight."

"Not before dark, you won't."

"Night flying doesn't bother me," Jake said.

The two men stood quietly for a moment; the snow dulled as the clouds pinched off the sunshine again. Jake felt the cold dig into his coat.

"Heard you lost one this morning," Moyer said.

It seemed so long ago since that morning when all the planes had left. "Yea, somebody screwed up. He had plenty of room to put down but chose a tree instead. Kinda stupid if you ask me." Jake tucked a wool scarf into his collar and pulled his helmet on buckling the strap.

"You would have done differently?" Moyer asked.

"Wouldn't have landed in a goddamn tree for Chrissake!" Jake answered.

The fuel and whiskey hoses were removed, and the heavy Standard was pushed around in a half circle from the pit. Jake swung his legs over the cockpit rim and dropped into the seat after brushing a thin layer of snow away. Feeling Moyer's eyes on him he pulled the goggles over his.

"You're not going to check your oil or anything?" Moyer asked.

"They already did it for me," Jake answered.

"Not like you to let someone else do those things," Moyer said. "But I guess everyone's going through changes these days."

"What's that supposed to mean?" Jake asked.

"Oh, I've just noticed how people change—myself included. Everyone's adapting to new ways, a new life."

"What are you talking about?" Jake asked.

"You, me, the Sioux Indians here. We've all changed some in the last few months. Hell, this place wasn't anything more than a watering hole last May, and these locals had never even seen an airplane. Now look at them. The twentieth century's dropped right into their laps, and they've taken a liking to it."

Jake glanced over at the cook shack as another pilot emerged picking his teeth. Moyer's dog, now fatter than ever, and the Sioux children danced around him yapping and holding out hands. Jake was convinced if the dog had a hand he too would have extended it.

"I haven't changed," Jake said.

"You probably the most," Moyer answered.

"Go to hell."

"Been there already."

Jake looked at him across the cockpit's rim. The snow had changed to a steady shower dusting the red beard. "I gotta go," Jake said.

"You've become a moody son-of-a-bitch," Moyer said.

"Not moody, just cold."

"Spend the night; only the short haulers going into Minnesota have to be off right away. There's reserve pilots inside. Let one of them take it. Henry's staying."

"Let him do what he wants. I get bored sitting around. I'm going, bye."

"We don't need more wrecks," Moyer said.

"I don't need advice. You might want to stand back, it gets awful cold when that thing starts blowing." He pointed at the propeller. A lineboy pulled the propeller through.

Moyer stepped back, the engine fired, sending a cloudy mixture of snow and mud flying past the tail.

"You're not the only pilot, Jake!" Moyer called through cupped hands. Jake shrugged pointing at his ears tucked under the padded helmet flaps. "You can't do it all!" Moyer waved, knowing Jake could not hear.

The huge biplane swung itself around in a cold blast of power and taxied quickly toward the runway, kicking up dirty snow. Jake held the brakes and ran the engine for several minutes without moving. When engine temperatures had risen enough, he opened the throttle wide and released the brakes.

Loaded full with gasoline and whiskey, the heavy biplane barely crawled over the rough surface. Only a lunatic could believe such a concoction of wood, steel and cotton, splattered with grease and mud, could fly. But it did. Using less than half the runway, it levitated, slowly moving over the trees, then turned south picking up speed as the wind pushed it along.

Moyer watched Jake disappear and listened for a few minutes to the changing pitch of the motor. The door to the cook shack slammed and another pilot faced him from across the tracks.

"Well?" Moyer hollered, again the sergeant on parade. "I'd been told you'd fallen into the crapper, and it would be spring thaw before we could fish you out..."

Jake was a thousand feet above the Minnesota woods in bright sunshine, having flown out of the snow shower almost immediately after leaving Canada. He squinted against the low sun to the southwest. His body was cold, but his mind seemed to have acquired a numbness that had nothing to do with temperature. Moyer's observation that everyone had changed, somehow, meant little to him. He realized that little of anything meant anything to him, except that flight at that moment.

He dropped lower into choppier air and watched his shadow race over the gray countryside. Experience allowed him to estimate his groundspeed at something close to 110 mph. He calculated his fuel and knew exactly how much ground he could cover before refueling.

"No sense sitting on the ground," he said and thrust his face into the icy slipstream. His body reacted to the jolt, but his mind still viewed everything with detachment.

Curious, he thought. Nothing seems to matter. "Hmmm," he muttered and shrugged. Abstractly, he replayed the early morning crash right down to the details of the doomed aircraft taxiing past the runway lights and the sight of the ruined pilot smashed in the wreckage. It seemed so far away, so removed.

"Such a waste," he said without conviction. He fumbled inside his pocket taking out a BABY RUTH. Flecks of red and white wrapper flew off in the wind, and he bit into the hard sweet lump. His mind ran to the remains of Billings' Travel Air.

"A waste," he said chewing.

Then he conjured up the last view he had of Sterns' biplane in the woods at Adair's.

"Silly of him," he said. The candy bar melted slowly in his mouth, and he picked a fleck of cold chocolate from his glove with his teeth.

"God, what a gorgeous day!" he said as loud as he could, watching the sun turn orange against the horizon.

Chapter
XVIII

The first round of Doc's autumn offensive was a success despite the one crash and the complete loss of an airplane. The ALMINCO LS-5 was quickly replaced by three new pilots in Curtiss JN-4 Jennies. They carried less than the Standard, so Doc reduced their salaries to compensate for the added expenses and longer times enroute. The other pilots complained when they heard about the new crews hiring on at reduced wages, fearing their own salaries could be cut as easily. The new pilots, however, were delighted just to be working. Any flying job that paid regularly was to be coveted. Doc ignored the grumblings from his crews, then considered the possibilities and put out the call for more pilots. They came.

"We couldn't have done better if we'd placed an ad in the New York Times, Arthur!" Doc said happily, chalking new figures to his over-crowded blackboards.

"Well, we had better be all packed up and moved out by Christmas," Art said from his desk in the hotel room. "Things are getting too big; too loose."

Art's constant sour observations irritated Doc. "Oh, you're a wet blanket. We are making money hand over fist! Look at our cost per gallon." Doc thumped a dusty column bulging with numbers. "Down 12 per cent!"

"Treasury man's still in town," Art said flatly.

"Can't touch us!"

"Yet. Have you seen how many pilots there are out at the field?"

Doc ignored the question and studied his boards.

"There's a boatload of 'em out there!" Art said.

A week into the campaign, the airfield outside Lincoln was swarming with pilots and an assortment of aircraft. Day or night someone was running an engine or trying to start one. Those with an airplane, made money. Those without, waited for a slot in one.

Trucks and wagons came and went at all hours, delivering gasoline and food and hauling away whiskey. No one left the farmhouse anymore without Doc's permission. Armed guards patrolled the farm's perimeter and the the dirt road leading in.

A week after the three Jennies arrived, one of them crashed in a snow shower in South Dakota. The pilot was unrecognizable after the fire contorted him into a charred lump with stubby arms and legs. The local paper carried a front page story about the mystery plane and its unknown pilot, but Doc never read the DELL RAPIDS SENTINEL and assumed the pilot had simply found other work and had left without giving notice. Without a forwarding address, Doc had nowhere to send the back pay. Someone flying an American Eagle biplane showed up looking for work and immediately filled the vacated position.

Jake, however, still flew the most, followed closely by Henry, who took it as a challenge to match Jake's record. Both men lived on the verge of exhaustion, catching sleep wherever they could—often in the cockpit, nodding until the engine's pitch howled violently as the plane's nose pitched over, then they would snap awake clutching at the stick. Jake tried tying a rope around the stick and to his seat, a primitive autopilot that seemed to work, until he awoke once with his compass heading almost a 180 degrees different from when he had last read it. His foggy brain was unable to calculate how many circles he might have flown while asleep, so he landed and passed out in his seat, waking an hour later with a cold drizzle soaking his face.

On any night, the bunkhouse in Canada would be crowded with snoring pilots curled up in corners; the damp stench of dirty bodies and old socks hung permanently in the air. The showers Moyer had installed beneath the water tank were exposed and frozen solid by the first week in November. No one seemed to notice the sour odors after flying 12 hours through rain and snow or pounding turbulence. They hit the bunks completely clothed and staggered out when someone would kick them muttering, "Time to go; get up!"

The cook shack ran continuously, churning out fried greasy meals at ever-increasing prices. A mound of garbage grew beside the building where the dogs and children rooted and fought over choice scraps. Thankfully, snow would cover the mess periodically, allowing Moyer to ignore it.

Doc was a gambler; not in the common sense with cards and dice but as a businessman. He took risks shunned by others and possessed a gambler's instinct for smelling odds— an innate talent, something that could never be learned.

He had never lived in the northern midwest and knew virtually nothing about the unpredictability of autumn weather, and its ability to range from warm and sunny one day to vicious winter the next. He was bored with details, and refused to let Art, or anyone else with less than an optimistic view, dwell on the possibilities for too long.

Doc knew, as only the gambler about to roll a pair of fours to hit eight the hard way, could know, that winter, true frozen winter, would hold off until December. It had to. He was strung out on a giant wager with every dollar reinvested, and by the end of the first week of November his luck was holding.

Reports from Moyer's showed tallies far out-stretching the entire inventory prior to the November push. The pilots no longer carried cash like so many independent bootleggers. The money was shuttled through a string of Canadian banks all under different names; all managed by Moyer and monitored by Arthur. Should a pilot crash and fall into unfriendly hands, he only had himself, the whiskey and an improbable tale. So far, the only ones to go down had the courtesy to die.

The activity at Moyer's went unmolested, until a Canadian customs official showed up one cold day swathed in a floor length seal skin coat with a large fur hat making his head look like a fuzzy black jack-o-lantern, except he never smiled.

He stepped from the train at sunset, looking extremely bored and put upon, as only the career bureaucrat can. Moyer had been warned of his arrival and had the field readied by scheduling the evening's arrivals to begin after the train's departure.

"Watch for a bonfire," Moyer had told the pilots. "If you don't see one burning, don't come in!"

The customs clerk poked lazily around the main office in Moyer's cabin, glancing only briefly at the bogus but impressive looking books Moyer kept updated and available. The customs man flipped through a few pages, briefly stopping to read entries about passengers transported between there and unreachable outposts in Ontario—outside his area of responsibility.

"Who would go there, eh?" he asked.

"Sportsmen," Moyer said. "Americans. They arrive by train, and someone flies them into the back lakes..." The official waved him into silence and flipped through the remaining pages, then slammed the ledger shut and sniffed at the air.

"Is that steak I smell?" he asked.

"Yes, it is!" Moyer answered cheerfully. "Care to stay to dinner? You're most welcome." Outside, the train whistle blew; schedules had to be met, customs official or no.

"Haven't the time, but I do admire the smell of those steaks, eh!" He looked straight at Moyer, eyes unblinking.

"Well, since you can't stay, do take some back with you." Moyer pointed toward the door, and the customs man started to move then said, "And, Mr. Moyer, if you could see clear of a little..." His voice trailed off, and he tapped a whiskey bottle Moyer had on the table with his finger nail.

"I was about to suggest just the thing for the ride home," Moyer said, and they walked outside. Passing the kitchen, Moyer stuck his head in the doorway briefly and ordered something unintelligible to the woman inside. They continued to the train, and as the official lifted his coat skirts to climb the steps, Moyer pointed to three men loading two cases of whiskey and two large wrapped packages onto the box car behind.

"Everything's in order here," the official said with a sniff. "Don't know what the complaints were about." He stepped inside the car.

"Do stop in again," Moyer called, and the door closed with a heavy click. The whistle blew, and with a blast of steam and black smoke, the locomotive pulled away.

"Light the damn fire!" Moyer called looking at his watch. "For Christ God Damn sake," he said to himself.

Minutes later the planes started to arrive, and a light snow fell.

The streets of Lincoln were cold and swept by a dry wind lifting swirls of dust and paper scraps high into the air as it puffed in ragged gusts between the buildings. Doc was on the phone most of the day arranging shipments and offering discounts to buyers with cash. He rarely left the apartment on the top floor of the Hotel Lindell, spending all day dressed in either his red or gold silk robes.

Art, too, was confined to the apartment reworking his books, keeping track of every gallon, every dollar running through the system. Sam Nyberg mostly sat around the stuffy apartment toying with his pistol and generally being bored. He left the running of his own outlet, Randall's old garage, to his cousin, who turned in a respectable profit without much effort. About the only chance Sam had to leave the hotel was to fetch cigarettes, a task he came to savor, watching the ashtrays fill until Doc would call, "Sam, run down to the corner and pick us up a carton of Luckys. Art, you smoke Luckys don't you?"

"No, Camels."

"And a carton of Camels for Art." Then he would return to the phone without another word.

Sam took these opportunities to stretch his legs; stretch them, that was, behind the wheel of the Lincoln. Slowly driving around town, he would pass any place that had a crowd, hoping to see the fingers point, the heads nod, all thinking, "...there goes an important man, that Sam Nyberg." Generally, however, he was ignored, even shunned, and after about 40 minutes of this, he would wander back to the hotel with the cigarettes. Doc never complained about the time it took, scarcely affording him a shrug.

Sam walked aimlessly around the room, briefly annoying Art working on his books. Then receiving no satisfaction there, he poured himself a tall drink from an open bottle and picked at the breakfast dishes still piled on the table in the center of the room. Like a patient confined to a hospital bed, Sam's favorite distractions throughout the long days were the three meals served by room service. Doc and Art rarely ate more than a few bites, preferring coffee and cigarettes instead. Sam obliged them in picking at the remains until his skinny frame sprouted a band around the waist.

"Well, sell everything...uh-huh, everything!" Doc said into the mouthpiece held in one hand, while his other held the receiver firm against his ear. He walked as he spoke, the silk robe rising in a swirl around him, the phone cord catching in his legs.

"I told you the damn stock market was headed down!" he now yelled. "Just sell whatever I got left; at whatever the price. Get rid of it!" He put the mouthpiece down long enough to twist a spent cigarette from the holder and press in another. "I know it's not much, damn it, but whose fault is that? If you'd listened to me..."

Art heard the thick southern drawl creeping into his voice, indicating the conversation was getting out of his control.

"Never mind, IF you can sell! Just sell!" The receiver slammed against the hook but missed, and the whole phone bounced to the floor in a tangle of loose parts and black wire. "Damn," he muttered retrieving it by the cord and thrusting it all at Sam to unwind.

"Bad day on Wall Street?" Art chirped from his corner.

"I know! I know! I should have listened to you a month ago," Doc said.

"You'll survive. You haven't lost much."

"Too much, and that jackass keeps saying things will turn around right after the new year." He paused a moment and asked, "What do you think, Art?"

"Sell."

"Now?"

"Yesterday, if at all possible," Art said.

"They're all crooks," Doc said. "Everyone of those sons-of-bitchin' stock brokers. Leeches! Parasites!" He watched Sam trying to stuff a pastry, oozing red jam and thick cream, into his mouth. Sam froze and pulled the pastry half way out.

"Did you wan' somfthin'?" he asked, the cream spitting. Doc shook his head slowly and turned away.

"You were right, Arthur," Doc said calming. "I should have gotten out when you did." He flicked his cigarette from the holder into an ashtray before it was half done and plucked another from the pack. "I'll bet Henry did all right, didn't he?"

"He did nicely," Art said.

"How much did he make?" Doc asked, but he knew Art would never tell. He had asked before, but some vague ethic kept Art from divulging anything about the personal finances of any of them.

"I'll bet he's walked off with 50 grand!" Doc said, hoping to catch some hint from Art through an unguarded reaction, but he never blinked.

"Jesus, what I could do with 50,000 in cash right now!" Doc said walking in a tight circle, the cigarette smoke following his head in a blue eddy.

"You've made a hell of a lot more than that already," Art said.

"But it's all tied up, Arthur. In airplanes, and inventory, and...and...Christ, you've got it floating all over the country for crying out loud. I can't say as I've got anything right now!"

"It'll settle itself out when we're through," Art said calmly.

Doc laughed. "If I didn't trust you so much, Arthur..." He laughed again, but it sounded almost forced. "You are a genius, Arthur. If anything was to ever happen to you, no one would ever unscramble this web." His voice was thick with southern drawl to the point of sounding backwoods. He glared at Art with an uncomfortably large smile.

Doc paced the room, puffing at his cigarette with irregular draws. "You okay?" Art asked.

"Certainly, why?" Doc answered quickly, pulling his robe tight and staring at Art with big questioning eyes.

"You're acting..."

"Tired," Doc said. "I'm tired. That's all." Doc rubbed his eyes. "So what's Henry doing with all that money? You can tell me that."

"Oh, he's not keeping it a secret. Most of it just goes into bank accounts."

"Where?"

"Iowa."

"Specifically?"

"None of your business."

"It is my business!" Doc snapped. "I...I'm concerned about him."

Art said nothing.

"All right! It's none of my business what he does with his money! Like buying fool airplanes. Why Iowa for Chris-sake?"

"I think he's sweet on that girl Jake has," Art said. He removed a pair of reading glasses and rubbed the bridge of his nose where they had sat for the last hour. "He tries to act coy about it, never really admitting it, but he mentions her name just a little too often. You know like a kid in school when..."

"Sam!" Doc called. "Get over here; look at this!"

Art realized Doc had no interest in Henry past his money and now stood at the window looking down at the street while waving Sam over. Sam pushed something into his mouth and washed it down with whiskey.

"What?" he asked.

"That car's back." Doc pointed. "The Chevrolet. You told me they weren't around anymore."

"They aren't! I mean, they weren't"

"Well, who's that?"

Sam pressed his forehead to the glass looking out. "They're back," he squeaked.

"Thank-you, you're a big help."

"I only see one inside," Sam said.

284

"No, there's two. See, in the back seat. And I'll bet the third one's in there, too. Just can't see him from up here."

"Who are they?"

"Capone. I know it," Doc said.

Art joined them at the window looking down at the parked Chevrolet in the street below. "Somehow, I just can't picture him sending out trouble in a green Chevy, just not the Chicago style."

"This isn't Chicago," Doc said with a growl. "It's Lincoln."

"You're right! If you want to hit someone in Lincoln, Nebraska a Chevy is definitely the car to use. The man takes no chances. Why I bet as soon as he discovered you were set up here in Lincoln, Mr. Capone turned to one of his torpedoes and said, 'Get a map; find Lincoln, Nebraska; load up the Chevy and rub this Doc outa business!' We are dealing with a very crafty enemy here."

"Shut up, Art," Doc said. "Sam, I want you to check them out—up close." He placed his hand on Sam's shoulder. "Find out what you can."

Sam felt his blood pump faster, and he reached under his jacket feeling for his pistol. This would be better than running for cigarettes.

Chapter
XIX

Jake was still tired, although he had slept for the last six hours, the first serious sleep he had since the autumn push began. He lit a cigarette off the one he had just finished. The wind had blown steadily since midnight out of the southwest bringing warmer air and with it moisture in the form of scattered white clouds. He unbuttoned his collar and ran his hand through his tangled hair trying to remember the last time he had bathed.

The Travel Air with the radial engine, an abomination someone had concocted out of stray parts, still refused to start. The pilot leaned over the cockpit's rim, swearing at the mechanic who stared blankly at the tangle of wires, hoses and cables between the firewall and cylinders. They were across the field from Jake, and their conversation was lost in the wind, but agitated gestures from the pilot conveyed the tone. Jake tried to think of the pilot's name, but quickly gave up. In fact, he had trouble remembering whether he had ever met him.

Lately, Jake's ability to remember names was slipping. He paid this little heed and compensated by repeating a name aloud before meeting the person. "Too many new pilots," he told himself, "impossible to keep all the damn names straight." He ignored the fact that even Henry's name or Lundy's were becoming difficult.

He took a long draw on the cigarette and scanned the horizon to the north for a sign of the returning LS-5. Only bare trees and a loose flock of crows hopping their way across the dead fields could be seen.

Suddenly, the Travel Air's motor sparked and gave a brief snort. Jake saw the pilot desperately pumping the throttle trying to keep the machine alive, but there followed a gasp and stone silence. The mechanic's posture changed from surprise over seeing the engine start to shrugging confusion

287

when it stopped. He shook his head, pointed at the motor and with a wave of disgust, at the pilot.

When he turned his back on the airplane, the pilot, apparently miffed, leaped from the cockpit and in full flying togs lunged for the mechanic with arms flailing. Jake put the cigarette in his mouth and folded his arms across his chest to watch the fight.

The mechanic had been caught by surprise with the first sucker punch from behind, but the pilot, swinging blindly, was soon flat on his back with the mechanic seated on his chest pummeling his face. The pilot waved his arms in useless defense like some leather-coated beetle being eaten by a sparrow. Jake giggled through the smoke and returned to watching the sky.

The clouds, gray and moist, were high and moving from the west. Their path, Jake noted, was at least 30 degrees off from the surface wind. He knew when the winds shifted from the northwest after dark, the temperature would drop. He knew this. He reworked the estimates in his head, and the truth was disquieting. A windshift meant weather, foul weather coming across the flat Dakotas from the Canadian Rockies.

"They probably grow this crap up in Siberia," he said looking at the clouds. He dropped the cigarette stub in the dirt and crushed it with his toe. He looked at his watch. The LS-5 was only slightly behind schedule. A quick unloading and smooth refuel would turn the plane around, with him at the controls, within 20 minutes of its arrival.

"Where are you?" he asked the sky.

Getting across most of Minnesota would be no problem, he thought. It was the empty stretches of woods that led into Canada that made his stomach constrict. There it would be cold; how cold mattered little. Anything below 40 was cold to Jake and he hated it. The hundreds of square miles of frozen nothing along the route meant hours of listening to a tired motor swinging an nine foot propeller in front of his face. It meant hours of straining out the usual hums and burps from the new ones that might tell of a valve sticking or a rod about to break. Even something as simple, and common, as dirt in the fuel line could drop him into the snow.

And, he pondered, should the Hisso engine somehow manage to deliver him again to Canada, he still had to encounter often vicious winds that could whip through the trees, making every approach a wild ride to a rutted and snowy runway almost certain to rip off his gear or catch a wingtip.

"Stupid way to make a living," he said and plunged his hands into his pockets to steady them.

He walked around the same wide circle he had been covering for the last half hour waiting for the LS-5. The wind blew a rough gust, throwing scraps of hay and corn stalks against the Travel Air.

The only other airplane on the field was a wrecked Jenny pushed against the barn. One wingtip rested in the dirt, the wing snapped somewhere about half way up the spar. The right landing gear leg was splintered as was the propeller. Two small girls, offspring of someone associated with the field, sat in the cockpits. One girl kept waving at an imaginary crowd below, while the one in the forward seat made spitting machine gun noises shooting down Germans, or bootleggers, or whoever it was little girls shot down, maybe little boys.

Jake looked at his watch again and said, "It's down somewhere. Write off another airplane." He pictured the LS-5 wedged in between two giant fir trees, the dead pilot half-covered with snow while timber wolves tugged the body from the wreckage.

Then, he saw the biplane before he heard its engine. It crept in from the northeast, crossed over the field, turned downwind, then back toward the runway descending toward the dead grass. Jake watched it bounce twice and roll out toward him, the propeller swinging in easy revolutions as the plane stopped.

"Beats working for a living," he said and ran to greet the pilot, a tall dark man named Martin with large teeth and no hair.

"Where were you?" Jake shouted.

"Up there," Martin answered, annoyed with the stupid question. "Where'd you think?"

Jake ignored him and waved at a truck to pull alongside. The crew in turn ignored him knowing better how to off load the whiskey and top the fuel tank. Martin killed the engine, and Jake popped open the cowling only to burn his knuckles checking the oil.

"Oil's low," Jake said.

"So add some," Martin answered.

"How much alcohol you running in the radiator?"

"I dunno, enough. You know." Martin's eyes were laced with red, and his whole body seemed ready to melt into the earth as he slid from the cockpit.

"How's the weather?" Jake asked.

Martin undid his flying suit in slow heavy movements while his mind digested the question. "Headwinds all the way," he finally said through a wide yawn. "That's good for you, of course."

"Yea, if it stays that way, which I doubt. What else?"

"Ah...damn cold, I know that; feels warmer here, though." He rubbed his fingers together as if feeling the air to see if it really was warm. He caught Jake's impatient stare. "Oh, well, the weather's about like this all the way up to Sioux City. Oh, don't land there at Rickenbaker Airport, there's some jerk from the Commerce Department snooping around bugging pilots, asking to see licenses and all that."

"You landed at a real airport?" Jake asked. "You can't do that—not loaded."

"Shit! I always do. Get good service, too."

"Did the Commerce Department guy see you?"

"Yea, came over and started snooping under the tarp, asking me for a license. I told him to keep his hands off the airplane, then I pulled his hat brim over his eyes and kicked him in the shins. He's only a little squirt, 'bout so high." He held his hand out at chest level, even with Jake's forehead. "So he run off to the office like he was going for help. I paid for the gas and skidattled. Flew east for a few miles to make him think I was going to Des Moines or something." He cocked his leg and loosed a noisy fart inside his leather suit. "Phwew!! Jeesus!" he said tugging at his collar waving away the fumes.

"So any other weather," Jake asked.

"Oh, ceiling's lower in Minnesota. Of course, that was a few hours ago. Probably changed by now. Lower, I expect."

"Any rain, snow, sleet?"

"No," Martin said popping his tired eyes wide open as if he had just awakened from a dream to find himself talking about the weather with a total stranger.

"Good," Jake said and turned.

"Well, nothing until you get about half way up the state. Then it, ah..." He yawned.

"It what?"

"It snows...and sleets. Sleet the first half; snow the second."

The Hisso engine was slow to restart, but finally did, and Jake was away, alone again in the only place where he felt truly at home. Alone, but uneasy, Jake felt a stranger to the sky. He wanted to ask someone if they knew something he did not, if, maybe, he should sit this one out.

With the wind at his back, he estimated he was doing better that 100 mph across western Iowa and into Minnesota. The clouds thickened over him but no sleet. He refueled at one of the usual stops outside a small town, where a squat man with closely trimmed gray hair knew how to fill the tank without asking questions.

Jake, however, had questions, "What do you think of the weather?"

"Could be worse," the answer came back flat. "I've seen worse."

"No, I mean, what do you think it's going to do? Do you think it'll get worse, snow or something?"

"Yes, snow or something. That's certain," he answered staring at the fuel sloshing into the tank.

"Thanks," Jake said.

"Ah, huh."

Jake paid for the gas and stuffed the receipt into his pocket. The engine balked again before starting, and Jake was aloft.

With white and gray prairie below, somewhere in Minnesota between Iowa and Canada, Jake picked up the first precipitation. The clouds hung low to his left, and he guided the plane to the right, but all ahead the sky drooped and turned a seawater gray.

The air was biting cold, his chin numb although wrapped in a wool scarf that kept flapping loose from his collar. The first drops struck like fat insects hitting the windshield, struts and wings. They splattered and streaked back leaving thick trails as the moisture, almost in crystal form, adhered and turned to ice.

Jake reached around the windshield and rubbed the glass from the outside. His heavy glove left a greasy smudge traced by the freezing rain. He pulled himself deeper into the cockpit away from the stinging cold wind and eased the plane's nose lower away from the clouds.

He talked to himself, "When this thing turns into an ice cube, I want to be near to the ground. I'm putting down in the first good field I see. Screw this!" He watched several good fields slide under the wings, but with each one he would look at the ice on the struts and say, "It'll take more." And it did growing steadily in clear mounds over anything poking into the wind.

"Not much of a weather briefing from that schmuck..." and he could not think of Martin's name. He had just seen him. The face he remembered, wide, almost comic, with that bald head. The name was gone. He shrugged it off.

Field after field—all good landing sites, but uninviting nonetheless—slid beneath. He measured the distance to Moyer's.

"About a hundred miles," he said to himself. "No sense risking a landing out in the middle of nowhere. Not when I've only got another hour or so to go."

"Right?" he asked.

"Absolutely," he answered. "You don't get paid to sit on the ground." A chunk of ice broke loose from the propeller and smacked into the wing struts before vanishing into the gray. Jake snapped his head to the left where the ice had struck and felt or imagined, he could not tell which, a sudden vibration from the uneven load on the propeller.

A tightness held his insides, and he forced himself to review the airplane's load capabilities. He tried to estimate the added weight from the ice and concluded that with the front cockpit empty and the fuel tank less than full he was still well below gross. A glance, however, at the wings rapidly losing their airfoil curves to the ice, only torqued further the knot in his guts.

He knew the other pilots had flown the Standards covered with ice, and he had done it himself with a full load up front including a dozen or so extra bottles tucked into empty spaces. Canadian whiskey sealed in its original bottle brought a healthy price back in Lincoln, so every pilot took to squeezing in the extra bottle wherever he could, including inside flying suits. Of course, these would often be half empty by the time they reached home.

Jake pulled his head deeper into the cockpit, his face practically touching the instrument panel. The airspeed indicator's needle pointed at 60, jerked a time or two and dropped below 40. He knew the pitot tube was almost clogged and pulled gently back on the stick more out of curiosity than to really prove anything.

"Oh beans!" he shouted as the biplane shuddered on the verge of a stall. He shoved the stick forward while adding more power. He swung his head from left to right, terrified and yet fascinated by the ice palace of coated wires and struts surrounding him. Edging his face over the cockpit's rim, he looked for a patch of flat dry ground, but the earth was lost to a cloud deck directly below. He had passed over it without notice and considered turning around for one of the empty fields behind. Gray mist swept past him, and he added power until the throttle hit the stop. The engine shook from the strain, and the water temperature nudged into the red.

"You've screwed yourself now, Jake old boy," he said with a calmness that scared him.

As quickly as panic had approached it fled, with the biplane popping into clear sky over a half-frozen lake surrounded by woods. Late afternoon sunlight poured through a gap in the clouds above reflecting off the icy flying wires, making them look like blown glass, beautiful but without strength.

Ice broke away from the propeller in brilliant crystals, striking the wires and breaking the fragile coat. The sun's blinding glow eased the tension in his body, although for the most part the ice remained on the wings and struts. He eased the power back slightly, and the plane still flew; the airspeed needle pegged at zero from the clogged pitot.

"Ten more minutes," Jake said. "Maybe 15." He studied the cold terrain of trees and frozen marshes trying to pick out familiar landmarks. Moyer's was somewhere near. He quessed he was already in Canada.

His fuel was good, his oil pressure normal. The water temperature was still running high, and he wondered if the radiator could have iced over. He listened to the motor and the wind through the frozen wires. More crystals flew past his head, and a sizable chunk broke loose from the lower wing.

"Maybe the air's warmer, above freezing," he said and shuddered violently from the cold.

The sunlight vanished into a thick layer of clouds above, and dry snowflakes skimmed off the ice on his wings.

"At least it won't stick," he mumbled, feeling thoroughly exhausted, his eyes heavy and muscles aching from tension and cold.

The earth was white now, and he checked the map.

"There's the lake!" he shouted to wake himself. "And...there's the the railroad track...I think..."

He strained to look over the cockpit's rim into the ever increasing swirl of snow.

"Yes, tracks! Got it now, you son-of-a-bitch!" He thumped the fuselage with a padded fist. The snow was steady now, but the visibility good. He followed the tracks across the marshes and into the trees, up the slight grade after Sprague, and—Moyer's!

The Indians heard the airplane's motor. Moyer ran toward them in his ankle length fur coat, his beaver hat pulled low over his ears. His breath puffed in a cloud of gray steam crystalizing in his red beard.

"Clear the runway!" he called, and the Indians whipped a team of horses dragging a crude plow. One driver rode the plow while the other led the team with an unrelenting stream of blows and vicious oaths he had learned from the pilots.

They had been plowing since the snow had begun to accumulate hours earlier, but despite their efforts had barely kept ahead of it. Now in the weak light of early evening, with Jake's biplane approaching, the field was blanketed in white with ragged windrows along the edges. Moyer poked a yardstick into the snow, noting three inches at the thinnest point. Off to the sides it would be over a foot.

"Good enough," he said to no one and slapped the stick against his boot.

Jake circled overhead. Moyer saw a face, swathed in cloth, peer over the cockpit's rim. He waved the yardstick calling, "C'mon in! Snow's not getting any less!"

Jake examined the strip watching the wind blow the snow into smooth mounds at both ends of the runway. The surrounding trees impeded the wind, keeping most of the drifts off to the sides. He saw a figure, he quessed to be Moyer, waving a stick. Jake turned away from the field and back into the wind. His turns were shallow, and he carried more power than normal, knowing the ice-coated machine would stall at a much higher speed than usual. Without the airspeed indicator he could only guess at his approach speed.

"Never relied on the damn instruments, anyhow," he said and turned onto final approach, feeling with all his senses for the correct speed.

Moyer plunged his hands deep into his coat pockets, leaving the yardstick in a snow bank. He watched the biplane, then studied the wind boiling the snowflakes into smoky puffs over the tree tops and across the runway. He estimated how long it would take to tug the biplane clear of the runway and fashion a new set of landing gear to it.

"Bound to bust something," he muttered.

The wind was steady across the trees, and Jake held an even but fast descent, until he dropped below the trees. He knew it was coming and corrected easily to the change in wind but still wallowed in the suddenly rough air close to the ground.

He hauled the ailerons first to the right, then with club-footed grace he stomped the rudder and jammed the stick to the opposite stop. The Standard, still encased in ice, responded like a mold of lead Jello.

"Good Christ!" Moyer shouted and recalculated the time needed to replace a shattered wing.

The Standard appeared to hover momentarily in a cloud of snowflakes. Its motor barked. Jake shoved the throttle full open, then with a sickening thud muffled by the snow, the whole mass of wires, fabric and ice dropped to the runway and came to an abrupt stop in the snow.

Jake opened his eyes.

"Well, I'm down."

He saw Moyer waving at the plow team while running through the snow toward the plane.

"Follow the plow!" Moyer called and pointed at the horses, their backs white with snow, being prodded into a position ahead of the spinning propeller. Their eyes were wide with fear; their nostrils puffing huge clouds of steam into the cold air. Both drivers slapped angry blows across their backs edging them forward. Moyer was shouting something at them, but his voice was mumbled by the wind and the terrified snorts and cries from the horses.

"Keep the motor running!" Moyer turned and shouted at Jake anticipating Jake's thought to kill it. Jake nodded and watched Moyer shuffle through the snow toward the horses and, grabbing his yardstick, join in thrashing them.

Suddenly, one reared, but the tangle of yoke and harness and a well-aimed blow from one of the drivers brought him back in place. Finally, the two horses were turned ahead of the biplane and the plow lurched forward, cutting a fresh path slightly wider than the plane's wheel base.

"All right!" Moyer shouted and waved Jake forward with his yardstick. Jake opened the throttle and blew a tremendous cloud of loose snow into the trees. Moyer kept waving him forward as if his gestures would help the airplane cut through the snow.

"Come on! Come on! That's it!" he yelled, but Jake heard none of it, only the howl of the Hisso motor as the Standard fishtailed behind the plow cutting through the windrows and catching in small drifts.

It took 30 minutes to bring the plane in and another half hour to tie it down, drain the oil, and add alcohol to the radiator. Jake and Moyer then trudged their way through the snow to the cabin, where a steady flow of blue smoke rose from a metal chimney. By then it was dark. Jake's feet were numb, and snowflakes clung to his face without melting.

Moyer slammed the door behind them as they stepped inside. Snow twisted in little eddies at his feet, and he dropped the latch in place and rolled a heavy blanket down covering the door.

"Keeps out drafts," he explained. Jake was already standing with his back to the iron stove, vigorously rubbing his backside and slapping his gloved hands together trying to press warmth back into his body.

"Take those clothes off!" Moyer commanded. "They only keep the cold in; there's no heat left in that fool body of yours." He gathered an armful of split firewood from a box by the wall and lifted the stove lid. Smoke rose slowly into the room. He stuffed the pieces in and slid the lid closed.

"I don't think I'll ever get warm," Jake said and shuddered violently.

"That stove really kicks up, and it'll blow you right out of here in a minute. I'll get some coffee on. I wasn't sure if anyone would be up yet this afternoon. Weather turned nasty awful quick. Of course, if anyone's going to make it in it would be you!"

Jake forced a smile at what he took for a compliment, but Moyer had turned his back to pull one of several clipboards off the wall. They were covered with receipts and schedules.

"This shows you and someone named Grieve," Moyer continued. "Yea, Wendall Grieve. Shows you two coming up tonight. I suspect this Grieve character is waiting out the weather somewhere. You ever heard of this Wendall Grieve?"

"No," Jake said, then added. "Yes, I did meet him once. He flies a Travel Air..."

"Yeah, that's what it says here—Travel Air." Moyer tapped the clipboard.

"But I don't think he'll be up today. Might not even be up tomorrow. When I left he couldn't even get the thing started. Just kept flooding it from what I could see. 'Course, the airplane looked like a real piece of shit to begin with."

Jake's teeth began to chatter uncontrollably even though the stove was heating rapidly. He rubbed his jaw and his whole body began to shiver.

"It...it looked like he made the th...thing out of a bunch of scraps." Jake shook from cold and beat his arms across his body. He unbuttoned his leather helmet and tossed it on the table. Snow melted off his boots in a pool. He worked his way out of his leather greatcoat and first layer of wool, dropping it all on the floor in a heap. The stove turned a dull red, pumping heat into the room.

"That big fellow, Henry, was the last to leave late this morning," Moyer said. "The weather was already turning sour."

"Hmmm," Jake grunted. "Hope he has enough sense to keep out of the ice."

"Is the weather bad coming up?"

Jake thought for a moment and said, "Naa!" He rubbed his hands together over the stove and thought of Henry loading up with ice somewhere south of there. With a shrug he pushed it out of his thoughts and into another part of his brain where he tossed everything unwanted.

Moyer pulled his hat back on and picked his mittens off the table.

"You stay here," he said. "I've got to make sure they get those horses put away properly." He reached for the blanket over the door. "You don't think that Grieve fellow will be up?"

"Seriously doubt it."

Moyer nodded. "Well, I'll light some smudge pots in case he tries."

"Whatever," Jake said, but Moyer was already through the door and had disappeared into the blowing snow. Jake pressed the blanket against the door again and returned to the stove.

"Florida," he said listening to the wind suck the smoke from the chimney.

Henry landed after dark when the first snow flurries had started to fly. Susan stood at the end of the pasture waving two oil lanterns, while her father stood at the other end in the darkness, swinging his lantern in a wide arc.

"See him yet?" Susan called.

"Huh?"

"What?"

"Can't hear you!"

"Never mind! I see him!" she yelled and waved her lanterns.

"Can't hear you!" her father shouted. "Here he comes!"

"Huh?"

Henry had circled their house several times in the dark, once quite low across the rooftop, until he saw the small lights appear on the back porch and head for the pasture. He still felt uneasy about night landings and hoped Doc would some-day spring for the cost of installing landing lights. He read somewhere that little red and green position lights were required on the planes for any flying after dark, but when he mentioned it to Doc without first thinking the rebuke was quick.

"Hell, boy! The object here is not to be seen!"

After that, he never bothered asking about landing lights.

Susan waved her lanterns until she heard the plane line up on final approach and drop slowly out of the darkness. Guess-ing its distance away, she stood at the edge of the field until the noise was uncomfortably close, then setting the lamps down she ran clear. Henry's Standard flew past in a soft hiss of singing wires and engine backfires. She grabbed the lanterns again and ran after him, tripping over a dirt mound and falling flat on her stomach.

The biplane headed toward the house following her father's light. She picked herself up and ran after him.

"Hello!" Mr. Bowers called. The engine stopped. "Oh! It's you, Henry!" He offered his hand over the cockpit. "I thought it would be Jake. I hope Susan's not too disappointed."

In the harsh glow of the lantern he saw Henry's face react with hurt.

"Oh, I didn't mean she would be disappointed to see you! It's just...well, you know, she probably thinks you're Jake." His voice trailed off.

"Oh, that's fine," Henry said with a forced laugh.

"I'm sure she'd be tickled to see you!" the father said, and was saved from further embarrassment by Susan running up.

"What a surprise!" she called rounding the tail meeting Henry as he stepped from the cockpit. "Oh! Henry!" Her voice made a poor effort at hiding her disappointment. "I...I thought you'd be Jake."

"No, I'm always Henry."

"No, no! You know! Oh never mind. Come in; you must be frozen." She took his arm guiding him toward the house being as cheerful as possible to cover her rude greeting.

"What brings you here?"

"Weather turned bad on me. I deviated."

"Oh?" Susan and her father stopped on the top stair to look at the sky, dark and still. The occasional snowflake blew through the glow from the kitchen.

"It's much worse up north!" Henry quickly injected.

"Oh," Susan said. "Are you heading north?"

"Well, no. Southwest actually. Lincoln."

"Weather's bad in Lincoln is it?" the father asked.

"I don't know. Could be. I didn't want to take any chances..."

300

"No, you shouldn't," Susan said. The three stood uncomfortably on the porch.

"I just thought it would be better to, you know, land here...If that's okay? I could head out!"

"No!" Susan protested and took his arm again.

"I could sleep in the barn."

"Nonsense! You come right inside. You must be starved."

"Truth is I am," Henry said.

"And you should get where it's warm." The door opened and in they went. "You were wise to put down here. You are always welcome here, Henry. Now take that suit off."

"Smells good in here," Henry said pulling the leather flying suit over his stocking feet. He sat on a kitchen chair, while Susan hovered nearby taking gloves, helmet and finally the suit. She gently folded it on the table, crossing the arms, giving them a final pat, before she set everything on a sideboard. The way she handled his clothes only made Henry long for her with a deep sorrow knowing there was always Jake.

"We've been baking," she said. "Bread's still warm; I'll get you some with supper."

"Thank-you," Henry said, and he inhaled the warm kitchen air with its mixture of warm bread, boiled potatoes and dish soap from the pan in the sink.

"I could stay here forever," he said too quickly. "It's awfully cold in the airplane. This is nice...Here, the kitchen...Is there anything I can do to help?" he asked springing to his feet.

"No, sit!" she answered and brushed by him toward the stove. He smelled of warm wool and motor oil; she tried not to look at him, but she noticed how much more attention he commanded in a room. She pressed the kitchen door shut.

"You don't have to make anything special," Henry said and yawned.

"Nothing special," she answered and shot a glance at him. He was so much larger than Jake in that kitchen, she thought. There was a healthier feeling about Henry that she only now realized was lacking in Jake. She caught a glimpse of her father leaning against the wall chewing on his pipe stem and grinning.

"So how is Jake?" she asked.

"He's in Canada tonight, I guess. I don't see him too much, but he's okay."

She paid little attention to the answer taking the opportunity to study him. She noted his voice was deeper than Jake's, less nasal.

"Does that suit really keep you warm," she asked.

"Not really, but it'll do," Henry answered.

"You ought to wear something underneath it. I mean, something more...of course I can see you've got something underneath." She had been looking at Henry as she spoke but quickly turned away. "Something more...heavy trousers, you know." She stirred vigorously at something on the stove and laughed. Her father settled himself into a chair and stared.

"I wear two pairs of pants, and long underwear, plus..."

"Well, that should be plenty," Susan said hoping to change the topic.

"The longjohns are plenty warm," Henry added.

"I'm sure," she said.

"Do they bind up on you from sittin' so much?" her father asked with a teasing smirk on his lips.

"What, the longjohns?"

"Daddy!"

"Yea," her father persisted. "Mine sorta' scrunch up between my legs—in the crotch."

"Soup!" Susan announced and dropped a large bowl onto the table in front of Henry and thrust a spoon at his nose.

"Thank-you," he said. Taking the spoon, his fingers briefly touched hers. The spoon was large enough to pass without touching, but they did. She patted his shoulder and said, "Eat." He did.

The room fell silent except for the occasional slurp from Henry and Susan fidgeting with pans near the stove. Mr. Bowers slowly packed and lit his pipe. He blew an even ring of smoke at the ceiling. It hung in the warm air above their heads like the abruptly terminated conversation.

Methodically, Henry scraped his bowl with the spoon and accepted a second from Susan. Her father slid his chair back and walked to the cellar. They listened to his footsteps descend the creaky stairs then return a minute later. He carried four beer bottles between his fingers and passed one to Henry.

"Thank-you."

Susan crowded baked chicken, green beans and a mound of potatoes onto a plate and slid it in front of Henry. He smiled without speaking, and she carved another hunk of warm bread for him. He nodded with a smile, and she took a seat at the table to watch him eat.

Not at all like Jake, she thought. Not finicky.

Slowly, he moved chicken and beans to his mouth, and with furtive glances at Susan he ate. Unlike Jake, he hardly touched the beer. Jake, she thought, would have been on his second already.

"Can I get you anything else?" she asked.

"Well, thank-you," he said and finished the chicken. Susan's father popped another bottle. White foam ran down the long neck soaking into the tablecloth.

"Oh, Daddy," she whined mopping at the foam.

"Must have over-primed that lot," he said. "You're not drinking, Henry. Don't you like it?"

"Oh, it's good. I'm just more hungry than thirsty I guess." To be polite, he hoisted the bottle and took a drink before returning to his plate.

"Not everyone has to get drunk, Daddy," Susan said.

"I'm just hungry, that's all," Henry repeated.

"All that food in your stomach soaks up the alcohol," Mr. Bowers declared and took a long draw, finishing with a prolonged belch. Susan hid her face and shook her head. "What? What'd I do?"

Susan rinsed the last plate, and Henry took it from her and dried it with a small damp towel. She wiped her hands quickly on her apron and started to fumble with the strings behind her back.

Henry rubbed the plate over and over watching her fiddle with the bow at the small of her back. She pulled at the wrong leads and instantly knotted the strings. Henry quickly set the plate on the counter and said, "Here, let me get that! You're in a knot."

She started to protest, but Henry was already behind her moving her hands away with gentle authority. She dropped her arms and stared straight ahead at the window over the sink. In the reflection she saw her face and shoulders, with Henry, a foot taller, head bowed, intently working the knot loose.

His features were distorted by the moisture on the cold pane, and without thinking she reached up with a towel and slowly rubbed the glass until he came into focus. What a handsome face, she admitted to herself, now that she could study him without being caught. She felt her body relax as his powerful hands rubbed clumsily at her back undoing the knot.

"How's it coming?" Her voice came out thin, and she reached behind touching his hands.

"Almost...There. Got the bastard!" he declared in triumph. "Oh, sorry."

"Silly, I've heard far worse around here!" She giggled and spun around. Henry remained planted only inches from her; his face slightly above hers staring down. She hesitated taking her apron off, holding it in a bunch close to his stomach. They stared at each other. Neither said anything, then as if on cue, they moved apart.

"Thank-you," she said, her voice husky, a little nervous. "Will you look at this!" She gathered four empty beer bottle from the table and set them near the cellar stairs. "He even finished yours!" She laughed. "He'll sleep well. Take a cannon to wake him."

Slowly, she walked across the room toward him, turned her head and looking through the glass in the door said, "Oh, look! It's snowing."

Henry moved beside her, and they gazed through the glass at a small area of the porch lit by the kitchen. Big wet snowflakes whirled past out of the darkness and collected on the railing.

They stood, shoulder to shoulder, noses touching the glass staring at the snow. Fog from their breath formed on the pane, and Henry wiped at it with his hand after first writing, SUSAN, in even letters. They laughed.

"Only one name?" she asked with a giggle.

"Need more steam," he said and breathed on the glass.

"Don't you just love snow?" she asked and inscribed a fancy H in the moisture.

Henry hesitated. "On the ground I love it," he said. "It's up there, it scares the hell out of me."

She had finished the H and was putting curls to an E when she stopped to look at him. "You don't like flying?"

"No, I love flying," he said. "And I even like the cold. Sitting in that open cockpit on a cold morning it...It's, well, I don't know. It lets you know you're alive, somehow. Sounds stupid, doesn't it?"

"No!"

"I just don't know how guys like Jake, and probably all the other pilots, can just plow their way through sleet and snow and fog. I get scared when I can't see where I'm going, and I know that's not right."

She turned from the window leaving his name unfinished.

"I don't think I'll ever fly like Jake Hollow," Henry said. "Nothing bothers him."

"No, nothing," Susan said absently. "Let's go outside!"

"Sure," he said, and she was away from him and into the next room gathering coats and gloves. Wrapped against the cold, Susan in soft wool, Henry in his leather flying suit, they ran down the back steps into the snow.

Henry slipped on the last step and tumbled to the ground amid giggles from Susan and theatrical howls from him. He lay on his back kicking like an overturned turtle, and Susan pulled him back to his feet with a strength that surprised him.

"Catch me!" she called and ran away, around his plane and into the darkness. He followed quickly, but she lost him in the shadows.

"Susan? Where are you?" he called.

FWAP! A loose snowball flew out of nowhere and caught him square on the chin. Cold fragments ran down his open collar, and he howled again. She ran off laughing toward the barn, and he took off after her catching her in a waist hugging tackle half way there.

Over they rolled in laughter and grunts, snow pressing into their faces and clothes until she was on her back looking straight up into Henry's face, all but shrouded by the darkness. He breathed heavily from the running, and his right knee pressed against the inside of her thigh. He gave a short laugh, and snowflakes settled onto her face melting against the warm skin.

"I'm glad it snowed," she said and felt more of his weight settle onto her.

Moyer stumbled back into the cabin amid shouts and oaths sworn against the cold and wind. The storm had increased, and snow had drifted to waist depth against the door. He pushed the door closed and shook the loose snow from his coat and hat.

"Your friend, Grieve's not coming in tonight. That's one thing I know for certain."

"Not my friend," Jake said huddled near the stove.

"That wind's getting miserable. Just hope the ropes keep your machine tied down."

"Not much we can do about that now," Jake said.

"Hey, that stove's about to melt!" Moyer exclaimed seeing the iron stove a bright cherry red. He kicked at the air vents with his boot. Clumps of snow hit the hot metal hissing and instantly turning into steam.

"It's cold in here!" Jake protested. "Got to burn more wood."

"You keep pumping that thing full, and we're likely to burn down! Don't want to sleep in the snow do you, eh?" Moyer shed his coat and hung it on a hook near the door. Beneath, he wore a red plaid shirt, and a layer of beige showed beneath that. He picked Jake's coat from the floor hanging it beside his.

"You need to warm up inside," he said and took a bottle from the shelf and uncorked it. "Here." He handed a glass to Jake and half filled it.

"I'll never be warm again," Jake said and drank.

"You complain too much." Moyer sat on the bed to remove his boots. A lump of blankets snarled, and he pulled them back exposing his yellow dog, curled with her nose poking under her paws.

"Lazy worthless hound," Moyer said. "You should be outside like any normal mutt." He rubbed the dog's head after pulling the covers back over her. The snarl turned to a soft groaning while he massaged her ears through the blanket.

He took the bottle and tipped it toward the ceiling. The wind blasted the shuttered windows with snow. Jake moved closer to the stove. He took the bottle from Moyer.

"Why can't we bootleg from Cuba for Chrissake?"

"Florida's got too many rum runners as it is," Moyer said. "Hey, did I ever tell you about the time I went with Doc down to Palm Beach, and we met Mae Murray..."

Chapter
XX

"Where is Jake Hollow?" Doc bellowed to no one in particular. The hotel room was bright with morning sunlight and cluttered with cigarette butts, empty bottles and papers and ledgers scattered on tables, chairs and floor.

"Still no word from Moyer," Art said calmly. He walked around the room picking up papers and sorting them into neat piles. Sam Nyberg stood quietly nearby watching Doc, afraid to speak. He knew enough to stay clear of his employer when the man was upset, or drunk, and it now appeared he was both.

"And Henry. Isn't he in yet?" Doc stood before the three chalk boards and poked a bony finger at Henry's name listed below one of the LS-5's. "According to this, he was supposed to be in last night. What the hell's going on around here? Doesn't anyone keep to schedule?" His cigarette holder was clenched between his teeth and bounced up and down with each syllable. Sam found this amusing, barely suppressing a grin.

"All the planes are ready to take-off for the morning run," Art said while reading a receipt stained with whiskey. "It's only a little after seven. That just puts us a few hours behind schedule."

"A few hours! Just a few hours?" Doc spat the words with sarcastic venom. "Do you realize how much money goes through this operation in 'just a few hours', as you so casually put it?"

"Of course I know," Art snapped. "I know everything about this operation." He held a fistful of wrinkled papers up to Doc and shook them. "Everything!"

"Well, what are we waiting on? Why aren't those airplanes heading north?" Doc slapped the chalkboard with his hand and left a dusty print over a column labeled, LOADS IN TRANSIT.

"The last message from Moyer was yesterday afternoon. He said the snow was increasing, and the 6 p.m. train was cancelled. Jake sent a telegram. Says the weather's bad all the way up—no sense anyone even starting out. The other pilots would never tell him to his face, but they've got a lot of respect for Jake. When he says don't fly, well, the smart money stays on the ground."

"Jake Hollow doesn't run this operation," Doc said.

"My guess is he wouldn't want to," Art said. "The man only functions inside an airplane. Outside of that, he's somewhat a fish out of water." Art checked his pockets. "Here's his telegram."

He pulled a yellow Western Union envelope from his jacket pocket and held it out to Doc who ignored it turning to the window. Sunshine washed over him as he looked into clear sky, his head turning from side to side in mock exaggeration.

"Doesn't look too bad to me," he said. "Tell those worthless pilots to get in the air. If they're waiting for weather to improve, they can wait enroute north. We've wasted enough time already."

He turned to the chalkboard, his back to Art and Sam. Art nodded to Sam to leave the room.

"Go tell them to take off," Art said. "All of them." Sam nodded and started to leave just as there was a knock. Sam looked to Art then opened the door. A bellboy rolled a cart covered with breakfast food past him into the room.

"Good morning!" the cheerful bellhop said. The aroma of bacon and coffee caught Sam who stood and watched him clear a table.

"I haven't had a chance to eat yet," Sam whined and instantly knew it was the wrong thing to say. Doc spun around and hurled a chalk eraser at him with uncanny accuracy. Before Sam could blink the eraser bounced off his forehead and ricocheted into a potted palm. Sam, eyes wide, bolted from the room and caught the elevator down. The front clerk thought it best not to mention the white powdery smudge across his face and simply bade him a, "good morning, sir," as Sam hurried through the lobby.

310

Nick Barchek sat in his government assigned Ford a half mile down the road from Doc's airfield. He gunned the engine in hopes of coaxing more heat from the anemic heater but knew it was useless. Since before sunrise he had sat alongside the road out of sight of the farmhouse. He had been seen there before, and some of Doc's drivers had even taken to waving and hitting their horns whenever they drove by.

Nick made little pretense of secrecy, and usually waved back. He never, however, moved any closer to the airfield than the woods along the runway. A handful of armed guards patrolled the perimeter on an irregular schedule, and Nick saw nothing to be gained by getting too close.

From where he was parked he could easily watch the planes arrive and depart, making note of aircraft types and schedules. Since his arrival, he had filled a notebook with airplane observing, all a waste of time, he decided. Without assistance, direction, or cooperation from the Omaha office—they refused to return his calls about the inoperative car heater—he was no more than some guy sitting in a car watching airplanes.

He skimmed through his log entries, countless airplanes, countless times. A separate section dealt with activity around the Hotel Lindell, but all his notes showed were the odd comings and goings of Sam Nyberg, and very occasionally the accountant, Arthur Reynolds. He concluded that Nyberg was the link between Doc and the actual working end of the enterprise. Nyberg collected and distributed payments. Doc never left the hotel anymore.

Nick sent reports to New York telling what he knew and expected to be recalled at anytime, but nothing happened. Out of boredom, he followed some of the delivery trucks, but without warrants or local assistance, he was just a guy who liked to watch trucks.

He pumped the accelerator and glanced at his watch. Ten more minutes, he decided. Ten more minutes, and he would leave.

"There's got to be a better way to make a living," he said through a sigh and opened the newspaper on the seat beside him.

A shadow passed over the car, and he looked at the sky. A large biplane slid low over the trees and vanished on final approach to the runway beyond. Nick opened his notebook and wrote: 7:15 A.M., STANDARD LS-5, REGISTRATION UNKNOWN, CARGO UNKNOWN, PILOT UNKNOWN, LANDED. He closed the book.

"Whoopee," he said flatly.

Until he had begun the Lincoln assignment Nick had never been closer to an airplane than the ones that occasionally flew low over the towers of downtown Manhattan on their way to Newark airport. Those he had ignored.

Beneath the newspaper on the seat beside him was EVERY BOY'S BOOK OF AEROPLANES. More out of need to fill the dull hours sitting around Lincoln, than out of true desire to learn about aircraft, Nick had become familiar with most of the book.

Spotting the Jennies and Standards was easy. Even Lucky Lundy's Curtiss Robin was simple to distinguish with its boxy cabin, high wing and square tail. The one airplane that initially stumped him, however, was an OX-5 powered Travel Air. Catching only a quick glimpse of the tail, he was indeed surprised to find bootleggers using Fokker D-7 fighter planes. For one entire evening he concocted a theory of a smuggler's air force running cargo planes with fighter escort to ward off competition. The ramifications in law enforcement were staggering. He envisioned the Treasury Department training agents to fly interceptors along the border and engaging in dogfights with underworld thugs.

Before his imagination got the better of him, however, he took a closer look at the D-7's and found them to be simple Travel Airs. Needless to say, the rough draft of his fighter theory never found its way to New York, but he did keep it, with an eye on fleshing it out for possible movie script. Each evening in his room he would take out the report and add to it writing in air battles and loose women all entwined in a netherworld of liquor and money.

SMUGGLERS AIR ARMADA, he titled it, then crossed that out. RUM FLIERS, and scratched that out, too. Finally,

BOOTLEG SKIES, but even that sounded too corny, and he tossed the whole thing back in his briefcase.

Henry landed as the other airplanes were warming their engines and making last minute preparations. Trucks moved between the planes delivering fuel and oil while mechanics tinkered with engines and shouted at pilots. Henry taxied between two uneven rows of planes.

Lucky Lundy's Robin sat in the chocks while Lundy ran the engine up to high rpm's, checking performance. Henry killed his motor and pulled his helmet off. He was cheated of the graceful silence that normally greets a pilot after the engines dies. Instead, the air was alive with OX-5 and OX-6 water cooled engines clacking and sputtering, along with several radials and the impressive Hispano Suizas of the larger Standards. In a back row, beside the derelict Jenny, he watched a pilot—he remembered his name as Grieve—furiously kicking a Travel Air.

"Mornin', Henry!" the black voice of Andrew, the mechanic, called. A truck backed up to the Standard to swap whiskey for gasoline. "We've been wonderin' where you might be." He climbed onto the wing dragging a hose in one hand.

"Oh, mornin', Andrew. I got stuck out east all night trying to stay out of the snow." Henry glanced around at the field virtually free of snow and added, "I guess you didn't get much of it here."

"No, but they got a whif of it up north."

A boy, about 14, silently pumped at a handle on the truck while Andrew filled the tank. "Doc sent out word you was to get yo' ass back up north just as soon as I get you turned around. You do somethin' to stir him up?"

"Not that I know of."

"Well, I hope you got a good night's sleep last night."

Henry smiled.

Suddenly, an engine roared, drowning out all other noise at the field. Henry and Andrew turned at the same time.

"Jesus H. Christ!" Henry cried scrambling from the cockpit. "Look at that nut! There's no one in the cockpit, and that throttle's wide open!"

He took off running toward Grieve and his Travel Air now roaring full power, the whole frame vibrating trying to shake loose the ropes and plow into the other planes and fuel trucks.

Grieve had apparently propped it and was headed for the cockpit when he slipped in the mud and fell with his head only inches from the propeller tip. Henry sprinted past gripping a wing and flung himself at the fuselage. In one lunge he grabbed the throttle and yanked back. The engine died.

"Hey, ya dumb bastard!" Grieve yelled. "What'd ya do that for? I just got it runnin'" He pushed himself from the mud and strode toward Henry, fists clenched, goggles pushed back on his forehead. He stood a full head shorter than Henry.

"You were about to kill someone!" Henry shouted.

"You mind your own business!"

"And you learn to start this damn thing or get someone who does." Henry started to leave. A small crowd of mechanics and drivers started to gather, a collection of blank faces and dumb grins awaiting a fight.

Henry avoided Wendall's eyes and brushed by him. A hand shot out and grabbed his arm as he passed. He felt the stubby fingers clench his bicep, and he tensed. Grieve, too, felt something, something unexpected. It was Henry's bicep instantly swell into something he could not possibly encircle with both hands. At the same time he felt Henry's left hand clasp his throat and lift him off the ground.

"You listen to me, you squeaky little fart, if you wanna' go around holdin' hands with someone, I suggest you learn to sweet talk them a little bit nicer. Understand?"

Wendall Grieve tried to nod his head and answer, but all that came out was a gurgling sound as he bobbed at the end of Henry's arm, his toes barely touching the ground.

Henry took a softer tone without relaxing his grip and led Grieve to the biplane. "Now, if you ever have any questions about starting this here airplane you just ask one of the lineboys." He twisted his neck toward the grinning crowd.

"And I'm certain any one of them would be happy to lend a hand. Is that understood?"

Again, Grieve grunted and sputtered under Henry's grasp. His eyes began to bulge out, and a panicked look crossed his face while he clawed at Henry's arm. The crowd began to sense that the lesson was going too far but did nothing to interfere. Sam Nyberg, who had been strutting around the line handing out assignments, kept to the back of the crowd until he saw Grieve being strangled, then pushed his way through for a better look.

"Now, if you promise to be good," Henry said, "and always have someone in the cockpit when starting your motor, I'll let you go." Henry spoke like he would to a small child. Grieve flailed wildly at his arm managing to push a sleeve up, but Henry's grip never slackened. His face was now swollen and purple, his tongue protruding like a steer hung in a slaughter-house. He nodded as best he could and kicked his feet in jerky motions.

Andrew started to call to Henry to let him loose, when Henry pulled Grieve's face close to his own and grinned. With that, he opened his fist, and his victim tumbled to the ground like a pile of wet laundry. No one moved to help.

"My plane refueled yet?" Henry asked Andrew.

"Almost...He looks bad, Henry."

"He'd look a lot worse if that plane had broke loose." Henry walked slowly through the crowd parting before him. Several minutes later the first plane took off.

Nick saw the first plane, a Curtiss Jenny, rise slowly from below the treeline bordering Doc's field. He noted the time, 8:03 a.m., and watched the biplane turn northeast, leveling several hundred feet above the ground. It was quickly followed by another Jenny, 8:05, that also banked gently in a right turn and leveled.

"Something's stirring," Nick said getting out of the car. He had been listening to the engines warming, on and off, for almost an hour. He placed his notebook on the car's roof and watched the plane's depart. A Standard LS-5 at 8:07, then another one at 8:09, then two more Jennies, almost in tandem. He noted, 'TWO JN-4'S AT 8:11 TURNED NORTHEAST.'

The air was cold, but Nick enjoyed watching the slow climbing biplanes lift one after the other above the trees and head out in a loose stream northbound. It was a beautiful sight, and he wondered what it would be like to be one of them, one of the pilots, crawling effortlessly through the cold morning sky on those frail-looking wings, laced together by a spider web of struts and wire.

The stream petered out after 8:30 and at 8:47 Nick recorded the last departure: 'ONE TRAVEL AIR (NOT A FOKKER D-7) WITH ROUND ENGINE, REGISTRATION UN-KNOWN, PILOT UNKNOWN, DEPARTED, HEADED NORTHEAST'. He waited a few more minutes until the sound of the airplanes faded, then folded his notebook, stuck the pencil back in his pocket and slid onto the driver's seat. The motor was running, but the car was cold. He mashed the gears, finding reverse.

He pulled out from the weeds onto the main highway leading to town. A truck loaded with hogs sped past. Looking over his shoulder, he saw the green Chevrolet parked in a grove of bare cottonwoods away from the highway. Three men, all in long winter overcoats, stood around the car stamping their feet trying to keep warm. Nick stopped to take a closer look.

He reached behind the seat for a pair of Army field glasses and turned them on the car. The figures were only slightly larger, but he recognized them as the three who had been lingering outside the Hotel Lindell. They too, he guessed, had been watching the morning departures.

One man leaned on the front right fender and smoked a cigarette. One of the others walked nervously back and forth behind the car, apparently explaining something to a third man who chewed on a short cigar.

Nick was twisted uncomfortably in the seat spying on the three. He turned momentarily, opening his door to step out. As he propped his elbows on the car's roof to brace the glasses, he saw the three men climb quickly into their car, two in the back, one up front driving. The Chevy pulled away from the thicket and lurched across the short field toward the highway. Through the binoculars Nick saw the man with the cigar seated in the rear point to his car and yell something at the driver. Nick, of course, heard none of this, but guessed its meaning and jumped into his Ford tossing the glasses onto the floor.

The gears complained bitterly as he jammed the stick forward into first. The government car took to the road in a backfiring cacophony of spitting gravel and whining gears. All four cylinders hammered away.

The speedometer needle passed 20, and he shifted gears. It edged over 35, and he looked in the rear mirror expecting to see the Chevy gain on him, but it hung back seemingly powered by the same gutless type motor.

"And I always thought gangsters drove Cadillacs," Nick said. He pressed the accelerator as hard as his wingtips could, and the Ford nudged to 45.

"C'mon, ya stupid piece o' junk!" he called in a clipped New York voice while pounding the steering wheel. The Chevy seemed to have found more power and grew in the rear view mirror.

Nick passed a sign reading, LINCOLN 6, then another sign: DOES YOUR HUSBAND. Nick shot a glance in the mirror; the Chevy grew larger and another sign approached: MISBEHAVE. Nick read the sign and checked the mirror. Another sign drew alongside and flew past, but Nick made a point to read: GRUNT AND GRUMBLE. He glanced again at the mirror where the Chevy slowly drew closer, the three heads clearly visible. Nick turned his attention back to the road where another sign slipped past: RANT AND RAVE. He could see the face of the cigar chewer leaning over the front seat shouting at the driver, but another sign was approaching and he turned to read: SHOOT THE BRUTE SOME. Nick kept his eyes on the roadside, waiting for the next sign completely ignoring the Chevy. Finally: BURMA SHAVE!

He took the fork that led to downtown along O street past the idle Standard factory. The Chevy filled his mirror. He could see the driver yelling back at the man with the cigar. The Ford would go no faster.

He started to panic, envisioning the Chevy pulling alongside on an empty stretch of road, the driver alternately watching the road and jockeying for position within inches of his door. Then the Chevy's rear window would roll down...

Wait a minute, Nick thought. Do Chevy's have rear windows?

Assuming it had a rear window, he envisioned it rolling down, and the muzzle of a Thompson submachine gun would protrude. The cigar-chewing thug behind the Tommy gun would squint and with an evil laugh...

Wait a minute, Nick thought. Maybe he would have to lean over the front seat and squint and laugh his evil laugh through the front window—that's if Chevy's don't have rear windows.

In either case, with the cigar clenched tightly in his teeth, the gunman would unloose a hundred rounds of .45 caliber slugs at close range into the government issue Ford, ripping through doors like they were made of balsa and mincing Nick into hamburger, practically within the Lincoln city limits.

The road was flat and empty. Nick felt he would push the gas pedal straight through the floor. He glanced from left to right looking for an out, a road, a farmhouse, anything that might offer a chance of escape, but all he saw was pasture and cornfields, all flat and barren. He checked the mirror again, and the Chevrolet was gone! He snapped his head left, and there it was inching up on him.

"It does have a rear window!" he shouted, and from the back seat the face with the cigar leaned out.

"Oh Shit!" Nick squealed, and he pummeled the steering wheel. A sign whizzed by. It said, LINCOLN, but he missed the mileage.

"C'mon, baby!" The speedometer read 55, and the transmission howled in protest. The front end wobbled and chattered, the steering wheel vibrating dangerously in his hands.

"Son-of-a-bitch," he hissed and looked over his shoulder. The Chevy approached his left rear fender, ready to pass. The driver, for some reason, sounded his horn. Nick looked. The driver waved at him to pull over and mouthed something, "Pull over" or "Your Mother" or "All over!" Nick could not distinguish.

"To hell with you, you son-of-a-bitch!" Nick saw the first buildings of Lincoln appear around a gradual bend in the road, the capital building poking above the distant trees. He approached a filling station and a lone house; not much cover, but at least it provided the illusion of security.

Can't shoot now, he thought. We're almost in town. See, a gas station. Hell, he then thought, they wouldn't think twice about killing me in a garage. Capone had just proven that was no obstacle in Chicago. Valentine's Day, he simply had a few of his boys walk into a garage, line up Mugs Moran's gang and...Nick remembered the photoggraghs circulating in the Department after the massacre. He envisioned himself in a black and white glossy, slumped against a chair, a pool of shiny blood beside what was left of his head.

"No, garages are out," he said and sped past the two buildings. "Gotta get downtown."

Suddenly, the Chevy was abeam him, and he swerved hard to the left. The Chevy skidded onto the dirt shoulder fishtailing in a cloud of gravel. The cigar chewer still leaned out the window, but now seemed to be flailing his arms grasping at air while the driver desperately tried to recover control of the car.

"Ah ha!" Nick shouted and laughed. "Take that, ya bastards!" He extended his arm, middle finger prominent, out the window in defiance while holding the wheel with the other hand.

The Chevy straightened and gained speed, closing again on his fender. It tried to pass, and Nick snapped the wheel again to block. Again, it swerved, and again Nick shouted and gestured. Several times the Chevy approached, and each time Nick countered, his confidence mounting.

The watertank atop the Lindell was visible over a rise when the Chevy again approached Nick's fender. The driver leaned on the horn almost continuously, and the cigarman hung out the window waving Nick over.

Nick turned the wheel to block, the Chevy backed, and he knew he was in trouble. The Ford skidded hard, the back end trying to enter Lincoln ahead of him. Nick countered with a hard right, and the car left the road at an almost perfect 90 degree angle, impacting the rim of a ditch. The Ford hit at a good speed mangling the spoke wheels and snapping the steering linkage. This mattered little, because the car cartwheeled sideways across the ditch, onto its back, then back onto its already demolished wheels, coming to rest finally in the corn stubble of a frozen field.

Nick still held the wheel, but his legs stuck at odd angles through the shattered windshield. All was quiet, and he saw a trickle of blood from a tear in his trousers near his left knee. His right leg was tucked under his left, and blood slowly oozed from the cuff. The world turned gray just as the pain flooded across his body. He moved in his seat lifting himself by the steering wheel. His vision was narrow, but he felt pain in both feet and decided that was a healthy sign.

The windshield held his legs fast. Any attempt to move them only made the jagged glass cut deeper into his skin. A thin line of blood dripped from his nose, and he wiped at it with his sleeve. His hands were also cut.

"Someone's going to be awfully upset when I return this thing," he said and rested his head against the seat amid the dull shards of broken glass. His vision blurred, and his ears hummed. He felt a strong desire to sleep.

A car stopped on the road, and Nick turned his head. It was the green Chevrolet. Three doors flew open, and three men jumped out. Nick saw them scramble through the ditch, one man slipped on the embankment and slid to the bottom, his long coat bunching under his arms. Nick snickered but made no effort to move.

A door opened, and Nick's right arm dropped off the seat. He tried to lift it in protest, but strength eluded him. A face with a cold cigar bent low and stared at him. Their voices were distant, vague. Someone moved his legs, and the pain blasted through him. He passed out.

Jake Hollow awoke to the sound of pounding on the cabin door. The room was dark and smelled of wood smoke and cold dirty socks. He heard Moyer snore in rasping irregular gasps somewhere in the corner behind the stove. The air was bitterly cold, the fire apparently having died sometime in the night. Jake poked a bare foot out from the covers and onto the wood floor.

"Jeesus sonuva...It's cold!" The pounding on the door continued and someone was shouting.

"Moyer!" the voice hollered, and the door shook from the blows. "You in there? You alive?" WHAM! WHAM! WHAM! The sound was muffled by the thick blanket hanging over the door. Jake's entire nervous system reacted painfully to his other organs metabolizing surplus ethanol into something resembling embalming fluid. In short, he was hungover—again.

WHAM! WHAM! "Moyer, get up, you worthless..." WHAM!

"All right," Jake yelled, but his voice came out a squeak. He gathered a blanket around himself and fell from bed.

"Middle of the damn night," he muttered. "Snowstorm..." WHAM! WHAM!

"I'm coming!" He stepped on an empty bottle in the dark, and it acted like a roller, sending him crashing to the floor, the blanket swirling over him in a loose pile.

"Ubbgh," he slobbered, his face brushing the floor.

WHAM! The violent knocking persisted.

"Coming!"

He pushed aside the hanging blanket, undid the latch, and an outside force opened the door. A blinding white light sent him crashing to the floor shielding his eyes and whimpering.

"Who are you?" a voice asked.

Jake slithered back into the shadows clutching his blanket.

"Where's Moyer?"

"Ohhh," Jake moaned and pointed weakly toward the stove. The stranger strode past him. Jake would have sworn he wore deep sea diving boots as the whole room shook under each step. A shutter was unlatched and pushed open allowing more cruel light and cold air to pour inside.

"My God, but it stinks in here."

Jake squinted at the voice, slowly discerning the railroad conductor in long overcoat and thick hat.

"He lives like a pig; like a hermit, and like a pig—like a hermit pig, eh!" The conductor grabbed a handful of blanket and uncovered the supine body of Moyer wedged into a corner, a disassembled pistol at his side.

"Wonder what he had in mind." The conductor prodded Moyer with the toe of his diving boot. "Get up. The day's begun."

Moyer made a sound like a boot being pulled from a muddy bog, then stood bolt upright, eyes wide, head erect.

"Morning!" he said. "Train in already?"

"It is, and we have quite a cargo for you, too. Are you with us, yet?" The conductor looked Moyer over from top to bottom as if trying to ascertain what might be holding him up.

"Be with you in a minute," he answered. "Tell the lads to start unloading; I only need a minute to freshen up a bit." Moyer stood absolutely motionless except for his mouth. It formed words with careful attention to diction.

"Good thing, Moyer. We can't hold the train more than 15 minutes." The conductor then nodded and left the room being careful not to step on Jake.

Moyer took one step and dropped to his knees recreating that noise like a muddy boot. Jake crawled to the window and pulled the glass portion shut. Outside, the world was brilliant in the clear dawn. The pine trees were white capped spires, motionless, heavy with snow. Occasionally, a bough would dip under the weight of the snow and loose its burden in a dusty avalanche down the tree.

The train sat outside the cabin billowing huge clouds of intensely white steam from under the locomotive. The Indian crew banged away at the ice encased water spout then swung it to the fireman on the tender. Winter had come to Canada, and Jake's first glimpse of it was from his knees, hungover, in a shack with a bear-sized, red-bearded whiskey smuggler who once knew Mae Murray.

He lifted himself onto the bed and watched Moyer button his trousers and pull two layers of plaid shirts over his torso. Strength seemed to flow into the man with each layer of clothing he donned. He took his long winter coat off a hook and worked his way into it. He then pulled the fur cap over his ears and wrapped a bright red scarf around his throat. In one smooth motion he clasped a whiskey bottle off a shelf, pulled the cork and took a long swallow. Returning the bottle, he puffed his chest, rubbed his hands together and called to Jake.

"Well then, Hollow. Better get a move on, eh! Day's a'wasting, eh!" He strode from the cabin letting in another chilled blast. Jake peered over the window ledge watching him march along the train shouting, "good mornings", while ordering Indians and train crewmen around in his cheery parade sergeant manner.

Jake shuffled around the room gathering socks and pants then crawled under the blankets to dress. After several minutes he emerged from the cabin, stooped over and lit a cigarette. The train was almost unloaded, and Moyer was directing two of his own crew to load whiskey kegs onto a nearby cart for the flight line. His hand signals were easy to interpret, and Jake envisioned him in uniform with stripes up his sleeve, knee deep in mud, ordering soldiers about with cases of ammunition while bombs exploded, and the world went crazy around him. He wondered briefly what made some men sergeants, and others soldiers and Indians. His reflections were shallow and easily interrupted by the smell of frying bacon from the mess shack. He turned for it after first stopping to retch in a snow drift.

The train pulled away to shrilling blasts on the whistle and mountains of black smoke and white steam. Behind the train

Moyer's ground crew scurried around like ants in the snow moving whiskey barrels and gasoline to the airfield.

Jake sat at the rough wooden table in the mess shack, drinking coffee and toying with a mound of bacon and slowly congealing gravy over hard dry biscuits. The quality of food had dropped off noticeably as the weather turned colder. Most pilots, however, were so tired and hungry by the time they sat down at the table they simply wolfed whatever was presented, chased with liberal amounts of liquor at 50 cents per glass.

The bitter coffee warmed Jake's chest, and he attempted a mouthful of greasy bacon. It threatened to return as soon as he swallowed, so he pushed the plate aside and placed two dollars on the table—cold weather prices were double the summer rates—and left. The cook scooped the two dollars up and scraped the leftovers into a frying pan for herself.

Jake's feet were clammy inside his heavy boots. He kicked several times at the side of the mess shack to improve circulation, but that did nothing. His socks were damp from sweat, and his feet would be cold all day. He trudged through the shallow snow across the tracks and along the smooth path leading to the Standard. Ground crews had most of the snow brushed from the wings, and gas and hooch was being pumped into the tanks at the same time. Further down the field Moyer directed snowplow operations. The same horses were being whipped by the same tired looking drivers.

Jake's was the only plane there, but still the field was alive with movement, all muffled by the new snow. Every now and then a voice would shout or one of the horses would complain, but overall, there was silence.

Jake unbuttoned the latches on the cowling, and a trace of snow fell past his glove and down his sleeve. One of the ground crew walked up with a metal can steaming with motor oil heated on the cook stove.

"Hello, Charlie," Jake called using the nickname the pilots had given him. His real name sounded like boiling oatmeal and Charlie was easier to remember. The Indian nodded with a straight smile and pointed at the motor.

"Oh, sure," Jake said and stepped aside letting Charlie add the warm oil. Jake watched over his shoulder. Above, the fueling was completed and the wagon moved. Charlie lit a handheld blowtorch and waved the loose flame back and forth over the stone cold cylinders. Once the motor was warm to touch they poured raw gas into the petcocks on the intake manifold.

While Charlie buttoned the cowling Jake climbed into the cockpit. The propeller was turned by hand until Charlie was satisfied it would start. He then removed a small metal can from under his coat, uncorked it, sniffed at it a moment and with a nod to Jake sprinkled a portion on the air filter.

Ether, Jake thought and wondered how well it would work.

"Pretty explosive stuff," he called to Charlie who smiled. Jake had heard of other pilots using it in cold weather but had never tried it himself. Of course, this was his first winter flying so far north.

Charlie signaled for Jake to start. The booster coil spun, and sparks shot to the gas filled cylinders, warmed by the blow torch and primed with ether.

Nothing.

Jake cranked and cranked and cranked, but the engine might as well have been filled with sea water for all the good it did. Charlie frowned and shouted at Jake, "Stop. Stop."

He approached the engine muttering to himself. He uncapped the ether and sprinkled a liberal amount on the filter and stepped back.

"Now. Now," he said confidently and waved at Jake to try again.

Jake took the booster coil handle and spun it round and round. He pictured the sparks showering through the leads into the cylinders where the explosive gases waited.

Nothing.

Jake leaned back in the seat and banged his cold feet against the sides trying to keep warm. He felt the damp cold settle in his butt, immobilized by the seat. Charlie scratched his head and took out the ether can again and sniffed at it. He wrinkled his brow—apparently it was real—and once again approached the machine.

Jake leaned over the fuselage and watched him splash away at the air filter saturating it with ether. Then without a word, he unbuttoned the cowling and climbing above the motor, poured ether straight into the petcocks.

"Now wait a minute," Jake called becoming apprehensive. "Do you know what you're doing? Isn't it possible to use too much? You know—Boom!"

"Yes, Boom," Charlie repeated and grinned. He sniffed the can again and held it up to Jake.

"No, thanks. Just get down, and button the cowling back up."

"Now. Now." Charlie bumped the propeller and waved for Jake to start. Jake obeyed with misgiving and cranked the booster. Again, nothing happened, when, suddenly, Jake noticed the main magneto switch was still turned OFF. Without that on, the sparks from the coil would never reach the engine.

Without thinking, and while still cranking, he snapped the switch ON.

KA POP! Great orange flames shot from the exhaust stack. COUGH-POO! Flames backfired through the carburetor, and Jake saw the propeller snap one way, then the other, then back again as flames and gray smoke blew from stacks and under the cowling.

Indians scattered, and Jake held onto the throttle wondering what to do. He looked over at Charlie who stood there grinning and staring at the propeller as it settled into a more regular rhythm and ran normally. Charlie grinned and turned away, sniffing at the ether can as he left.

Once the motor idled on its own, Jake climbed out and ran back to the cabin. Moyer was filling in receipts and making entries on clipboards. A cigarette burned itself out in the ashtray at his elbow. Jake stuffed a wrinkled shirt into his bag and closed it.

"Thanks for the place to stay," he said.

Moyer looked up from the ledger, and Jake noticed for the first time that he wore glasses, half glasses used for reading.

"Leaving, are you?"

"Plane's warming up outside. I've got eight cold hours of flying ahead of me. See you next time."

"Here's your receipts." Moyer pressed several carbon printed forms on him and stood. "Did I tell you about Adair?" he asked.

Jake spun around at the name of a woman he had made a conscious effort to forget, although whenever he closed his eyes she seemed to appear.

"What?" he asked. "What about her?"

"She's dead."

Nick's mind tried to push through the confusing figures and voices that swam in the thick fog surrounding him. He saw bright lights and faces. Someone asked if he could hear them. He answered, "yes", and they asked the question again and shook him. He thought his legs hurt but wondered why he never felt the hands that moved him onto the table below the bright lights. Then the lights faded as did the voices.

His mind was in New York. The woman in the apartment across the hall from his, opened her door and handed him a large drink. It was too heavy to carry or bring to his lips, so he set it on a steering wheel in her living room.

She laughed and scooped meatloaf repeatedly onto a chipped blue platter. All the while she reminded him to, "save the receipts; save the receipts!"

Someone puffed on a cigar. Someone sneezed.

"Can you hear me?" There were more lights, and voices around him in a room.

Nick answered, "yes, yes."

"He doesn't hear us," a voice said. Everyone seemed to be wearing white. No, Nick realized, not everyone. Someone wore gray, or brown and smoked a cigar. Someone ordered the cigar put out.

Nick tried to scream, "Get out! Get out!" But the man with the cigar lurked somewhere over him brandishing the

Thompson sub machine gun. He saw the barrel pointing directly at him, and the green Chevrolet pulling alongside.

A hundred rounds of .45 caliber ammunition would spray him at any minute. He smelled the cigar, and still they complained to the man who smoked it.

"Yes, get away!" Nick screamed, but the words never left his mind. A mask slid over his face, and he smelled black rubber. The car seat—his face was on the seat, and someone moved his legs. There was broken glass everywhere and blood, his own blood. He could no longer smell the cigar.

He pushed the rubber mask away, and a voice said, "I need help here; he's strong."

Three men in long coats pulled him from the Ford. Oh God, the Ford! I'm going to catch hell about that, he thought. He laughed still pushing away the mask.

"I remember walking to the Chevy," he said. "At least I think I remember walking." He turned his eyes to one of the figures in white. "Is my voice working okay?"

"Hold his arms down!" a white voice said.

"I tried to get in the front seat, and the man with the cigar said, 'no, that seat's broken. Get in the back.'"

The white voice said softly, "You'll be all right; just breathe normally." The rubber mask formed a seal around his face, and the lights went out—Poof!

Chapter
XXI

The Hotel Lindell received no advance notice about Doc's plans to move out. In fact, no one knew about it until Doc casually announced his intention to relocate his headquarters to the farm to be closer to the operation. It was late on a Wednesday night, a week after they had heard the news of Adair's death. Sam was picking at a plate of cold meat and cream pie. Art was scratching away at his ledgers under a harsh yellow lamp in the sitting room.

Doc stood at the chalk boards rearranging names in what appeared to be no particular scheme. It was almost midnight. Sam had asked to go home earlier and had met with such a sharp rebuke from Doc, that he now sat moodily by himself eating. He had gained more weight in the last week, and his acne had spread to previously unscarred regions of his face and neck.

"Where's Hollow?" Doc asked. For hours he had been working the boards every now and then shooting an odd question to the room.

Art glanced at Sam then at Doc's back trying to determine just who was expected to answer the question. "Must be my turn," he muttered and flipped through a list of active pilots running his finger along the column until he found Jake's.

"He's flying Standard number three, and is due in at nine a.m.." He ran his finger along the horizontal line beside Jake's name. "Carrying 120 gallons. That reflects the increased load orders you issued..."

"I know what I ordered," Doc said curtly, barely glancing at Art over his shoulder. His eyes were set deep into his tired face. The artificial light only deepened the creases in his skin. He smoked a cigarette without the holder. "I only asked when he was due." He tapped a smudged line on his board. "Just double checking."

Art slid his schedule back into a neat folder and dropped it into a file cabinet at his side.

"There's not too many good flying days left before winter, you know, Arthur."

"Um Hmm," Art hummed.

"Once winter's in we're shut down for good." Doc erased the smudge beside Jake's name using the flat of his hand. He tried printing in something else, but his hand shook distorting the letters, and he rubbed it out again with his palm.

"Any new complaints from those pilots?" he asked.

"No more than the usual. They say night landings are dangerous. We have had a number of accidents; had to replace quite a few wing tips and gear legs. They want landing lights on their planes."

"Then let them buy them," Doc said.

"Nobody has any money. That's another complaint. They want to know when they get paid. Some of those pilots are owed sizable amounts."

"They'll get paid. I've told you that. We pay them now, and half of them will be off before we finish the season. The other half will tear up Lincoln in one night, pissing it all away and bringing the roof down on us. They don't understand how this works you know! They're only pilots."

"They want real airplane mechanics out there..."

"Well, why don't we invite Henry Ford to bring us some Tri-Motors, and we'll build a terminal..."

"Hey, you asked. I'm only telling you what I hear." Art was the only one who could talk back to Doc, but even he invoked this privilege rarely.

"I know," Doc said rubbing his eyes. A long ash dropped from his cigarette onto the carpet. "Any other problems?"

"I think they suspect you've substituted cheaper gas in the tanks. They say the engines have been running like shit."

"Sam," Doc turned. "I thought you said they wouldn't notice the difference in octane if we dyed it!"

"Well, nobody seemed to notice it when Randall did it. Saves you a nickel a gallon."

"Maybe airplane engines are different," Doc mused. "Anyhow, they can live with it.

"They say the food stinks."

Doc waved this complaint away without comment.

"They want more sleep between turn arounds."

"They can sleep in the airplanes—IN THE AIR!"

"I think they do," Art said. "Moyer wired us that whoever flew that American Eagle came in last week, turned over the airport and just kept heading north. Plowed into the trees somewhere north of there."

"Did they find him?"

"Didn't bother to look."

Doc blinked. "He was empty, I assume."

"Oh, yes."

"Good."

Art shifted in his chair. "The biggest complaint, and I think this one might prove the most troublesome, is the weather. It's getting colder..."

"Always gets colder in winter," Doc said.

"Canada cold," Art said. "Remember Lake Placid? Apparently this is worse. The planes won't start. Oil turns to tar if left out, and sitting immobile in those open cockpits for hours the pilots freeze. Most of these guys aren't experienced with cold weather."

Doc snickered and turned from the boards. "The cold will keep their minds off the increased loads they're going to carry."

"Another increase?"

"Why not?"

"Oh, I could probably think of a reason; so could they, I suppose."

"Who's our best pilot, Art?" Doc asked looking straight at him.

"Hollow, Jake Hollow."

"By how much?"

"Oh, I'd have to add up the numbers, but I'd say he's an easy 20 percent ahead of the nearest competitor."

"Makes money for us, right?"

"Yes."

"We pay him good money, right?"

"Well, lately we haven't paid anyone, but we owe him good money."

"Does he complain?"

Art thought a minute. "Not openly, no."

"Sam," Doc called. "Had any complaints from Jake Hollow about flying conditions—new loads and all?"

Sam was stuffing a glob of cream pie into his mouth but managed, "Not about flying."

"There! You see, Arthur. Jake Hollow is our benchmark. Everyone else should measure up to him."

"But Hollow's crazy!"

"Hell, Arthur, they're all crazy, or they wouldn't be pilots!"

"No," Art protested. "I mean Jake Hollow is really going nuts, truly crackers! I dunno, it's like shell shock or something. He keeps to himself, never rests, something's driving him..."

"Good!" Doc said. "It should drive them all."

"It's going to drive him right into the ground."

"Not my concern," Doc said. "Although, I hate to lose a good airplane."

The room fell silent for several minutes while Sam scraped his plate and Doc lit another cigarette. Arthur poured a short drink for himself, then poured one for Doc.

"Thank-you," Doc said.

Art rolled his glass between his fingers, staring at the light reflecting through the whiskey. Finally he said, "You know, one of our best pilots, after Hollow, is Henry. He's turned into quite the flyer."

"Hollow earns more."

"True, but Henry's final take will be much higher when he gets his share of profit."

Doc set his drink down turning on Arthur. "Henry's a pilot. Pilot's don't get shares, just wages."

"Henry's a partner," Art said. "A 20 per cent partner. We promised him that when we started. You can't screw him out of that! That's wrong."

"Don't give me right and wrong." Doc spoke softly with a sideglance at Sam who left the table and settled into a wing chair. Within seconds he was asleep.

Doc leaned over Arthur. "Henry quit being a partner when he took up flying. I didn't need another pilot, especially a green one."

"But he proved himself," Art said.

"Coincidentally. What I really needed was a right arm, a side man. Someone to run the physical aspects of the operation."

"Like him?" Art jerked a thumb at Sam.

"He's adequate...damn adequate," Doc said and turned his back. "Plus he runs a fair volume of hooch through that garage of his. He deserves to be a partner more than Henry. I plan to give him Henry's share. I will not give it to some damn pilot, that's for certain!" Doc swept from the room into his bedroom.

Art thumbed through the pages of his books and slowly added the figures calculating a 20 percent slice of the enterprise. He pulled out Henry's personal file and mentally added that sum to Henry's already respectable earnings from flying and brief turn in the stock market.

"A person could live nicely on that," he said to himself and looked over at Sam asleep in the chair, his mouth open. "Sure as hell not giving it to that!"

Suddenly, Doc breezed into the room again and flicked on the overhead lights. He charged over to Sam and vigorously shook his head.

"Get up! Get up! We are moving," he announced.

"Wha..." Sam mumbled rubbing his eyes.

"What are you talking about?" Art asked.

"Out to the field; get the trucks. I want all this packed and moved out right now. Tonight." He pushed Sam toward the door, planting a hat on the boy's head. Sam stumbled from the room, and Doc closed the door behind him.

"Where are we moving to, and why for Chrissake?" Art asked.

"Out to the airfield—the farmhouse. We need to be closer to the troops. Direct supervision, that's what's needed. Be in the center of things." He was charged with sudden energy, moving quickly about the room picking up papers.

"There's no room at the farm!"

"Sounds like a line out of a Christmas play."

"Well, there is no room," Art protested. "They're already sleeping in every conceivable nook and cranny out there."

"Well, then a few pilots shall sleep with the camels and sheep in the manger, eh?"

Less than an hour later Sam returned with two trucks and four sleepy pilots shanghaied to move baggage. He ordered them roughly about while doing nothing himself. They loaded boxes and suitcases into the hallway, down the elevator and through the front lobby where a drowsy night clerk ran frantically around questioning everyone passing his desk including a tired bellhop, who was soon carrying boxes himself.

The chalkboards were the last items to leave, followed by Doc, Art and Sam. The clerk, a thin mousy type with long strands of greasy hair combed across his bald spot ran in front of Doc blocking his exit.

"Are we checking out, sir?"

"Not unless you plan to join us. We, however," and he indicated his entourage with an imperious wave of his cigarette holder, "shall be leaving."

"Do you wish to settle the bill?"

"Certainly. Samuel, take care of that, please." Doc led Arthur through the front doors and climbed into the awaiting car. The chalk boards were hoisted onto one of the trucks in front of them.

Inside, the clerk squirmed his way behind the desk and rummaged through a file box until he found the unpaid bills for Doc's suite.

Ding! Sam plunked the brass bell with his palm.

"Please don't do that," the clerk snapped without looking up. He sat behind a small partition and entered a long string of charges into the adding machine wrenching the handle with

334

each entry. The front door opened again, and the sound of the trucks flowed in momentarily. He tapped at the machine. He finished one list and called to Sam.

"It will take me a minute to figure the phone charges. This is very unusual, you know. Imagine, checking out in the middle of the night. Did you have any calls this evening? Never mind, I'll check the log myself." He stood to get the receipts from the switchboard and noticed Sam had left. Dashing to the counter he caught a glimpse of the Lincoln pulling away.

"What?" he huffed and started for the stwichboard, when he saw a small lump on the counter top. It was a bullet, set neatly where his bell had been. He dropped it quickly in the waste basket and stood alone in the dark lobby and shivered.

Sam drove the car with one hand on the wheel while he inspected the bell from the clerk's desk, slowly turning it, wondering what possible use he might have for it. Deciding he had none, he rolled down his window and tossed it into the darkness.

"Close your window, Sam," Doc growled from the back seat.

"Yes, sir," Sam answered. The long automobile moved through the deserted streets of Lincoln and entered the highway leading to the farm.

Doc sat quietly in the rear seat and watched the trucks ahead moving slowly in the windy night. A light snow fell, caught in the glare of the headlamps—white fireflies in the otherwise black night. He was comfortable and excited.

"You know, Art, we're close now," he said.

"Oh?" Art answered and peered out the window.

"No, I don't mean close to the farm. I mean we are almost there—to our goal. It's almost December, and the money's still rolling in. The planes are still flying, and so far the law's stayed off our backs."

"They should; we pay them well enough."

"I just can't figure why the competition hasn't moved on us yet," Doc said. "I thought we might attract more attention." He sounded disappointed.

335

"The Treasury Department's been watching us. They've made no secret about that."

"It'll take them at least a year to decide what they want to do."

They rode in silence another mile and Doc spoke. "That green car's been prowling around the farm, I hear. If you ask me, that's one of Capone's squads—out of Chicago." Doc was proud to imagine himself threatening Al Capone. "I'll bet he controlled this territory, and we've cut into his take. We were bound to ruffle a few feathers somewhere."

"Capone?" Art laughed while looking out the window at the snow. "No, I don't think so. We're a two bit operation compared with anything Capone's got going. He doesn't know us from a hole in the ground. No, whoever ran the small amount of booze around here before was strictly mom and pop stuff—Clara and Randall types.

"Hell, this is William Jennings Bryan country; got temperance written all over it. We buffaloed them, and they haven't figured out how to get rid of us yet, but given enough time they will. We simply created our own market and moved in. It's about time to move out, too."

"Arthur, you might be misreading the situation," Doc said. "There's room for market expansion here. As it is we can't handle all the orders."

"You're not thinking of staying through the winter are you?"

Doc was silent a moment. "And why not?"

"Doc!"

"We are making money, Arthur." His voice took on its thicker southern accent.

"And becoming too big without protecting ourselves."

"We did it in New York."

"In New York we had a support staff that stretched up and down the East Coast. Rothstein's influence stretched a lot further than Lake Placid Village..."

"Don't give me Rothstein."

"Here we have no protection beyond a few pay-offs local-ly." Art turned sideways facing Doc. "The beauty of this

336

whole deal is we move in fast; make a killing, and while everybody starts getting interested in us, we've already packed up and left. That was the plan."

"Plans change, Arthur. One must be flexible to take advantage of opportunity, or someone else will."

"Let them! We've made a small bundle, now we leave. Set up somewhere else. Like Cuba; it's warmer there."

"Cold weather didn't chase Capone out of Chicago!"

"This isn't Chicago, and you are not Al Capone." Arthur's voice struck cold and sharp. They rode in silence for another mile or so. "Al Capone in Nebraska." Art laughed. "For Chrissake."

Doc laughed, too, and poked Art in the shoulder. "This town would shit if Capone moved in."

"It's already started to shit, because you're here!" They both laughed, and Sam turned in his seat.

"Can I pass these guys? They aren't doing any better than 30, and I have to keep down shifting."

"Stick behind them, Samuel. There's no rush."

Sam turned away with a sigh and moodily shifted gears again.

"Why do you suppose the Treasury Department really sent someone out here?" Doc asked.

"Simple. Rothstein left behind quite a string of unpaid bills. I suspect your name's flagged somewhere, and somehow they heard you were out here, so they fired off some nobody to check you out."

"Those were some times, weren't they, Art?"

"You mean New York?"

"Yes. With Arnie Rothstein and the whole East Coast kissing his rear end..."

"Remember Mae?" Art asked.

"Mae Murray? Oh yes!" They looked at each other with conspiratorial grins. "Do you think Moyer ever...?"

"Moyer?" Art laughed. "Well, who knows, maybe."

Doc spoke softly gazing at the ceiling, "Mae Murray..."

"You knew Mae Murray?" Sam's young voice popped into the conversation, his head completely turned, eyes off the road.

"Oh yes," Doc answered. "Drive the car! Drive!"

Art leaned closer to Doc, so Sam could not hear. "I wonder if Mae ever thinks about us. Or if she ever thinks about Adair. Weren't those two a pair?"

"Arnie thought so, anyhow," Doc said. Sam kept turning his head, straining to catch part of the conversation. Doc flicked a finger at his ear to keep him looking forward.

"Who do you think Arnie liked better, Mae or Adair?" Doc asked.

"Mae was closer to his age..."

"Yes, and I think that's why he went for Adair!" The two laughed again sharing familiar secrets, then fell silent, both reliving memories. The clock ticked methodically from second to second.

"Arnold Rothstein," Doc spoke the name, his tone implying a respect for it. "Who do you think killed him?"

"Same person who killed Adair," Art said flatly. "Vince."

The two men sat quietly, both staring into the dark snow. The truck ahead turned off the highway onto the dirt road leading to the farmhouse. The chalkboards swayed from side to side in the wind.

"Did Vince and Adair owe us any money?" Doc asked.

"No, Vince paid everything off ahead of time. He must have been planning it for some while," Art said. "How very thorough of him."

"Damn considerate I'd say. The man never let his personal life interfere with business." Doc watched the clock tick off one full minute then said, "Do you have their records with you?"

"No, burned them when the account was closed."

"Fine."

Chapter
XXII

Saturday nights, regardless of the weather, meant business for Clara—big business. Aside from her downtown eatery now passing more than the occasional snort under the table to the locals, Clara ran three roadhouses outside city limits. All three did a healthy business with the university crowd, transients waiting between trains, and, as Clara liked to call them, "our colored clientele." Any way it came in, the money from her speaks continued to flow.

It was early Saturday evening, and she was in the back room of her diner, supervising the unloading of a shipment from Doc. The walls were lined to the ceiling with unlabeled bottles in wooden cases. This was the cut whiskey, mostly corn liquor flavored with Canadian. Doc had warned her that shipments would taper off after the first of December, and he combined this with an attractive discount if she paid cash in advance for larger orders. Clara took advantage of the deal and stockpiled for winter, a time when she perceived prices doubling.

"Put the straight Canadian over in the corner and set a case aside for Mayor Twimms." She barked orders to four black men who shuttled unmarked crates from the truck parked in the alley.

"Clara!" the cook called from the kitchen. "I think you'd better come out here." His voice was strained, but she knew Lionel to be the nervous type and decided it could wait.

"In a minute," she answered. A dull booming vibrated distantly through the walls.

"I don't know," Lionel persisted, "I think you'd better get out here."

"Can't you take care of it? I'm busy." The booming grew in intensity, thumping at a regular beat. "What-the-hell.."

"Clara!" Lionel screeched, and a sharp tearing crash exploded in the dining room.

"Lock that door!" Clara hollered to the driver standing near the back door. "Get the truck out of here..." She was halfway through the door to the kitchen, and the back door was being latched, when there was a rattling crash, and the back door flew open slamming the driver to the floor. He curled into a tight ball covering his head to ward off the blows from a mob, mostly women, plowing through the breech.

Clara glanced over her shoulder and saw about 20 women and a handful of men wielding clubs, baseball bats, axes and just about anything threatening they could find—one man carried a mop—charge into the storeroom and set about smashing open whiskey cases and bashing drivers.

"Lionel, bolt the front door," she cried running through the small kitchen. Instantly, she saw it was too late. A crowd, easily twice the size of the one in back, was busy smashing windows and axing tables in a frenzy of righteous vigor. One man, someone she had sold cut whiskey to on a regular basis, stood near the counter. As a woman set bottles on the counter he would obliterate them one at a time with a bat. Seeing Clara, he grinned sheepishly and nodded toward the woman setting up the bottles. He took another swing. SMASH! Glass and booze filled the air. In the middle of the room, amid the axers and bashers, stood a tall man pounding a bass drum with, LINCOLN ANTI-SALOON LEAGUE painted crudely on the skin.

BOOM! BOOM! He pounded the rhythm propelling the mob on to more destructive abandon. BOOM! BOOM! While he drummed, a woman climbed onto a chair and tried to make a speech. Reading from a prepared text, she shouted above the crowd, but Clara could only catch snippets of her tirade, something about Satan and youth and Jesus. Clara wondered idly if the woman was for Satan and against youth, or for youth and merely invoking the name of Jesus to silence the crowd. Either way, the crowd ignored her and continued to rip Clara's place to shreds.

Bottles crashed, and chairs broke—a scene familiar to her roadhouses on Saturday nights with the regular crowd. Clara noticed several men slip in with the mob, grab what they could from the unsmashed stock and slide out again while the drum boomed. Suddenly, the speechmaker fell off her chair when someone smashed it with an axe.

Outside, a handful of uniformed policemen appeared and kept their distance without interfering. Clara recognized them as recipients of free lunches and campaigners for various police benevolent associations, all to which she contributed generously.

The crowd soon tired of breaking fixtures and turned on Clara with an ugly howl. Hands groped at her legs and face. Her left shoe vanished. The drum pounded continuously, and she felt the weight of the mob pressing her to the floor. Her dress was pulled over her face and torn. Anonymous hands pummeled her head with weak blows. A hand grasped her breast. A small woman, hair tucked neatly behind her head and smelling of lilac, repeatedly smacked her tiny fist into Clara's face. Clara broke an arm free, and with the strength that came from 20 years of slinging hash, she back-handed the woman, catching her square on the nose with her knotted fist. She drew blood.

"Ha!" she shouted, and the woman disappeared into the crowd, just as police whistles squealed and blue uniforms waded through the crowd swinging nightsticks.

Clara rolled painfully on the floor littered with broken glass and wood splinters. Slowly, the mob was herded from the diner, and the shattered doors barred. Clara sat up against the wall behind the counter and checked her teeth, all there.

"Pretty rough on you, eh, Clara?" A sad looking sergeant peered down at her and shook his head. "This was bound to happen. You can't flaunt success forever."

Clara pushed herself to her feet accepting help from the sergeant. He led her to the back room, where she surveyed a scene almost identical to the front, except the delivery truck was overturned in the alleyway. Whiskey ran in a shiny stream to the gutter.

"We got to take you in, Clara. Come on." The sergeant guided her outside to a car. The crowd stood well clear of the entrance and cheered as Clara was driven away. Inside, policemen sifted through the rubble and placed undamaged bottles into several cases for processing as evidence.

The county attorney refused to press charges against Clara when it was discovered that no usable evidence made it to the station. Clara was released after midnight and drove out to check her roadhouses. They were untouched, and business was brisk.

The following morning, while airplanes departed for Canada, she stood in Doc's new office at the airfield, decrying the lack of police protection and pleading with him to restock her business.

"I can sympathize, Clara, but you must realize—here please take my hankerchief; that is a nasty bruise! You must realize I have other orders to fill.."

"Oh please! Those animals destroyed almost my entire stock! I don't know how I can make do..." She blew her nose. "I called my insurance man, Carl Bentz, he's out there now making the adjustment, but that will only cover fixtures, not inventory. You understand I couldn't possibly have insured..." She blew her nose again. "...that."

"Clara, I'll—no, keep the hankerchief—I'll see what I can do, but you realize you may have to pay a bit more. This is terribly sudden."

"Oh thank-you!" She sprang out of her chair and planted a string of wet kisses on his mouth. "Thank-you, thank..."

"Yes, that's fine..."

She blew her nose again with vigor.

"Wire Moyer," Doc said to Art after Clara had left. "We have ten planes due to arrive before nightfall." He paced in front of the chalkboards. "Each plane is to carry 10 percent more; that's like getting a planeload free, isn't it?" He smiled.

"They won't like this," Art said.

"Add onto that telegram there's a 50 dollar bonus to the pilot who makes it back first."

"50?"

"All right! Make it a hundred, a hundred bucks to the first pilot back with the higher load. I haven't seen a pilot yet who'd turn down that kind of dough."

Art scribbled the message on a piece of paper, folded it and called for Sam to run to the Western Union Office at the train station.

Chapter
XXIII

Sam, dressed in his camel's hair coat and galoshes, walked across the empty field behind the house. He held his new .38 revolver in his right hand and moved toward a flock of grackles picking through the harvested corn rows. Snow had fallen steadily since dawn, and the black birds stood in sharp contrast against the white. The air was still. The only sound came from his boots crunching the snow. He moved within 50 feet of the flock and stopped. Snowflakes collected on his coat, and he raised the pistol.

The birds pecked at the frozen earth scratching out bits of corn. The front sight rested on a black body, and he clicked the hammer back. As he did, the bird shifted but, immediately, another took its place.

No matter, Sam thought, and squeezed the trigger. BAM! The shot barked in the still air, followed quickly by five others. The flock lifted in great excitement, snow kicking in exploding puffs as lead slugs passed through feathers and tiny organs before impacting the earth.

He tracked the flock, emptying his pistol in wild shots until there were only dry clicks, and the birds settled on the far side of the field to renew gleaning. He snapped open the cylinder and dropped the warm empty shells into the snow. While rummaging through his coat pocket for more bullets, one of the mechanics ran toward him from the house. Sam ignored him, walking to the dead birds.

"Sam!" the mechanic called, the weak voice coming out of a steam cloud. "Sam!"

Sam kicked at the black and scarlet remains of a grackle and cursed his marksmanship. "Two out of six," he murmured and pressed new bullets onto the cylinder.

"Sam, Doc wants you to get to town."

Sam clicked the pistol closed and sighted on the indistinct center of the flock. BAM! BAM! BAM! The three shots rang out in even procession, but the flock continued its progress across the ground uninterupted.

"Ahhh," Sam whined.

"Sam! Doc wants you!"

"Ah, I'm coming. You went and made me miss that shot, you know!" He brushed past the messenger, stepping on the grackles.

"Sorry."

Sam shoved the pistol into its holster through a flap sewn into his overcoat. The flap was unnoticeable and allowed him to get to the gun without going through a lot of buttons. He had read in a magazine that Machine-gun Jack McGurn had the same sort of coat made for him.

"What's he want?"

"I don't know. Just said he wants you to go into town for something. I don't know."

Sam acted annoyed, more out of principal than anything. He had looked forward to shooting birds all morning, but a trip to town offered a promising diversion.

"That old goat can't spend two minutes without my help," he said without looking at the mechanic. The two shuffled through the snow, faces down against the flakes, toward the house.

Inside, Art handed him the telegram to send and reminded him to bring back a receipt.

"I know, I know," Sam said heading out the door.

Snow had piled on the Lincoln Brougham in a thick layer. Sam brushed a small hole clear on the windshield and roared away. Driving the long powerful car alone was infinitely more fun than chauffeuring Art and Doc around at old lady speeds.

Once on the highway, he opened the throttle wide and sped past the Burma Shave signs pointing his pistol at each one and reciting them by heart.

"Does your husband—Pow!" he said.

"Misbehave—Boom!"

"Grunt and grumble—Ka Plooey!"

"Rant and rave—Bang!"

"Shoot the brute—Pow! Pow!"

"Some Burma Shave!" He laughed almost hysterically, always finding the signs amusing.

The Lincoln flew past the lone filling station outside Lincoln, the speedometer reading over 70. He passed a snow plow and slowed entering town. As always he took his time with the errand. Lincoln was his town, his birthplace. It was where he grew up, and he desperately wanted everyone to see him.

Despite the cold he rolled down the windows on both sides of the car and cruised through the streets with his left arm out and right hand casually gliding the long automobile—longer than anyone else's in town, except possibly Lyon's Funeral Home on Q street. Then, their passengers rode lying down.

Sam waved to several girls across from Woolworth. They turned their noses up with dramatic disdain, but the one, a red head, slightly plump, glanced over her shoulder as Sam turned the corner and thumbed his nose in return.

He waited impatiently for a team of dray horses to pull a wagonload of coal through the intersection by the library. Gunning the engine, he hoped to spook the animals, but the street-wise horses, too, ignored him and plodded on. A small boy throwing snowballs at the library stopped long enough to stare at the car, then as Sam deemed him worthy of a wave, he hurled a snowball through the side window, splattering against the walnut door panel.

"You little bastard!" Sam screamed and opened the door forgetting the car was in gear. His foot slipped off the clutch, and the machine lurched forward, bounced through the intersection and onto the sidewalk crushing a dormant shrub. He grabbed the wheel. Faces appeared at the steamy windows in the library, and he backed away.

Delivering the telegram took less than five minutes. Sam folded the receipt into his inside pocket and backed the Lincoln away from the depot. The ride through town was uneventful. Mostly the streets were empty, only the odd car and occasional pedestrian moving slowly through the snow.

He gazed at the flimsy Christmas decorations strung from the street lamps. The business districts were putting on a jolly face to ward off mounting financial blues. Sam squelched an urge to take a shot at a pink faced Santa staring down at him from atop Baker's Hardware store on North 9th Street.

Someone stepped around the corner, and he recognized her as Darla Hendersen, the girl who had sat next to him in Algebra in the ninth grade, the year she quit school. She walked quickly along the street with the gait of a thoroughbred horse, her butt swinging in quick exaggerated movements, and her feet stepping high through the slush.

Sam followed her for half a block watching. He remembered her as an incredibly stupid girl, flat-chested and smelling of wet wool. He also remembered how hard she tried to make friends without success.

"Darla, hey Darla!" he called leaning out the window. The Lincoln slowed to a pace matching hers. She kept walking but turned. She wore round glasses and squinted to see who called.

"It's me, Sam." It meant nothing. "Sam Nyberg—I used to sit next to you in Algebra, Mr. Garner's class." She stopped. but said nothing, only stared. "I put the mud turtle in your bookbag..."

"Nyberg?" She giggled and walked toward the car, slowly, but still kicking high her feet. "Sam Nyberg? Is that you?— giggle, giggle."

"Yeah," Sam said. "Whatcha doin'?" He tried to sound casual.

"Where'd you get the car? Giggle, giggle."

"Oh this? I just use this for business." He gunned the engine. "It gets me around. You doing anything?"

"Yes," she answered blankly.

"Well, you wanna do something else? Like go for a ride with me?"

She shrugged and wandered around the car climbing into the front seat. She was still flat-chested and smelled of wool.

"Gee, swell car!" she said without looking at Sam.

"Where would you like to go?"

"Can we put the windows up?" She had hers rolled up before he answered.

"I know where we can get us a drink—whiskey—good stuff, too."

She giggled and looked out the window. "Gee, swell!" was her reply.

Sam slipped the car into gear, and off they went toward his filling station. It was time he checked on his cousin, anyhow.

"So, Darla, whatcha been up to since last I saw you?" Sam imitated Art's clipped city accent trying to sound worldly. He drove the car with one hand on the wheel while the other slithered across the seatback toward her shoulder.

"Working," she answered, her own voice flat, tell-tale midwestern. A sinus problem made every word sound as if she was preparing to spit. "You've got more pimples than you had in Algebra!"

His right arm stopped mid-slither, and fell to the seat. They rounded a corner and drove by the hospital. She stared with a dull gaze out the window and occasionally snorted to clear her head. Sam racked his brain for something to inject into the conversation, when he saw the green Chevrolet parked outside the hospital. It was empty.

He pulled over to the curb. "Wait here a minute," he said and left her while he ran to the Chevy. It looked like any other car, except the front right seat was torn, and a spring poked through. "You'd think Al Capone could afford something better," he said.

He looked around quickly and saw that he was alone. The Chevy was only visible from a brief stretch of the road, and the road was quiet. Several trees and a long wall leading to the service entrance to the hospital provided good cover a short distance away. If he parked the Lincoln along the wall, away from the street, he could wait out of sight a few feet from the Chevy, taking the owners by surprise as they approached. He would only be exposed to the street for an instant, then he could disappear behind the wall and drive off like nothing happened. The hospital, he decided, was an excellent place to kill people.

He ran back to the car where Darla waited, a stale grin on her face. He backed the Lincoln beside the wall.

"Where'd you go?" she asked.

"Never mind," he said. "Do you want to make some money?" He flashed a roll of bills. Her eyes opened wide. "Say, 20 bucks?"

"Certainly! But can't we find a better place than this?" She looked around at the empty back seat.

"I want you to sit here in my seat," he said, and Darla giggled. "I'm going to meet someone around the corner. Now you watch out here." He pointed down the wall toward the hospital's back entrance. "And if anyone comes along, you hit the horn. See?" He demonstrated.

"That's it? I don't get it."

"Well, you see, I'm waiting for someone over there, and they might accidentally come from this way, and I wouldn't want to miss them. Can you do it?"

"For 20 bucks?"

"All right, 25! Now, sit tight, and hit the horn if anyone comes!" He jumped out again and ran down the wall toward the Chevy.

The snow fell harder now, and the world was silent. Only the occasional car or wagon moved along the street. Sam squatted behind the wall and peered through a bush at the Chevy. Snow piled on his hat, and his feet turned cold. He removed the pistol from its holster and held it inside his coat pocket. Being brand new, he was anxious to try it on something other than grackles.

He reviewed the plan in his head. The three men would walk down the steps toward the car. One would open the front door to drive. The front right seat was broken, so the other two would climb in the back, but not at the same time, of course. Whoever was last to get in would get it first—POW! Then he would lunge through the open door, and deliver one shot to the driver's head—POW! And whoever was last in the back seat would be caught with no way out—POW! One, two three; a piece of cake.

HONK! HONK! Sam turned. It was Darla back at the car. He ran back toward the car. Darla pressed the horn, and he opened the door.

"What?" he asked, out of breath.

"Is it a man or a woman I'm supposed to watch for?"

Nick woke. His head was throbbing and swathed in cotton gauze. His eyes slowly focused, and he realized he was in a hospital room. The events prior to losing consciousness drifted back to him. He remembered the car chase and realized he had rolled the Ford. After that things became fuzzy.

"You awake?" a male voice asked from a corner of the room behind him.

Nick eased himself onto his elbows and turned. Someone sat on a straight back chair, a newspaper at his side.

"You've been sleeping a long time, pal. We were beginning to wonder if you'd ever get up. How do you feel?" The man had moved to the bed. The door opened and the smell of cigar preceded the next visitor. Nick flinched.

"How is he?"

"Awake, I think."

"You with us?" the cigar smoker asked and leaned on the bed.

Nick had no idea what to do but stare. "Who are you?" he asked slowly, his mouth dry, the words distant and soft.

The cigar smoker reached inside his jacket and produced a wallet. With a deft, practiced motion he flipped it open, and a badge, large and silver with an eagle and laurel fringe, flashed at Nick. "Federal Prohibition Officers, Special Agent McFarland. This is Ron Towers." A third man with black wavy hair entered the room joining them bedside. "And this is Meyers, Frank Meyers." Meyers nodded, hello, and Nick dropped back on the pillow.

"Then you're not gangsters, mobsters, bad guys..."

"No, don't get paid enough to be on the other side. Now, tell me something, Mr. Barchek." McFarland sat on the bed and leaned close to Nick's face. "What the hell are you doing here? Isn't this a long way from New York?"

351

"You tell me," Nick said. "For weeks I've been doing nothing but watching airplanes and hanging around hotels. I'm supposed to be watching bootleggers."

"With airplanes," McFarland said. "We know all about that. They run a field outside town. A small operation, but enough to bring us down from Omaha."

"So you're after him, too?"

"Not really," McFarland said. "Mostly we deal with the local moonshiners—corn whiskey, beer making. That sort of game. But this fellow, Doc, has been getting some notice. We just came down to see what's what. State boys don't like us here. Locals aren't too cooperative, either."

"What have you found?"

"Mostly we've been keeping an eye on you. Heard you checked out a car. Didn't like the idea of another district snooping around ours. Turns out you're pretty dull company."

"Thanks," Nick said.

"Don't mention it," McFarland answered. "As for this Doc, we haven't found a lick about him neither." He struck a match against the wall to relight his cigar. "Damn things never stay lit."

"They stay stinky, though," Meyers chirped.

"You never mind!" McFarland snapped with a grin and turned back to Nick. "You find anything worthwhile?"

"Nothing usable," Nick said. "I know he must be doing well, always see lots of trucks and airplanes moving through there."

"Seen any booze?"

"No, never got close enough. Don't want to, really. This Doc used to be a big time operator on the east coast. Worked for Rothstein, Arnold Rothstein. Some people even think he might have killed old Arnie."

"Did he?"

Nick pushed himself into a sitting position, and his legs ignited with burning pain. "Jeez, what the hell did I do to myself?" He lifted the covers and saw his two legs wrapped in gauze damp with blood. He dropped the covers.

"Did he kill Rothstein?" McFarland persisted.

"Nobody knows," Nick said. "Our office was closing in on him when it happened, so we lost out on a lucrative catch. My boss spent a lot of field time investigating him and lost a boatload of credibility when it fell through and Doc vanished.

"The Washington office likes headlines whenever overtime's used. They like results they can take to Congress at appropriation time. That's where my boss fell down. A year's investigation costs a bundle, and when it turns up dry...well, makes everybody look bad, especially the man in charge."

"So they sent you out here to finish up on the guy?"

"That's the cover, anyway," Nick said. "I think they just sent me out to stir the old man a little; let him know we're still watching. Personal vendetta." Nick adjusted his weight, and the pain raced through him again. "Actually, I think they just wanted me out of the office. I harbor no delusions about my worth in the overall scheme of things, organizationally speaking."

"I noticed you had no gun," McFarland said. "Did they issue you one?"

"Sure, but I don't like them, so I left it home. It's in my desk at the office." Nick touched his face, carefully taking stock of damage. "I'm an accountant, anyhow. I can't do anything until I see their books, and they all keep books, detailed books. Once I get them, I join with IRS and can do a hell of a lot more damage than you gunfighters."

"Don't get excited," McFarland said. "They give us guns, but I don't think the three of us together could hit the wall. We mostly go around checking pharmacies for illegal prescriptions for beer and wine. We keep a low profile; leave the tough stuff to the FBI."

"So why the interest in Doc's bunch?" Nick asked.

"We've been getting a lot of phone calls about him. We tried to ignore them, but after a while we figured we'd better make an appearance, especially with you on the scene. So here we are."

"Do you plan to do anything?" Nick asked.

"Not really," McFarland answered. "We're not heroes. Unless this fellow walks up and surrenders himself, which isn't too damn likely, we intend to observe, record and head home after a couple more weeks to fill out reports." His cigar had gone out again, and he struck a match against the wall and relit it puffing hard. Blue smoke washed past Nick who waved at it with a bandaged hand.

"You plan to ignore him forever? He can't hide a bootlegging operation that size indefinitely." Nick said.

"Don't intend to," McFarland said. "Word has it he'll be done by the first of December. Then he folds up and pushes south."

"What makes you think that?" Nick asked.

"He told us so!" McFarland said, and the other two men laughed.

Nick looked incredulous. He stared at the three men in dumpy gray suits standing around his bed. "He told you? Doc told you?"

"Yes," McFarland said. "He called us one day and said, hey, if you guys just lay off for two or three weeks 'til about the first, then, he said, he'll wind up business and scram leaving behind enough of the operation, and a few patsies to boot, for us to move in and make a splashy arrest with lots of headlines."

Nick could not believe what he was hearing. He knew the local police were on the take, but that was common enough. He somehow expected the federal agents to be immune. McFarland noted his reaction, and struck another match against the wall.

"And what's your end of the take?" Nick asked. "Secret bank accounts somewhere, or just cash stuffed in an envelope shoved under your door?"

"Now wait a minute, Barchek! Don't get all high and mighty on me! We haven't taken a thing here and won't either. We leave that to the locals." The cigar refused to stay lit, and he thrust it cold into his mouth working it into the corner while he spoke.

"The way I figure it, this Prohibition thing won't last another ten years, maybe not even that long. Then what happens to us? They won't need Prohibition agents if there isn't anything left to prohibit. So I figure, the folks we're trying to beat now are going to be the legitimate liquor dealers in the future, and we will somehow get swallowed into the regulation end of the business. Why get all hard-nosed when it'll all blow over in a few years. Hell, I like working for the Feds! It doesn't pay much, but it's secure, and none of us exactly works too hard."

"But why cut deals with bootleggers?" Nick asked. "You really don't expect them to help you somehow after Prohibition ends, do you?"

McFarland leaned closer to Nick. "Why? Because if I don't, I don't get shit." He stood. "There's no way the three of us are going to raid Doc's operation and haul the whole mob off to Leavenworth in leg irons. He knows that. He's no dummy. But he also knows that we can be a nuisance, a thorn in his side, so he sizes us up, and one day he makes a phone call. We talk.

"Now, he's got 'til the first, then we move in. He doesn't let on to anyone in his operation what the deal is, and to them it's business as usual. Then on the first, he tiptoes off, and we scoop up a handful of niggers and pilots holding a couple hundred bottles of hooch. Makes good copy for the newspapers—lots of airplanes confiscated, some guns, and a small amount of cash. Everyone gets their picture taken handcuffing genuine bootleggers, and we three have job security guaranteed."

"And Doc?" Nick asked.

"He's gone! Hell, if we spent the next four years working up a case, and if he, for some dumb reason, stayed long enough for us to drag him in, he'd cut some deal with the prosecutor and walk. This way we cut the deal in advance. There's less paperwork, less trouble, and the case runs our way, not his."

Nick slid under the sheets and gazed at the ceiling. He thought of Fitz, his supervisor, doomed to rot in his present job in New York, no chance of moving up, because he turned up empty handed after a lengthy investigation.

He looked at McFarland—forty years old, fat, balding and chewing on a cold stub of a cigar. McFarland would advance in this business. He would have the portfolio, the newspaper clippings. Nick closed his eyes. He hurt all over. The cigar smell nauseated him.

"I'm tired," he said.

"I can imagine," McFarland said. "We'll just leave you alone now. Take care, and get better. Try to get yourself on your feet by the first, and we'll let you in on the raid. Hell, even if you're still down, we'll pencil you in as an advisor. It'd be good publicity for you." McFarland patted his shoulder and waved for the others to leave with him. "Let's get some lunch," he said leaving the room.

A nurse glided into the room after they left. Nick was in growing pain, and he asked when he could see the doctor.

"Tomorrow, or maybe tonight if he comes back," she said and injected something into his buttocks. After fluffing his pillow so it was lodged uncomfortably under his neck, she left. Nick turned his head and looked out the window watching the snow mount on the window sill. Shortly before he dropped into sleep he heard three sharp pops, followed by silence and three more pops, then silence.

Sam crouched behind the wall in the snow, his head poked now and then through the bushes to watch the Chevy. His feet were painfully cold, and he wondered how long Darla would wait in the car.

The snow increased and soon covered his shoulders in a thick white shawl. He fondled the pistol in his coat pocket and watched the parking lot. A shovel scraped along a sidewalk somewhere behind him in short rasping strokes.

He reviewed the plan for the twentieth time in his head: the three men walking down the steps, the car door opening, the last one in, getting shot first—POP; then the driver—POP; and finally, the remaining one in the rear seat. He reworked the steps back to the Lincoln and the casual getaway through

356

town. He thought he would take Darla out for coffee and cake at the roadhouse where he had gone the night he killed Billings.

"Wait 'til Doc hears about this," he murmured.

He waited. He debated whether to stand up to urinate against the wall behind the bush or abandon his ambush and go back to Darla. The snow fell, and the scraping kept up its tireless pace in the distance.

Suddenly, the doors at the top of the stairs opened, and three men stepped into the snow. One chewed a cigar and pulled his collar up against the cold. One slipped on the steps. They all laughed and walked toward the car. Sam's heart pounded against his chest. His hands pumped sweat against the pistol's grip. He pulled his hat brim low over his eyes and crouched closer to the wall.

Their voices drew closer.

"Where to, Mac?" one asked.

"Shit," the one called, Mac, answered, "I'm tired of meat-loaf. Let's try that Dago place near the train station."

A car door opened. Sam leaned around the corner, pistol drawn, hammer cocked. His throat was tight, mouth dry.

"You wanna drive?" one of the men asked and held the door open.

"No, it's your turn, I drove last time."

The cigar chewer started to climb into the back seat and said, "I don't give a rat's ass who drives; just get going. I'm starved!"

The one holding the driver's door open put a hand on the other man's shoulder and laughing said, "You drive."

"Balls! I always gotta drive!" He broke away, and they struggled amid grunts and laughter for the back seat. Sam watched in astonishment, as his plan snagged in the childlike antics of his intended victims. He hunched back against the wall.

"Hell, I'll drive!" the cigarman called and climbed out again and into the front seat. Immediately, the other two headed for the back.

"After you, my dear Alfonse!"

357

"No, no, you first, my dear Gastone!"

Meyers was about to follow Towers into the car, when he saw someone rush across the snowy lot toward him. McFarland had already started the motor and was trying to light his cigar, when Sam aimed his pistol at Meyers and pulled the trigger.

"Oh," was all Meyers uttered, seeing the revolver pointed at his face only ten paces away.

Sam waited for the gun's report, but nothing happened. Instantly, he knew what went wrong and lowered the pistol to snap the safety off.

"What's the..." McFarland started to say and looked over his shoulder.

Meyers groped madly for his gun, ripping buttons off his coat when Sam raised the pistol again and fired.

BAM!

Meyers fell back in the snow clutching his stomach. Towers heard the shot, but somehow thought it was connected with the general clowning around about who drove.

Sam slipped as he ran to the car, his gun held straight ahead of him. McFarland pulled his pistol from under his coat and raised it at Sam, but Sam's went off first.

BAM!

The shot entered McFarland's face two inches above and to the right of his cigar and passed through his brain exiting behind his ear. It lodged in the wood dashboard above the glove box.

Sam spun around to shoot the remaining back seat passenger when he slipped again in the snow.

BAM!

The gun went off, the bullet passing harmlessly through the car's roof and vanishing in the falling snow. Towers had his .38 revolver unholstered and clawed his way out the door as Sam recovered his balance. The two men stood almost face to face when Towers thrust his pistol at Sam's throat and fired.

POW!

The bullet drove through his esophagus and missed his spinal cord, tearing a chunk out of his neck in the back, then dug into a small maple tree behind.

As Sam staggered back, more in surprise than from pain, Towers fired again. The second slug punctured his camel's hair coat and drove straight into his heart, exploding it like a water balloon before it lodged into his back bone. The third shot, Sam never felt. It burrowed into his brain as he lay on the snow bank beside the moaning and bleeding Meyers.

Towers held the gun with both hands pointed at the prostrate assailant. A scarlet slush of blood formed in the snow under the body. A larger and faster growing puddle grew under Meyers who lifted his head and stared pleadingly at Towers. Towers leaned against the car and felt his own blood drain from his head, and Meyer's low wailing faded in and out.

Still gripping his pistol he sat on the running board and vomited. From the hospital several persons came running through the snow. A stretcher appeared, and Meyers was whisked off amid shouts and questions.

Slowly, Towers recovered his strength and pushed away from the car.

"Get back," he said weakly, and everyone ignored him. "Get back! Get away!" he shouted waving his pistol. "Don't touch anything!" He reached inside the car past McFarland's lifeless body and switched off the motor. A cold cigar stump was still clenched between the deadman's teeth.

Darla became impatient waiting in the car. She heard the gunshots, but they were muffled by the wall and the snow, so she thought nothing of them. Twenty minutes later, thoroughly bored, she left the car and headed around the wall where Sam had crouched. Finding him gone, she looked around the corner at the small crowd gathered around the green car. Two police cars were parked in the street nearby.

"Hmmm," she giggled and walked home in the snow.

359

Chapter
XXIV

Nobody saw Wendall Grieve crash, but everyone at the farm heard it. To most, it sounded like a tree falling, a quick crunch then silence. By the time the ground crew reached the wreckage in the gully at the end of the runway he was dead.

They backed a truck alongside and drained the unused gasoline and all the whiskey from the plane, then they argued among themselves, deciding who would remove the corpse. The ground was frozen, but with a couple of pick axes, and a lot of swearing, they had the body buried in about an hour. Nobody wanted his clothes, and his wallet was turned over to Arthur. The plane still possessed salvageable parts, so they towed it to the barn and left it beside the wrecked Jenny and the remnants of an LS-5.

By the time Henry landed that day, snow had thoroughly covered the gravesite, and he never noticed the smashed Travel Air. He stomped into the farmhouse brushing snow from his shoulders. Inside the air was hot and smelled of coffee. He poured himself a cup in the kitchen and stooped down to rub the ears of Lundy's dog. Lundy was in Canada.

"Hiya, girl! Ohh, that's a good girl, yes, a good girl!" The dog jumped at Henry's face and licked his nose. Henry rolled onto the floor, and the dog dropped onto his chest pinning him down. Art limped into the kitchen, his left hand leaning heavily on a cane.

"You keep fooling around with that bitch, and Lundy's going to get jealous," he said. "Thank-you for pouring me a cup," he added taking Henry's coffee from the table.

"Hello, Art!" Henry pushed the dog aside and stood. "Is there anything to eat around here?"

"There was some stew," Art said and lifted a lid off a cold pot. A foul odor assaulted his nostrils, and he slammed the lid back. "Jesus, maybe that was last week we had stew."

"Nevermind, I'll find something." Henry rummaged through the cabinets until he found a tin of Ritz crackers and stuffed four into his mouth while Art spoke.

"There's been some bad news, Henry," he said and sat at the table. He lit a cigarette and looked for an ashtray. The room was littered with dirty dishes, cigarette butts and empty bottles. Finally, he shrugged and dropped the cold match on the floor. The dog licked it up and chewed it into a mushy paste.

"Business bad?" Henry asked and took a seat across from Art.

"No, that's been good." Art reached into his jacket pocket and took out a wallet. He dropped it on the table and slid it over to Henry who took it. "It belonged to Grieve; he's dead."

Henry, of course, cared little for Grieve, but the fat leather wallet of the dead pilot unsettled him. "Crashed?" he asked.

Art nodded. "Several hours before you came in; right at the north end of the runway. That's where he's buried."

"So why tell me?" Henry leaned back in the chair. He took another cracker and popped it in his mouth. It disappeared like a communion wafer.

"Take the wallet, Henry."

"Why, I don't need any money."

"No, there's no money." Art pressed it at him. "I've got an idea, and I don't want to tell you everything just yet...in case it falls through. Just hang onto the wallet." Henry picked it up and thumbed through the folds. It contained a few pieces of paper with phone numbers, a Utah driver's license and a birth certificate. Henry spread the last on the table.

"Why the hell would he have this?" he asked.

"Chances are, everything Grieve owned was with him in the airplane. Look at the driver's license; the address."

Henry read aloud, "General Delivery, Ogden, Utah. So?"

"People don't live at general deliveries. I don't think Mr. Grieve had a home."

"Most of these pilots here don't"

"And," Art continued, "I don't think he had a family either."

362

"Excuse me, Art, but I don't have any idea what this has to do with anything. And I don't think I care either." He tossed the papers onto the table

"There's discharge papers from the US Army in there, too and a pilot's license."

"Be fun to have one of them," Henry mused.

Art brought his cigatette to his lips and took a long draw. He held the smoke in his lungs and looked at Henry saying nothing. Outside, a truck pulled up, its brakes squealing. They both heard the truck door open then slam. Footsteps rushed up the back stairs.

"Henry, there's a person in that pile of papers. An unaccounted for person. Nobody knew Wendall Grieve lived, and no one knows he died, no one important."

Henry stared. The back door slammed. Lundy's dog moved closer to Henry's chair.

"Arthur!" Doc's voice boomed through the house. "Arthur! Oh, there you are. There's trouble...Hello, Henry. Art, it was Sam all right. He's dead."

"Sam Nyberg?" Henry asked.

"Yes," Doc said and looked around. A few pilots moved through the house, seemingly without purpose. "Let's go upstairs. Henry, come with us!"

The three left the kitchen and reassembled in a bedroom at the top of the stairs where Doc had made his office. The chalkboards leaned against the wall beside a small bed covered in papers. Clothing lay scattered about the room; his scarlet robe bunched in a wad in the corner.

"Close the door," Doc said in a hushed voice. "Henry, you're staying down here until further notice. No more flying!" Doc wiped a pale hand across his face drawing his features down.

"Sam tried to kill some federal agents in town..."

"Oh, shit!" Art sputtered.

"Now wait," Doc said. "He did manage to get one..."

"Dead?"

"Yes..."

"Oh royal shit!" Art cussed.

"Another's in the hospital apparently going to live."

"Big deal," Art said.

Doc's voice changed from hushed to angered. "What the hell did that idiot have in mind?"

"What federal agents?" Henry asked.

"Was it the jerk from the Treasury..."

"No, his name is Barchek," Doc said. "Seems he's been in an accident and was in the hospital. The ones Sam hit were just visiting, I guess."

"Where'd they come from?" Art asked.

"Remember that green car that's been watching us?"

"The Chevy?" Henry asked.

"Yes," Doc said. "Those were Feds."

"Not Capone after all, huh?" Art said.

"I guess not, although none of that matters right now." Doc picked an eraser from the board's ledge and rubbed Sam's name from an independent slot at the edge of the board. "I talked to the mayor."

"He saw you?"

"No, I called. He wasn't too anxious to talk. He says he can stall the police, his own police, but there's no way the Feds are going to sit quietly when word spreads about one of their own getting killed."

"So we pack up and leave, right?" Art asked.

"No," Doc said. "I think I bought us some time. We've still got, what?" He glanced at the boards. "There's at least nine planeloads still due in."

"So what!" Art said.

"Look, it's going to work out. First of all, they said they would treat Sam like an independent, just some local punk working on his own. The mayor's going to lead the newspapers down to Sam's garage and oversee a big raid. That'll keep everyone happy for a couple of days."

"What about the Fed who's still alive?" Art asked.

"Actually there's two," Doc said. "One wounded and one unhurt; he's the one who killed Sam."

"He knows who we are I assume," Art said.

A note of panic and irritation had crept into his voice.

"True," Doc said and looked at the two for a moment. "But I don't think he'll do anything."

Art and Henry exchanged confused looks and waited for Doc to explain.

"I made some phone calls a while back," Doc said. "I talked to the Feds...in Omaha. I said if they backed off until the first of December we would be through, and they could move in and make a few juicy arrests when we pulled out. It was a sound business deal."

There was silence, broken finally by Henry, "Arrest who?"

"Oh, not us, of course," Doc said with a laugh. "Not you, Henry! I would certainly keep you clean. We'd have been gone before they ever got here."

"Why didn't you tell me?" Art asked slowly.

"I would have, if I saw a need to. Don't be so nervous."

Art and Henry were silent. Doc stepped between them putting his hand on Henry's shoulder, an awkward reach.

"Now, Henry, you stay close to me, and as the planes return from Canada today, I want you to personally get them turned around and back out. No time wasted. Understood?"

Henry bobbed slightly shifting his weight from one foot to the other.

"Good!" Doc said and slapped his back. "Arthur, you get one of the ground crew to take you into town. Take a truck, the car's too conspicuous. Let's keep a low profile for a while. Telegram Moyer and tell him what I told Henry here: Get those airplanes turned around in no time at all. No rest period. Every pilot turns around and pushes back, and cram everything they can in those machines. No one should be taking off with more than enough empty space to sit his fanny in the seat."

"How'd Grieve crash?" Henry asked.

Doc turned. Henry had slumped into a chair and was toying with a brown wallet.

"I don't know," Doc said. "I understand he was a piss-poor pilot to begin with; must have screwed up."

"I think he was flying much too far over gross," Henry muttered.

"No one else has had any trouble," Doc snapped. "Have you?"

"It's not comfortable."

"No one said this was a comfortable job. They want comfort? Tell them to go back to their circus acts, hauling fat old ladies around all day for ten bucks. They can quit any time they want. I'm not holding anyone here!" Doc's voice had the usual southern resonance he put on whenever he meant to lecture or bully, but for some reason, Henry noted, it lacked the venom or true threat of implied authority that was needed to make it work. Doc seemed to be going through the motions without the conviction.

Henry said nothing and stuffed the wallet into his pocket and stood. He looked straight at Doc. "You were planning to leave the pilots to the cops, weren't you?"

Doc looked up at Henry and saw a tired man in a bearskin flying suit. He looked harmless. "No, I would have passed the word, given everybody notice. True, a few may have found themselves caught up in the raid, but, hell, they'd be released as first offenders and sent packing. Small price to pay for the kind of money I've been paying them."

"Nobody's been paid much lately," Henry said.

"It's coming!" Doc said defensively.

"What about me?" Henry asked. "Would you have told me, or was I supposed to be caught up accidentally in the raid, too? I've got a police record, you know, and I could be looking at hard time somewhere."

Doc said nothing, his mouth starting to form a reply, but Henry had left the room.

"Moody little schmuck," he said.

"You wouldn't have left him, would you?" Art asked.

Doc turned to his chalkboards and studied the names completely ignoring Art. With a careful even sweep, he erased Henry's name and replaced it with a reserve pilot named, Levandowski.

"I'll get that telegram off to Moyer," Art said reaching for his cane. He glanced once at Doc's back and left the room.

Outside, a plane landed quietly in the snow and, with gentle prods of the throttle, taxied toward the house. Doc peered through the frosty window and watched Jake Hollow climb stiffly from the cockpit while a truck pulled alongside. He saw him pull a whiskey bottle from his coat pocket, take a quick swallow and toss it empty into the snow. Doc wrote Jake's arrival time under his name.

"Reliable man, that Mr. Hollow," he muttered.

Henry stood near the kitchen stove. The lid was removed and orange flames licked at the rim. He held Grieve's wallet in one hand and read the birth certificate holding it with the other hand. Jake came through the door stamping his feet and seeking warmth.

"Oh good," he said and pulled his gauntlets off dropping them to the floor while he held his hands over the open flame.

"Florida, Henry! I keep telling myself: Two more weeks, then off to Florida; sunshine and orange groves!" He shuddered violently and unbuttoned his long leather coat. "What's that?"

"You going to Florida?" Henry asked.

"As soon as I can," Jake answered. "What's that? Your wallet? Just trying to heat it up so you can have something warm to sit on?"

"Is Susan going with you?"

"I haven't really asked her. Not sure I will."

"Do you like her?" Henry asked still staring at the wallet and the flames.

"Yes."

"Do you love her?"

Jake thought a moment, a moment too long, and he knew the answer, "no."

"I do," Henry said. "Leave her alone." There was no malice in his voice, no threat, only command. Jake felt awkward. The hair on his hands began to curl holding them too close to the flame. Henry handed Jake the wallet after folding the birth certificate back inside.

"Grieve is dead," he said. "His identity is up for grabs. You want it? Could come in handy."

367

"What's wrong with who I am?" Jake asked.

"I haven't decided who you are," Henry replied.

Little time was wasted, and Jake was airborne again. From horizon to horizon the landscape was a gray expanse of frozen earth. The sun hid behind clouds, making the snow as dull as the leaden sky above him. The air was smooth. Jake flew between weather systems, wet to the south, bitterly cold and drier the further north he went.

He pulled himself into the fetal position, shoulders hunched, face barely able to see over the cockpit's rim. Hours of northern flying had taught him this pose. It kept his face from the icy wind and what little body heat he could produce, seated for mile after mile, seemed to stay inside his leather clothes better. He glanced at the airspeed.

Instead of a number, he saw a face. Faint at first, it made him start. He blinked. The face sharpened. Features appeared, and he looked away. Below, Minnesota was as uninviting as he had yet seen it. A desolate expanse of prairie and slate-gray lakes. Jake felt the fear reach up from the empty world. He looked back into his cockpit, pulling himself tighter into his crouch.

There it was again, staring at him with one eye, one dead but probing eye. Jake shook his head and stamped his feet against the floor. He rocked the ailerons, then looked back at the airspeed indicator.

"No," he said. "Go away!"

The one eye stared at him. It grew in size, almost reaching beyond the instrument's glass. Jake stared back. He knew who it was. "You're dead," he said. "You screwed up, Hirsh-whatever-your-name was. You are dead!" He covered the instrument with his hand and looked away.

"Where am I?" he asked. Looking over the side, he saw nothing familiar, although he had been across this same terrain a hundred times. Someone beckoned from a lake. Someone distant, small.

"What the hell?" Jake said and dipped a wing. He was far above the earth, flying in the cold, hard air. He looked across

368

the landscape where not a road, nor a house could be seen. Yet, there on the edge of a small lake someone waved.

His hand slipped from the airspeed face, and he kept his gaze over the cockpit, knowing the eye of the dead pilot still watched him. Jake stared at the figure below.

Despite his altitude, he could make out details of the person. Slender, it wore a simple fedora and loose clothing, far too light for this far north. Another pilot, he thought. Someone just down, maybe. One of Doc's?

He knew the truth. The figure below waved an arm in a steady motion, beckoning Jake to come to it. Jake shuddered. He looked away. The eye stared at him, and he shifted his gaze to the other side. Sterns was directly ahead. Jake banked, hit left rudder and tried to avoid smacking into the vision.

"Sterns!" he bellowed. "Sterns!"

He looked back at the lake. The figure still waved, and he knew it was Adair. As dead as the landscape, yet he could make out the tight curves of her legs beneath the loose trousers. His desire for her turned to revulsion, and he shook a finger over the rim. "Go!" he shouted, then turned to Sterns. "Go to her! She's there!" He pointed over his shoulder. "She's waiting there! She never loved you but go anyhow! You two deserve each other!" He had to laugh at himself. He thought Sterns laughed with him, so he banked the airplane back to hit him. "You were really a dumb-shit of a pilot, Sterns. You had potential, but you screwed up somewhere." He pointed to the airspeed indicator. "Like him, another fuck-up!" He felt Sterns' image move closer, almost surrounding him, and he banked away.

"Leave me!" he screamed. "Leave me, you sons-of-bitches!" He looked over the rim. Without realizing it, he was circling the lake where Adair still waved. "Go to her! Leave me be!"

He circled the lake, the eye of dead Hirsh watching, critical of his flying abilities. Sterns stood on his right wing, and Adair still called from the frozen water.

Jake held the joystick with both hands. He pulled himself so far inside the cockpit the stick almost touched his chin. He closed his eyes.

The others appeared.

Billings, smashed in the night fog, then dead in the hospital. Wendall Grieve, barely gone, hidden in the mud at the end of the runway. "They'll find you," Jake told him. "Just because you're a screw-up doesn't mean they won't find you. Oh, they'll find you all right." He laughed and opened his eyes. The biplane still held its long circle around the lake.

"You know you were nuts, don't you?" Jake called to Adair. "You're crazy, and that no-face hubby was nuts, too! You were made for each other." He looked around, barely relaxing his crouch. "Where is he?" He looked. "You can't hide from me just because you haven't got a face, you know!"

When Vince did appear, Jake experienced terror like nothing that had ever gripped him before. The apparition settled onto the airplane like an immense weight, an amorphous horror beyond anything Jake's flying world could have done to him.

It came all at once. Jake heard the thunder of the stallion come through the heavy clouds above him, then when he thought the beast would trample both himself and the machine, the being, the memory of Vince was everywhere. Jake pulled himself into the cockpit. His eyes could not close tight enough to keep out the fear. "How can I see a man with no face?" he whimpered. "You can't be," he said, his voice was almost pleading.

He looked up. The airplane had lost altitude. He tried to level and pull back on the stick, but his arms were frozen, holding the plane in its spiral toward the lake.

Adair was more distinct, her arms waving in mockery. Sterns lounged on the wing, amused with Jake's fear. The dead and mangled visions of Billings and Grieve, even though he had never actually seen their ruined bodies, danced like evil cherubim around his spiraling airplane.

"I won't join you," Jake said. His arms tried to pull against the slow descent. "You won't force me out of my sky." But the weight of Vince and the others, pressed him lower.

He passed across the lake. Adair was abeam him. She stopped waving and stared. Jake managed to level the wings in time to avoid colliding with a stand of trees on the bank. His arms were heavy, his body frozen. He fought against the weight of the dead ones. "Get off!" he shouted and felt the terror slacken but only a trace. Vince laughed. The others joined in.

"I know what you think," Jake said. His arms strained against the heavy controls. He skimmed across the trees again, approaching the lake. Passing Adair, he looked her straight in the face. Her scar was an ugly worm across the gray, dead skin. "You can't have me!" he snapped. His arms were tired to the point he could no longer think of flying. His only thoughts were of getting on the ground. It was the first time he had ever sought sanctuary on the ground. Always he had found it in the sky. Even against thunder and vicious wind, his solution had been to get airborne, to outrun the threat. Now, he weakened under the crushing weight of the dead.

A field beyond the lake offered a smooth approach. Jake pulled himself from the fetal position and stuck his head into the slipstream. Cold air bit at his cheeks. He smelled the exhaust and felt himself regain a speck of control, still Vince and Sterns pressed him to the ground.

He pulled the throttle all the way back. His feet were clumsy against the rudders, and his hands barely moved the stick. When the wheels hit, his speed was too great, and he ballooned.

"Laugh!" he screamed. "Go and laugh!"

The airplane lost speed and settled onto the frozen earth with a bone-jarring rumble. The shock threw Jake from one side to the other. His feet slipped off the rudders, and he heard his teeth snap together, missing his tongue.

The biplane rolled to a stop at the crest of the hill. Jake left the engine running. It ticked over, kicking bits of ice crystal

371

from the ground and sending them in a blur around the tail. Jake closed his eyes.

One by one the images wove through his darkness. Billings, his face warm, a distant pilot, now gone. He faded. Grieve was next. "I never liked you," Jake said. "You weren't worth a shit, not as a pilot and certainly not as anything else. That really doesn't matter, now, does it? I know I could out-fly you...but that doesn't matter much anymore, does it?"

Jake let the memory of Wendall Grieve linger for a while. It was like staring into a mirror, a mirror that held back nothing, and the image coming back hurt, hurt terribly. "You don't think I'm worth a shit, either, do you?" Then Wendall Grieve faded.

Richard-Robert Hirsh-Whatever was next, or at least what was left of him. "You had no right," Jake said. "It doesn't matter that I didn't know you, any of you, really. You just didn't have the right to go killing yourselves in front of me. I've got better things to do."

He thought of Susan. Her presence was a smile in this horrible bag of memories. He wanted her, and at the same time knew she was gone from him. Not like the others, dead, but gone nonetheless. "I can't have her, not her." He looked back at the lake. Adair's image was almost gone. "And not you, either," he called. There was tenderness in his voice. "I can't belong to you, either. Please go. Please, all of you, please go." He heard his voice crack, the first time that had happened since he had been told his own mother was dying.

He made no effort to suppress the tears. They came in a flood, collecting inside his goggles, and he pushed them onto his forehead, and the tears flowed across his frozen cheeks.

He cried.

He cried.

His body shook, and he rolled his head against the padded backrest. His arms fell limp to his side, his legs dropped as though all the muscles had died. He felt the weight of the dead raise from his shoulders and chest. He felt the cold settle in their place.

How long he slept, Jake would never know. When he did awake, the engine was still ticking over, as though it would run at idle forever, waiting for the pilot to stir from his sleep and take them out of there.

Jake looked around. He looked up. Only sky, still leaden, but void of faces. He looked at the airspeed, and the dead eye was gone. He moved himself. Although stiff from cold, it felt as though his body had been drained of the weariness that had plagued him for the last few months. He inhaled the sharp cold of the north and pulled the goggles across his eyes. He was ready to go back up. The sky was no longer solely his. He could still be master of how he flew, of how he lived, but he knew there were others. He would never forget them. Somehow they could all share that same sky.

Chapter
XXV

Art opened his briefcase and set it on the table beside the phone in the Western Union office at the train station. A passenger train waited at the siding, hissing small tufts of white steam in the cold air. Snowflakes bounced off the ROCK ISLAND painted on the cars. Porters trudged along the platform pushing carts, while the few passengers waved to family or smoked cigarettes, waiting to board.

"The operator's on the line, sir. Just let me know when you're through, and I'll figure the charges. Do you have a lot of calls to make?" The telegraph agent was friendly and slid wood into the stove while Art settled into a chair.

"I'll be calling quite a few people, yes. I think it's warm enough in here, thank-you." Art held the receiver and waited for the agent to leave.

"Well, if you need anything else, you tell me."

"I will...Oh, here." Art rummaged through his case and handed the agent a message. "Send this, name's on the top. I'll pay for the telegram along with the long distance charges." The agent read the message quickly and left.

"Operator?" Art said. "Yes, I need to make a few calls. The first one's in Iowa; Boone, Iowa. The First Community Bank of Iowa, I have the number..." Art waited for his call and watched the train hiss and grunt pulling away from the platform. Someone from the bank came on the line, and silence returned to the station. A thin dusting of snow soon covered the tracks.

"Hello, First Community Bank, Theodore Dwyer, may I help you?" The voice was thin and full of static. Art almost had to shout to be heard.

"Hello, yes, my name is Grieve, Wendall Grieve. I'm calling from Utah, and I'd like to open an account."

"From Utah, sir. That's quite a long ways. Why would you want to open an account with us?"

"It is a long way, and why is not important. Can I open an account?"

"Well, normally we do this in person. Can you come in?"

Art looked at the receiver and shook his head. "Look, Bud, do you want my dough or not. I can't get out there for a couple of days, but I need to open an account—today—right now. Interested?"

The voice at the other end huffed and babbled for a minute about procedures and regulations. Art said he could give him the information over the phone, and the funds would be wired to the bank shortly. The banker agreed and took the information, including Grieve's mother's first name, which Art had copied off Wendall's birth certificate. An obituary attached to the certificate indicated the woman had died in 1919 of influenza.

"Well, Mr. Grieve, I have an account waiting for you. The number is 4-1-3-7, and as soon as you can, come in please and sign the papers."

"Yes, I will. In a day or so, maybe. Thank-you. Good-bye." Art hung up. "Idiot."

He read the name of a second Iowa bank from the list in his briefcase. The operator rang him through, and a similar conversation led to a second account being opened in Wendall Grieve's name. Art sat in the Western Union office for almost an hour calling banks and opening accounts. The telegraph agent entered once to stoke the fire, but a curt wave from Art sent him out again.

Most of the bootlegging money from the operation was laundered through a handful of banks in Florida and Texas where it mingled with the vast liquid fortunes made during the last decade in oil and land speculation. Art's close contacts with the banks from the Rothstein days made switching funds easy. The banks were used to the shell games of moving money to keep investigatory noses away. Art's machinations paled in comparison to what they had seen. Art and Doc were now small potatoes in the underworld economy, not worth the money it would take to trace.

Art shouted into the mouthpiece, the long distance connection was full of static.

"Hello, can you hear me? Ah huh, good...I wish to close an account, yes close...ah huh, look you'll have to speak up, I can hardly hear you...yes, that's better."

Art stood, moving closer to the hot stove. "The account number is G-6-7-3-4-4-4...yes, a four...no, three fours; four, four, four...that's it!" He poked a cigarette into his mouth and patted his pockets in search of a match. Not finding one, he glanced around the room while the line was silent except for the long distance hum.

"What? No, a Gee, as in garbage; 6-7-3-4-4-4...yes, I'll hold. Christ," he muttered the last and opened the door to the stove and stuck his cigarette inside. Flames danced toward the sleeve of his coat and singed the wool.

"Goddam...Hello, yes, that's me," he yelled into the mouthpiece again. "Now, close the account, and I want the money wired to...huh? Well, get another damn pencil! Christ! Christ!" He puffed on the cigarette.

"You do?" he asked. "Good! Send the money to First Community Bank, Boone, Iowa. To be deposited to Wendall Grieve, number 4-1-3-7...no, that's the full address; it'll get there...no! Iowa! Iowa! Idaho is a whole different place!"

After five more calls with varying degrees of clarity, Art had shuffled a good portion of the organization's money into Iowa banks under the name, Wendall Grieve, a man whose body was planted in the frozen Nebraska soil.

By dusk, the smudge pots along the runway were burning with angry little flames, whipped by a southerly wind that brought warmer air turning the snow to rain. The atmosphere at the airfield was confused and tense. Throughout the afternoon and evening, the planes returned from Canada. Their cargoes were quickly off loaded, and the planes serviced and turned around. Henry met each pilot with harsh instructions to forget about resting.

"Grab a quick bite," he would call. "There's something in the kitchen. Take yourself a leak and get back in the air. Doc's orders!"

Two of the LS-5 pilots were almost asleep at the controls when they taxied in, so Doc ordered them out and two reserve pilots were substituted.

"Every one of these pilots is well behind schedule," Doc complained somewhere around dawn. His chalkboard was a dusty scrawl of crossed out names and numbers. It was apparent to Henry that Doc really had no idea what was going on or how many planes were unaccounted for.

"Weather's been a big problem. They're having to fly long detours. I know some of the pilots are just plain sitting on the ground in places waiting it out."

"On the ground?" Doc scowled. "They're paid to fly, not sit."

The sky outside his second floor bedroom was a pale gray; the airfield motionless below except for the flickers from the smudge pots. Doc shuffled in a horseshoe arc around his boards. Each time he passed the window he would look outside as if someone had just landed. Henry fell asleep in the chair by the door.

While Henry slept the mayor of Lincoln leading four squad cars of police armed with axes and shotguns, and accompanied by two reporters, raided Sam's garage outside town, catching his cousin unaware sleeping in the backroom.

In the gunbattle that followed, as reported by the morning paper, the cousin was killed. Luckily, the paper went on, no law enforcement officers were wounded. The cousin's body, however, was taken to the county morgue with over 50 bullet holes and countless shotgun pellets throughout.

The mayor supervised the dismantling of the "bootleggers hideout" and the destruction of over 40 cases of "pure Canadian hooch". The stock was actually a cheap mixture of local moonshine and Doc's whiskey. The paper reported the seizure of an undisclosed quantity of cash and numerous firearms. The mayor was photographed holding the sawed off shotgun used in the gunbattle, while standing in front of the riddled body of Sam's cousin, the face tactfully covered to avoid viewer discomfort. The grainy black and white photograph made it look as though the cousin was laying in a puddle of motor oil.

Another photograph appeared on page two of the afternoon edition showing the mayor being congratulated by Federal prohibition agent, Towers. A detailed account of Towers' "desperate duel" with the "mad dog" Nyberg ran in two columns below the photo. An artist's rendition of the battle ran in eight frames along the bottom of the page. It portrayed Towers and the wounded Meyers dodging machine-gun bullets while returning their own withering fusillade. Numerous eye witnesses supported the story with their own accounts. The article concluded with a statement from Mayor Twimms: "The raid today concludes a three month investigation into illegal liquor trading in the Lincoln area. The investigation was a combined effort of local, state and federal authorities, through whose efforts illegal trafficking in bootleg alcohol has been eradicated!"

Doc read the newspaper and smiled. "The mayor's bought us time. We're safe for at least another week."

Art shook his head. "The mayor's saving his own ass and hoping we disappear; which we should."

"Nonsense!"

By late evening, the temperature had risen above 40, and the rain melted the snow turning the field into a bog of mud and slush. Wendall Grieve's grave was now a brown soupy puddle at the end of the runway. The planes had to land over it, but no one paid it any mind. The pilots, exhausted from the endless flying, no longer cared about anything. Planes would hit the runway in a bouncing spray of mud; the pilots just satisfied to survive.

Gear legs bent, tires popped, and no one cared. The mechanics, too, stumbled around in exhausted stupor trying to patch together the abused machines. One biplane touched down, and as it rolled toward the ramp, it lifted its tail and stuck its nose in the mud. The propeller shattered, and the engine came to an abrupt halt. The pilot awoke finding himself pointed at the mud. He slowly crawled out, dragged himself into the house and without comment fell onto the floor, where he slept completely wrapped in flying togs.

The biplane's tail was lassoed and pulled down. The propeller was unbolted, and another substituted. Gasoline was pumped into the tank, and someone else crawled into the cockpit.

"Contact!"

"Contact!" The new propeller was swung, and the engine restarted before it had a chance to cool. The pilot ran the engine up until the vibration seemed ready to shake the plane apart. He shrugged and departed. Two Curtiss Jennies stood on the field at 3 a.m. in a cold steady rain. The engine idled gently on one, while the pilot listened to a last minute briefing from Henry. Doc watched from his second floor office. By the glow of a bonfire, he saw Henry point to a map spread across a lower wing out of the rain. His finger traced a route, but it was too far away for Doc to see. The Jenny pilot looked at the map, said something to Henry, shrugged and climbed into his plane. Within minutes, it had disappeared into the night.

The second Jenny was being refueled while two mechanics argued over how to patch a tear in the fabric on the left wing tip. They heated the dope over an open flame and slapped it onto a wet cotton patch. The whole thing slid about in a gooey mush without really adhering.

"It ain't gonna hold, Henry!"

"I don't care!" he hollered. "Just fix it and get this son-of-a-bitch back in the air. It'll hold long enough."

"I wouldn't want my ass in this thing," one mechanic snapped and brushed warm dope in crude strokes across the cotton skin. From the farmhouse, a drowsy pilot slogged through the half melted snow toward the biplane. Head down, he walked like a soldier asked once too often to risk all against an unbeatable foe. Henry rushed over to meet him.

"Levandowski, isn't it?" he asked cheerfully while scrutinizing his clipboard, the names smudged by the rain.

"No, Wong. Julius T. Wong." The pilot plodded by him and dropped himself into the cockpit and fell asleep, rain dripping off his helmet. Henry checked down his list and penciled WONG over a smeared entry.

The mechanics finished with their fabric patch and turned the plane into the wind. Wong was shaken awake and the engine started. Henry stared at the biplane being swallowed into the night before it lifted from the runway. He checked his clipboard and glanced at his watch.

"That should be the last one in," he said to himself.

With the clipboard tucked under his arm, he walked slowly to the back porch stopping at one of the smudge pots. The rain hissed on the hot cannister, the flame twisting as if to avoid getting wet.

"Put the runway lights out," he called. "That's it for tonight." His voice was strained from lack of sleep; his neck ached. Two ghost-like figures moved away from the warmth of the bonfire and began to snuff out the oil lamps one-by-one. Henry stood in the rain and watched until the last orange flame vanished, then he walked inside and collapsed on a stuffed chair beside the fireplace. The room was full of snoring pilots. Lundy's dog sneezed in the kitchen and walked to Henry's side. His hand flicked weakly at the dog's head, and he watched the bonfire fade through the window, its yellow glow dancing on the walls. He was asleep before it died.

POP!

"That's it! Now, another!"

POP!

"Ah huh, and one with just you and the mayor. Mayor, if you'll move closer to the bed, and you, sir, if you could sit up straighter. You tend to slouch and slide down under the covers."

Nick Barchek struggled to move himself further up the headboard without tearing open the raw wounds on his legs. A nurse, young and pretty, lent a hand and Mayor Twimms supervised with a beaming smile.

"That's it," the photographer called and aimed his camera. "Shake hands! Hold that telegram from the Treasury Secretary just a little higher...and smile..."

POP! The flash went off again. The smoke from the burnt magnesium powder billowed to the ceiling like a small bomb blast.

The photographer moved quickly around the bed rearranging subjects. "I'd like to get one of Agent Towers and Agent Barchek together reading the letter of commendation."

"It's a telegram," Nick said.

"Whatever," the photographer quipped and pushed an unsmiling Towers toward the bed. "Now, each take hold of the letter. Mayor, if you'll just step aside for this one...yes, just for this one. I'll take some more of you in a moment."

The mayor, a little flustered, moved out of camera range and continued his beaming.

POP! The flash fired again.

"Proud day, boys; proud day!" the mayor declared. "Say, can we get that other young man, ah agent Mayers..."

"Meyers," Towers corrected him.

"Yes, Agent Meyers. Is he awake, yet?"

The nurse spoke. "He's still unconscious. There's a good deal of in..."

"I'll answer that!" A doctor—the hospital administrator— said. "Mr. Meyers has suffered severe abdominal trauma and is currently fighting the first stages of peritonitis. Our staff is doing everything possible to rectify this..."

POP!

The camera caught him with his mouth open, and his right index finger touching the tip of his nose. "...ah, this situation. A full and complete recovery is anticipated." He turned to the camera with a big smile, arms folded, but the photographer was busy sliding film plates from the camera to the bag.

Nick became dizzy, maybe from the throbbing pain in both legs or just from the crush of newsmen and spectators there for the ceremonies. He and Towers had been handed a key to the city, and congratulated by the mayor and each department head including water and sewage.

The telegram from the Secretary of the Treasury had arrived that morning, and the mayor had made a big play of reading it to the two agents while they nodded politely. Towers nodding because he was on the verge of sobbing; Barchek because he had just been injected again with morphine.

The story of the gunbattle between federal agents and the assassin, Nyberg, had been picked up by the news wire service and Towers and Barchek found themselves heroes.

Nick pressed his hand against his thigh to stem the pulsating dull pain running up his leg.

"Could we wind up now?" he asked. The mayor looked around wondering who had spoken. "I'm tired. Please leave." A stab of pain shot to his groin, and he winced.

"Now, everyone please leave the room!" the hospital administrator called. He and the mayor vied for the honor of directing the crowd outside. Nick watched the nurse's backside as she was ushered out with the others.

"Towers," Nick called. "Would you stay a minute?" Towers turned and nodded. The mayor announced that the agents needed to be alone for a sensitive conference, and that he, the mayor, would keep everyone informed. Towers closed the door. The room was silent, and the two men said nothing for a moment. Fast moving clouds allowed a shaft of bright mid-day sunlight into the room.

Nick spoke. "What happened?"

"We walked into a trap." Towers sat on an empty bed across from Nick. He looked worn out. "The guy was waiting for us. Mac got it in the car, and Meyers took one in the gut." He closed his eyes.

"The papers said there was quite a gunfight."

Towers lay back on the bed and snickered without opening his eyes. "Shit! Ole Mac was sitting at the wheel, and Meyers—his first name is Aaron, did you know that? He used Frank for a business name, so people wouldn't think him too Jewish."

He rolled his head toward Nick and opened his eyes. "So Frank took one shot in the belly without his gun ever leaving the holster. They took it off him inside the hospital." He pushed himself up on one elbow. "And me?" He looked straight at Nick. "I was scared shitless and just started shooting. Puked my guts out later. They didn't photograph me doing that, not good press you know! AGENT PUKES OUT GUTS AFTER BLASTING GANGSTER!" He pronounced the last like a headline.

"You got him."

"Yea," Towers laughed.

"What now?"

"Now?" He sat all the way up dropping his feet over the edge. "Nothing. Nothing now. We do absolutely nothing, and the whole world comes to us. Mac would love it." He stood. "Hell, the damn Secretary of the goddam Treasury sent us a telegram. Congratulations! Your careers are secured forever and ever, amen. Sorry about your friend, love, the Big Man."

"So we leave it? You know the garage they raided was only an outlet. Doc's still running his booze through the farm. We leave it?"

"Look, Barchek, you and I haven't got anything in common, except we got caught up together in this stupid mess. And you! Shit. All you did was roll a car and sit your ass in the hospital while Frank and Mac get shot up! I don't give a goddamn what you do. But me? I'm done! I paid my dues, and I'm done! Time to pick up the prizes and go home." He turned and stood at the door. Nick was silent.

Towers spoke without turning. "I suggest you get better in New York. Leave this alone." He left.

Nick never saw Ron Towers again. Towers returned to Omaha, where he was promoted to field supervisor, a desk job. Meyers died of peritonitis two days after the hospital administrator declared him on the road to full recovery.

On July 4th, 1930, a plaque was dedicated inside the main lobby of the Federal Building in Omaha. It read: TO THE MEMORY OF AGENTS JOSEPH P. McFARLAND AND FRANK AARON MAYERS...and something about devotion and sacrifice. Towers pointed out the misspelling of Frank's name, but he was told the plaque was already paid for and too bad. Ron Towers retired from federal service in 1948 having never risen above his field supervisor job. In 1951, the Federal Building in Omaha was renovated, and the plaque lost.

It was dark. Doc paced the one-room office on the second floor of the farmhouse. He paced like an animal, more like one in a pound than a zoo. A single oil lamp illuminated the room, spreading harsh grotesque shadows on the walls exaggerating features. Art entered quietly, unnoticed. He watched Doc scribble a name on the chalkboard with an unsteady hand then erase it immediately with vengeance.

Art looked closely at the many names on the board, and saw, sprinkled throughout the names of the pilots, other names from the past—names from Lake Placid and New York City, and a few he did not recognize. Atop one column was the name VINCE and a column of dollar amounts below this which Art quickly estimated to add up to almost a quarter of a million dollars. He had no idea what the figure could mean. Moyer's name appeared beside Vince's and beneath his, a host of names from the Rothstein empire, mostly pimps and gamblers, many of whom were now dead.

Doc spun around. The oil lamp illuminated one side of his face, leaving the other half in shadow. The lines and creases in his skin were deepened making him look years older. His eyes were sunk deep inside his head. He smoked a cigarette without the holder. Ashes had fallen in dirty tufts onto his gold robe. He seemed embarrassed finding Arthur behind him, and he slowly erased the board.

"Arthur," he said, and that was all.

"Have you slept?" Art hobbled to a small desk and gathered papers scattered there. "Henry is asleep downstairs. I'm going to bed now, myself..."

"There seems to be undue concern with sleeping around here! Isn't anyone running the show?" He strode to the window and opened it. Cold damp air flowed in, but he ignored it. "Where are the guards? Are they out? Or have they, too, gone to bed?"

"They're out," Art said. "Twice as many as usual."

Doc looked into the dying bonfire below. The rain had turned to a cold drizzle, almost fog. And why isn't anyone out there working on the airplanes?" he asked.

"All the planes have gone. It's time to sleep."

"Where are the airplanes?" Doc stuck his head through the open window so far that Arthur thought he might tip over. "Where are they?" he hollered into the fog.

"Come inside," Art said taking his arm. "Everything's taken care of. Nobody will be in for a while. Come inside and go to bed."

Doc turned from the window and smiled toward Art, but he was looking past him at the chalkboards. "I have a plan, Arthur. We can make ourselves a fortune!" He ran to the boards and tapped them.

"We double the planes, and double the loads. Uh huh? You see? Then we set up a whole separate fleet of airplanes right here on the field. Planes that don't go to Canada. They stay right in the U.S. and deliver to new markets, like, ah, like Cheyenne and Des Moines or Kansas City. We open new markets, Arthur."

"All those markets are already covered," Art said.

"By truckers, Art. Everybody uses trucks, and trucks get bumped off. We, on the other hand, are successful because we fly over the competition—faster and safer. Can't touch an airplane once it's in the sky!"

"It's time to close up shop, Doc. It's all over here. We are through." Art spoke softly, then his voice became cold, clipped with impatience. "We've made our money, and it's time to pull out."

"No!" Doc shouted and fumbled for a magazine on his bed. "Look!" He thumbed through the pages until he came to an article on the Ford Tri-Motors, a three-engine airliner pictured at the head of the page.

"We buy one of these, and suddenly we are carrying the same load that ALL of our planes carry right now in just one trip! Think of it, Arthur. One airplane replacing all of those rag and wire contraptions down there." He pointed toward the open window, at the empty ramp in the fog.

Art glanced briefly at the article and tossed the magazine on the desk. "No," he said. "We're not in the business anymore. Go to bed." He tried to leave, but Doc blocked him and picked the magazine off the desk.

"You haven't even read it yet. These things can make it up to Canada in one trip. No stops for gas! That means no exposure. No fuel trucks along the way to charge us double for gas. And only two pilots instead of the ten or fifteen we need now." He pushed the magazine close to Art's nose. "Read! Read! If we purchased two of these Tri-Motors and always kept one in the air while the other is being serviced. Then put the second one up when the first one gets back...why we'd have a never ending supply. The profit potential is immense, Arthur, immense!"

Art snatched the magazine from his hand and tossed it through the window. "It's over! Go to bed!" He slammed the window shut and left the room.

Doc glowered at the closed door. He looked around the room and slowly walked to the chalkboards. Carefully, he printed the name, ARTHUR, on the board and just as carefully erased it.

He heard Art climb into a squeaky bed in the adjoining room, and there was silence. Downstairs Lundy's dog began to scratch at herself, and in the lonely silence it seemed as though the whole house shook with her. Doc settled onto his bed and fell asleep in his gold robe with a piece of chalk in his hand.

Chapter
XXVI

Susan's truck whined, and the gears chattered as she shifted from first to reverse, rocking it free from yet another mud hole. The morning sun was brilliant and still low on the horizon. The air was cold, somehow adding to a vague loneliness found in the mud and empty road. The truck lurched forward, the rear wheels biting into hard dirt, and she drove away, mud dripping from the running boards and fenders.

Through the cold air, she could hear the train leave the station in Luther. The distant whistle bounced across the flat countryside the way it did every morning, the way it had for as long as she could remember. She loved the sound and often toyed with the dream of listening to that sound from inside the train, from a passenger car leaving the farms and the mud. The whistle now, however, only made her think of Jake Hollow, and that made her sad and confused.

Her eyes stared blankly at the road while she listened to the whistle fade. In a flash she noticed the hard surface had once again turned to mush.

"Eeee," was the thin noise from her throat. The front tires clipped the mud, and she strained to turn the wheel. A brown geyser shot past the fender splattering the edge of the windshield and hitting her through the open sides.

"Bluhh," she stammered. The truck bounced off the road and onto the dead grass and brambles along the embankment. Her feet slid off the pedals, and the truck cut through four strands of barbed wire before it settled in a thicket of low trees.

"Blast all!" she swore and struck the steering wheel with her fist. She walked around the truck deciding the best way to back out. The rear wheels were bogged almost to the axle, and rusty wire looped in coils around the bumper. She reached for a shovel and a couple of planks in the rear and for the next 30 minutes worked on extricating herself.

"Stupid, stupid..." she muttered, prying the last board under the wheels. By now she was thick with m. ud, and her patience severely tried.

"Bastard!" she howled, invoking the strength found only in certain words, to force the board under the tire. "There."

Before she stepped into the cab, she heard the faint sound of a motor, far in the distance, like a tractor moving across a field. She looked to the sky. A handful of crows winged from one hedge row to another. The sun was higher, its rays warm against her face. A dog barked, and she saw it.

Two wings, thin and frail looking, and as it drew closer the wires came into focus along with the bony landing gear beneath the square fuselage. It was a Standard.

She started to wave before the pilot could possibly have seen her, and continued waving after the biplane passed over in the dull rumble of its powerful motor.

She threw the shovel onto the truck's bed, but it slid right across landing in the mud on the far side.

"Bastard," she said having decided the word could be useful in a variety of circumstances. She ran around and retrieved the shovel placing it carefully back on the truck. She climbed inside.

Back and forth she rocked the truck mashing gears and using the word, "bastard," more liberally than may have been called for. But it seemed to work. The truck worked free and in a searing, rattling cacophony of flying mud and springing barbed wire she drove down the embankment, keeping to the firm grass beside the road, then taking the road when it hardened a hundred feet beyond.

By now, the biplane was out of sight, and she was still two miles from home.

Was it Jake? She asked herself. Henry?

The usual wave of confused emotion followed her thoughts.

"What good is either one of them?" she asked aloud. "Both unreliable; both involved in God knows what illegal goings on." Rather than dissuade her from either one, the illicit aura of her two suitors excited her, and she pressed the accelerator pedal all the harder, although it was already flat against the floorboards.

She turned the the final corner leading to the house when a second plane glided overhead obviously aimed toward the pasture. Another LS-5.

"Jake and Henry!" she exclaimed and pulled into the driveway. She braked to a stop beside the old Buick her father had recently purchased. She jumped from the truck and ran toward the Standard already parked near the house. Her father stood talking with the pilot.

"Jake?" she called on the run. The pilot tugged off his helmet, and a sad tired face turned toward her. She stopped short.

"Oh, I'm sorry," she said breathing hard. "I thought you were..."

"Jake Hollow?" the pilot asked. His words came slowly from a tired and frozen brain.

"Yes," she said.

"He's behind me somewhere. That might be him now."

All three watched the biplane descend slowly on final approach. Sunlight caught the wings with a glint, and a shadow raced along the ground trying to meet the airplane.

"Looks kinda slow," was all the pilot said, and they watched the Standard LS-5 flutter momentarily in the air above the pasture then drop like a shoe onto the turf.

"Oomph!" her father said. "That's gotta hurt!"

The biplane lifted back into the air but only briefly. It touched again, and its right gear leg crumpled under the weight, sending the whole plane into a ground loop to the right. Bits of dead grass and mud flew into the air. The plane slid on its wing tip, off the landing strip and into the barbed wire fence. The propeller chewed once into the earth and froze.

"Well, he's down, anyhow," the pilot announced cheerfully. Susan was already running toward the wreck.

"Please don't be Jake!" she prayed. "Please don't be Jake! Please don't be..." Before she rounded the undamaged wing tip she saw the pilot drop from the cockpit and remove his helmet and scarf.

"Hello!" he called with an uplifted voice and wide grin. A thin trickle of blood ran from his nose. He wiped at it with his sleeve. "I can explain this."

"You're not Jake!" she said. "You're bleeding."

"Right on both counts," he said. "I don't think I've blocked the runway; the others should make it okay."

Before Susan could say, "others?", another biplane circled overhead, apparently reviewing the strip. The pilot with the bloody nose waved and pointed further down the field.

"Over there!" he called, as if the pilot above could hear. "Land over there!" Satisfied his message had been conveyed, he stood back to watch, wiping at his nose.

"Might want to stand over here, little girl." He patted the fuselage where he stood and grinned while holding his sleeve to one nostril. Susan nodded and moved off the field, away from the stranger.

The pilot flew his approach much too fast and had to circle again.

"What's wrong with him?" Susan asked.

"Nothing, he's just tired—like me—like everyone. We've been flying steady for two weeks, catching sleep here and there, but never enough to recharge the old batteries." He made the last comment about batteries sound lewd, but she ignored him. "Who are you, anyhow?" he asked.

"Susan," she answered without turning.

"Jim, I'm Jim...that's what you can call me—Jim." His words dropped flat against her cool reception. She watched the arrival. Again, the plane was too fast, but the pilot seemed determined to put the machine on the ground, and that he did with a wallop. The wheels hit and bounced.

"Keep at it!" Jim cheered. The wheels hit again, and the biplane drove along the ground past Jim's plane clipping its wing.

Bits of wood and fabric exploded into the air, and the landing plane swerved down the strip racing toward the house. Susan saw her father and the other pilot run for the porch and crouch behind a pillar. Out of control, it plowed into the parked LS-5 with a sheering crunching splatter of wood and metal.

"Whoa! Good landing! What a pro!" Jim called and hooted a rebel yell. "That Jake sure can make an entrance."

Susan heard the name and was off again running toward the house. From the porch, her father cautiously approached the wreckage as if it would burst into flames with a false step. Susan, on the other hand, ran full speed and clambered onto the remnants of a wing and reached for him.

"Jake!" she called, almost screamed, and he turned a lazy grin and waved.

"Hello, Susan. I seem to be getting quite a collection of airplanes here."

She reached over and hugged him, pressing his face deep into her shoulder, so he had to push her away to breathe.

"Hello, Jake," her father said. "You can just park there if you like."

"Won't be in the way, now will it?"

"Naw!"

Susan listened to their dry exchange, reviewing the twisted mess of wires and broken wings. Jake's plane was implanted into the side of the other Standard. His propeller, even at idle power, had chewed away the cockpit area and spit the remains 30 feet in all directions. The wings were wedded together in such a knot that it was hard to tell what belonged where.

"You all right?" the father asked.

"Just fine," Jake answered. "Can I offer you a drink?"

"What've you got?"

"Oh, about a hundred and twenty gallons of Canada's finest."

"I'll get a glass."

"Make it two," Jake said.

"Three," the first pilot added.

Her father nodded and went inside.

"What are you doing here?" Susan asked. "All of you?"

Jake pushed himself from the cockpit, and taking her hand they both slipped off the wing and tumbled to the ground.

"Excuse me," he said without his eyes meeting hers. He was friendly, cheerful but distant.

"I don't mind," she said with a teasing giggle, but he was already lifting himself off her, and her flirtations were lost. Her father stepped off the back porch with an armful of glasses and coffee mugs.

"Where's the whiskey?" he called. "What are you doing on the ground, Suzie? Get up; have a glass."

From somewhere a hose appeared, and the cap was removed from Jake's whiskey hopper. With the hose in his mouth, he started the siphon and filled the glasses. Liquor splashed on the fuselage. They laughed.

"To Doc!" Jake called and held his glass high.

"To Doc! To his whiskey!"

"To our money!"

"It's best to drink his whiskey," Jake said after finishing off his first glass and siphoning another. "We may never see our money."

The hose was passed around filling glasses. Susan's father took a drink straight off the hose, and the others laughed deep male laughter. A great tension seemed to be released with the unchecked flow of the liquor and the sight of the ruined biplanes. Without actually saying it, they knew it was the end. Whether they were to ever see their money or not, it was over.

Susan pretended to sip from her glass, but stood silently at the edge of the circle listening to the raw laughter and mostly watching Jake, who never looked at her.

Other planes arrived, sometimes alone, and occasionally in pairs. There were more bad landings, cheered on by drunken pilots lining the runway, and more broken wing tips and buckled gear. None, however, approached the disasterous consequences of Jake's landing.

"Where the hell is this place?" each new arrival would ask as he was handed a tumbler of whiskey.

By noon, Lundy's Curtiss Robin had landed, without incident, to a hail of boos and raspberries from the crowd. Jake met him with a half-filled mason jar.

"What the devil, Jake? What are we doing here? Thank-you."

"Never mind that, just drink." Jake lifted his own glass and saluted the sky. It was the only thing he could readily identify after dropping onto his back.

When Doc awoke the sun was already high, the light having passed his window and warmed the opposite side of the house. It took more than a glance at the clock to convince himself he had slept for over ten hours. He remembered nothing from the sleep—no dreams, no stirring.

The room was cold, and the old farmhouse silent. The wind blew gray smoke from the chimney in a quick swirl by his window then carried it away. He shed his gold robe, letting it drop to the floor. Passing a full length mirror, partially hidden by the chalkboards, he saw an old man, thin and frail with dark puffy flesh under watery eyes. The shoulders protruded in sharp points beneath his undershirt. His legs were white, hairless and so weak he could not imagine how they held him up.

"Where did you come from, old man? he asked the mirror and saw the image smile weakly back at him.

He tried sucking in the slight pouch over his stomach and turned sideways to admire the result—an old man with his gut pulled in. He reached for his red robe and chanced a look out the window. The yard was empty except for the old Packard parked near the house and Henry's own airplane, the Standard J-1.

Normally, it was parked and covered behind the barn. He never used it for business. Now it sat unattended at the end of the runway, tied down, facing into the wind. The canvas tarpaulins were removed from the cockpits as if readied for flight. Doc ran barefoot from his room and called at the top of the stairs.

"Henry! Henry! Where are you?"

He ran down the stairs and into the front room, cluttered with papers, cold cigarette butts and various empty bottles and dirty dishes. The bed where Lundy's dog slept was empty.

"Arthur!" Doc called feeling a lonely wave of panic surge over him. "Henry!" He poked his head into the kitchen. A fire burned in the stove and an empty box sat on the floor beside it. The room was warm, but he left quickly running to the back porch.

"Henry!" he called, almost screeching. The sunlight was brilliant and stung his eyes. Shading them with one hand, he called, "Hen-ry! Henn-ry!"

He ran to the end of the porch and stubbed his toe on a discarded propeller. "Shit!"

He called, "Henry!" His voice bounced across the empty fields and melted into the woods beyond. The only sound returning was the rustling of dead limbs in the cottonwood behind the house.

Doc slapped the railing and walked back into the house avoiding the propeller. The door refused to close properly, so he left it ajar. Art appeared from a back room carrying two boxes overflowing with papers. He limped badly.

"Where the hell is everyone?" Doc asked and tried to slam the door. It merely bounced back at him.

"Good morning, Doc. You slept quite a while. How are you feeling this morning?" Art limped by him without stopping, a friendly smile on his face.

"It's almost three o'clock, Arthur. Where's the first shipment. Has anyone come in yet?"

"No."

Doc stared at the retreating back of his accountant carrying the boxes into the kitchen. He followed.

"No?" he asked.

Art slid the lid back on the stove and, without stopping to explain, pulled a handful of papers from the top file and dropped them into the fire. Flames lapped the rim devouring them.

"What are you burning there? Vince and Adair's records?" He tried desperately to grasp what was happening.

"No, I burnt them long ago, when they died." He dropped another fistful of records into the firebox. "These are our records."

Doc knew his mouth dropped open but did nothing to correct it. Art methodically fed the enterprise's records, handful at a time, into the flames. Never once did he look at Doc. As if to emphasize his lack of concern, he removed a cigarette from his pocket and holding a flaming ledger page lit it.

"Where's Henry?" Doc asked, his voice soft.

"Went to town," Art said and glanced at his watch. "He should be along soon."

"Who said he could leave the farm?"

"Nobody said whether he could or could not do anything. He left. I asked him to run the pilots into town." Art dropped a whole sheaf of papers into the fire letting the pages fall through his fingers like loose snow.

Doc felt his authority vanish. "What pilots?"

Art turned, his look hard. "Not all the pilots had their own planes. Those still on reserve this morning were paid off, and Henry took them to the train station. They're gone."

He read a page, grunted to himself and dropped it into the stove. "I also had Henry send a telegram to Moyer." He recited the text from memory: "Last flight out today. Will be last ever. Payment will follow. Good job. Take the winter off. And I signed your name."

You had no right." The voice was low and strained. "We still had time, lots of time."

"We had nothing! Barely enough time to leave town. I told you that last night."

"You are wrong!"

"No, Doc you're wrong. Why can't you understand. It's all over here. We had a good thing, but now it's time to pull out. Get out while we still can."

"Bull!" Doc spat the word, but it did nothing.

Art returned to dropping papers into the fire. Outside, a truck pulled into the driveway and squeaked to a stop. Doc heard Henry's voice ordering someone about, but the commands were lost through the walls. Doc left the kitchen.

"Good morning, Doc," Henry greeted him coming through the door from the porch. "You aren't ready to go? You'd better get dressed."

Doc, in bare feet, was dwarfed by the man in the leather jacket and knee high boots. "Where am I going?" he asked.

A smile froze on Henry's face, and he looked around for help. Art entered and said, "I haven't told him yet, Henry."

Doc turned, his expression alone asked, "told me what?"

"You're going to the train station, Doc." Somehow, Art seemed larger than Doc. "There's a train that leaves at 4:30, 4:36 actually. You have a Pullman compartment reserved under the name of Whitfield, Wayne Whitfield. You're booked through Kansas City and further south to Houston, Texas."

Doc turned his back, and feeling the blood drain from his face, he sat down on a straight backed chair. "Houston? Why Houston?"

Art continued. "There's a bank account waiting there for you, I'll give you the number later. It's Texas Trust Bank. We did business with them from Lake Placid. They'll be expecting you. Everything's taken care of. Taxes are all paid through some old payroll fronts..."

"And you?" Doc waved at both of them.

"I'm not going with you; neither is Henry."

Doc nodded and licked his cracked lips. He stood quietly to leave, but with one foot on the bottom stair he said, "You are traitors, you know." It was a calm declaration.

"Oh, Doc!"

"No, you are, Henry, and don't deny it." He drew himself up. How much money do you have? Right now. How much?"

"I don't know..."

"A lot. Far more than you could've made when I found you playing sandlot baseball down south. No, Henry, you are a traitor." He climbed another step and stopped.

"How much do I get, Arthur?"

"One third."

Doc shook his head and smiled. "That wasn't our original agreement, was it?" He looked into the two faces and saw the cracks of doubt and guilt form. "Henry was in for 20 per cent wasn't he? Plus I paid him mighty handsomely for flying, too!"

"That's not exactly true..." Art said, but was cut off.

"No? No? Henry, did I ever say I would cut you out?"

"Well,..."

"No. There you have it! And all you had to do was drive a car for me," Doc said. "Who ran this operation, gentlemen? Whose idea was it? Who smoothed over every wrinkle along the way?"

"Who went to the Feds?" Henry asked.

"Who made you rich?" Doc shouted.

Doc took two more steps. Henry spoke. "You were going to leave us."

Doc shook his head as if he were listening to an ungrateful child complain.

"This is an equitable split," Art said.

Again, Doc only shook his head. He slowly and with great dignity climbed the remaining steps. They heard the door latch click behind him.

Art and Henry, alone in the front room amid the cigarette butts and debris from weeks of pilots living in close quarters, looked at each other. Art, knowing Henry would weaken, said, "Forget him, Henry. He gets a third. It's fair. Get the car loaded."

A bonfire burned rapidly in the yard near the house sending a column of blue smoke into the cloudless winter sky. Doc stood at the open window in his room, silent, occasionally rubbing his thin hands together behind his back. He was dressed in his finest pastel gray three piece suit, a matching homburg hat perched like a crown on his head. His spats were spotless white; his out of style high button shoes were polished to a brilliant black shine.

He stared down into the flames. A blackman, indistinguishable to him from any of a number of the mechanics and drivers on the field, walked to the flames and without ceremony took a sledge hammer and smashed a chalkboard standing beside the fire. The black slate crumbled into the mud, and the wooden frame was tipped onto the heap and consumed.

Doc's face remained blank, and he watched another chalkboard, still covered with names and aircraft numbers, carried to the edge of the bonfire. The executioner's hammer crashed into the scribblings, and, again, the frame was dropped on the fire where it crackled and vanished.

"This the last of it?" a voice asked from behind him.

"Yes," Doc said. "Thank-you." He waved the man away and stood alone in the cold empty room, barely lit by the late afternoon sun. The wallpaper, he noticed for the first time, was stained from water leaks and peeled away from the wall near the ceiling. He shivered and tapped his hat tighter onto his head. Outside, a truck started, and the driver gunned the accelerator impatiently awaiting someone. Doc left the room and descended the stairs.

"There's not much time, Doc," Henry spoke from the bottom of the stairway where he held a winter coat out for him.

Doc wiggled into the heavy coat. "Thank-you, Henry. Are you driving?"

"No. Andrew will. Art will ride with you to the station."

Doc fastened the last button and tucked a beige scarf around his neck. He then pulled a pair of chamois skin gloves over his hands and flexed his fingers. Without a word or a glance back, he left the house and entered the Packard. Alone in the back seat, he stared straight ahead, conscious of the eyes upon him. From the corners of his vision he saw the bonfire flicker, and ahead he watched Henry help two men load the last of the salvageable airplane parts onto the truck. A canvas flap was dropped, and with a snort of exhaust smoke and crunch of gears the truck departed.

The Packard's door opened on the passenger's side, front seat. Art stepped in.

"We'll have to hurry," he said, but Doc only stared.

The driver's door opened, and Andrew slid in and started the motor. They left the farmstead behind and drove the short road to the main highway.

"Look," Andrew said.

"Just pull up and stop," Art said.

Ahead, at the intersection, traffic was blocked by two police cars, three unmarked cars and a pair of motorcycles with uniformed riders. The Packard slowed and Doc saw Mayor Twimms poke his head from one of the plain cars. Andrew rolled the window down, and a policeman approached.

"Follow us. Understand?"

"Yes, sir," Andrew answered without making eye contact.

After several minutes, while cars jockeyed for position around the Packard, the convoy pulled away. Approaching town, they turned off the main route.

"It's 4:36 now," Doc said looking at his watch. "What time was my train, Arthur?"

"It might be late."

"Hmmm," Doc said and slid his watch back into its pocket.

The convoy skirted the business and residential districts and approached the station from behind the stockyards and slaughterhouse west of town. Doc looked at the pens where the cattle stood unemotional in the black mud waiting. Occasionally, one would stir in the tight quarters, and the whole herd would shift a few steps and stop.

"Looks like they've held the train for you," Andrew called turning into the gravel parking lot beside the station.

Car doors opened, and policemen deployed themselves along the platform blocking curious pedestrians. The engineer peered down from the locomotive, an expression on his face similar to the cattle in the pens—waiting, unquestioning. A conductor paced impatiently on the platform.

"Seven minutes behind schedule!" he said, but no one cared.

Mayor Twimms sat in his car a discrete distance from the station. Doc's eyes caught his briefly, and he nodded politely, but the mayor looked away lest someone should notice.

"Your things are already on board," Art said from Doc's side standing on the platform.

"I should think so." There was a pause while neither man said anything. The only sounds came from the hissing of the steam engine and the sporadic coughs from the many policemen surrounding them.

"Good-bye, Doc."

"Good-bye, Arthur." He took one step onto the train's stairs then stopped. He looked at Art with a trace of a grin on his thin lips. "You do good work." And he was gone.

Chapter
XXVII

Jake Hollow could hardly turn around without finding Susan Bowers close by his elbow, her round eyes full of longing, her mouth ready with a smile—too ready. She held a glass in her hand but drank sparingly and only when she knew he was watching. Whiskey to her was a burning sour drink which made her giggle and always left her nauseated. Jake, on the other hand, drank without so much as a grimace, the way Susan's father drank.

By nightfall all the planes were in, and all the pilots were drunk. The Bowers' house was a smelly roar of crude songs and overly embellished flying stories. Lucky Lundy sat in a corner alternately snoozing and pining over the loss of his dog. Several pilots tried, but in vain, to explain that the dog was still alive and had not been kidnapped or diced into the stew pot as Lundy charged. Lundy drank himself unconscious and was still that way when the first truck from Lincoln arrived with the dog. After licking her master's face, and finding it still warm, the dog set out working the room for handouts.

"There's plenty of stew left!" Susan called over the racket in the kitchen. Her father, his eyes heavy with drink, scooped wet dollars off the beer soaked table while Jake and two other pilots slapped fresh money down for the next poker hand. No one was particularly good at the game, and the pot migrated from one player to the next without anyone gaining an advantage.

"More stew, Jake?"

"Huh?" Jake barely glanced at Susan, but treated her with a distant politeness, worse than ignoring her entirely. He felt his efforts to avoid her might border on cruelty, but he made no effort to change.

"There's plenty..."

"No. No, thank-you, Susan," he said and poured a drink.

"Susie, leave the poor man alone!" her father bellowed. "Christ, you been pesterin' him all day."

She felt the searing embarrassment that only a parent can inflict.

"I only thought he might be hungry..."

"Hell's bells, girl! You've stuffed enough potatoes into him to turn him into an Irishman! Now whose deal is it?"

"Yours," one of the players muttered, his face on the table. "The cards are in your hands."

"So they are. Well, ante up! How's the liquor holding up?"

"Fine, fine!"

"There's more beer. Isn't there more beer, Susie?"

She nodded absently and stood behind Jake.

"Oh, now she's pouting," her father said, and Jake turned his head. Susan was indeed mid-pout. Her lower lip and chin stuck out from her face, and her eyes cast a killer frown across the table.

"I have to go see a man about a horse," Jake said and stood uneasily from the table.

"At this hour?" Susan asked.

"Ahhh..." Jake had no idea how to respond, and he stumbled through the back door and off the porch. Susan was right behind him following him to a wide tree in the shadows.

"Why don't you wait here?" he said to her.

She stopped abruptly. Jake rounded the tree. She toyed with the remnants of a snow pile with her toe until she noticed the urine stain etchings and stepped away.

"What-the-hell am I going to do with her?" Jake asked himself facing the tree. He finally buttoned his trousers again and stood alone in the dark. A warm damp wind blew from the southwest, and he breathed in deeply. Tomorrow it would swing out of the north, he thought. And that meant more snow, more winter. He desperately wanted to collect his money and head south. He figured Doc owed him about $10,000, and although he had spent most of his previous earnings or loaned it to Susan's father, he figured $10,000 would be plenty to start out again.

"I could buy another plane in Alabama, or somewhere," he spoke to the tree as if it could listen. "Christ, Florida's full of money! Five bucks a ride, no ten, at least! Fifteen bucks a ride and I'll be busy all winter."

He leaned his forehead against the trunk and closed his eyes. The alcohol was already burning off in his system and a wave of guilt and depression flooded over him.

"Susan," he muttered, he thought quietly. He pressed his forehead firmly against the shag bark hardly feeling it dig into his skin. "What do I do?" He slowly rotated his head grinding the bark into his skin. "She can't come with me. I don't love her!" He jumped, hearing his own voice echo off the house.

Fearing she might have heard, he suddenly longed for her, longed to apologize and make new promises.

Then, "No! She can't come!"

He pushed his head away from the tree feeling pain where the bark had rubbed. "What'll I tell her? What'll she think?"

"She'll think I'm taking the world's longest piss for one thing!"

He stepped from behind the tree determined to confront her. Luckily, she was gone. The door to the kitchen was closed. The muffled rumble of male voices vibrated into the darkness. The door swung open and a pilot fell out, catching the pillar by the stairs and swinging himself onto the ground beside Jake.

He undid his fly over the snow pile, and Jake walked past him up the stairs.

"Hey, ever write your name in the snow?"

Jake stepped back inside.

"Thought you'd fallen asleep out there," Susan's father said. "What did you say to Susan?"

"Nothing, why?"

"Oh, just thought you two might have had words. She wandered through here like she just lost her best friend. Went to bed." He tried to shuffle the wet deck, and finally resorted to simply mashing the beer-soaked cards into a disorderly pile. "You want in the game?"

"No." Jake started to leave.

"You might want to leave her be." The father spoke carefully, his tone momentarily sober, protective.

"I'm just going to catch some sleep."

"The couch is your best bet, downstairs."

Jake nodded and left the room. In the living room, pilots had staked out all the furniture, and while most were dropping into sleep, the stories continued although enthusiasm was waning. The couch was taken by someone dressed in full leather flying coat and boots. He snored with his mouth wide open. Lundy was curled in a corner, his dog asleep at his side. Someone had draped a blanket over his shoulders and head. Jake, without hesitation, removed it for himself. The dog stirred briefly, and, showing no interest in Jake, moaned and settled itself against Lundy's stomach.

Jake picked his way over sleeping forms to a spot below the window and curled with his face to the wall on the hardwood floor. He turned once to gaze at the staircase leading to Susan's room. While he tried to think about her, his mind shut down, and he was asleep.

Jake Hollow had no dreams that night, and only woke once long enough to shuffle outside to the tree with the blanket over his shoulders. By then, the house was cold, and everyone asleep. His head pounded, and his mouth was cotton dry, but he curled back onto the hard floor and slept past dawn.

"Jake!" The voice calling his name broke through the blackness inside his brain, and he tried to ignore it. "Jake!" Something shook him. "Get up!"

Life flowed slowly through his veins and into his tortured head. He woke and for a moment, a brief moment, felt no pain, then with a thick bump the agony of last night's alcohol throbbed in his skull.

"You look like a bag of wet underwear!"

It was Art, and he stood over Jake leaning on a cane. He was dressed in a clean suit and had shaved recently. He smelled of Bay Rum.

"Where'd you come from?"

"Got in this morning, about two hours ago."

"Bring any money?"

"Don't you want breakfast first?"

Jake felt his stomach lurch at the mention of food. The headache he could handle; it was the nausea that truly plagued him.

"Just want my money."

Art reached inside his suit pocket and took out a stack of bills wrapped with a rubberband and a slip of paper with Jake's name atop the first bill.

"$8,236." He dropped the bundle on Jake. It bounced off the corner of the blanket and onto the floor. Slowly, Jake picked the stack off the floor and stared at it. "Count it."

Jake turned the bills over several times before he spoke. "Eight thousand?"

"And 236."

"That's all?"

"You were by far the highest paid pilot in the lot. Count it."

Jake thumbed through the bills, mostly fifties. "It looks like about eight grand. I thought I had more coming."

"If you're satisfied the full amount's there, just sign the slip." Art handed him a fountain pen after unscrewing the cap.

"Are you sure I only get eight grand?"

"I can show you the books. Every flight is logged, along with every bonus, and expenses." Art stared at him knowing the books were now ashes.

Jake waved Art away and scribbled his name on the slip then handed it and the pen back.

"Thank-you, Jake Hollow." Art slipped the paper into his jacket and looked for a moment at the worn out hungover pilot at his feet. "Good luck to you," he said. "You know..." He paused and looked around at the empty room. Bright sunlight poured through the dirty windows. Jake was the last one asleep on the floor. The rest of the pilots had been paid off and were being shuttled to the train station in Luther or were outside preparing their airplanes.

"You know, Jake," Art continued. "You're a good employee—a good pilot."

"I know I'm a good pilot," he said. "Or at least I used to think I was a good pilot. I could care less about being a good employee—anybody's employee!"

"We made good money off you. I understand you plan to move south."

"Florida," Jake said.

"I have connections down there, if you need any help."

Jake waved the bundle of money weakly in the air. "This is help enough...although I did expect a little more."

Art laughed. "You're an employee. You only get paid for what you do. Dollars for work units. That's no way to get rich."

"I wasn't asking to get rich."

"But you could, Jake."

Art leaned heavily on his cane and removed a silver cigarette case from his pocket. He offered one to Jake who took it, then slipped one into his own mouth. After lighting both from the same match he said, "Jake, Florida's not a bad choice. Cuba's just a 90 mile skip across the water, and it's full of rum."

Jake blew a smoke ring and poked his finger through it.

"You get yourself settled down there, and maybe in a couple of months when that's gone..." He nodded at the money on Jake's lap.

"This?" Jake asked. "Hell, this'll last me forever."

"Yes? Well, if it should run out, you call me." The accountant took a calling card from his billfold and held it out to Jake.

CONSUELA'S IMPORTS
CUBAN GIFT ITEMS
MIAMI, FLA.
TELEPHONE: MI4-4167

"Consuela?" Jake asked, amused.

"Just mention my name, first name will be adequate, and they'll get a message to me." Art leaned over and held out his hand. "Shall we say in a few months, Mr. Hollow?"

Jake grinned at Art and took his hand. The grip was firm, not what he expected from an accountant. "We'll see, Art. We'll see."

"Where's Susan?" Jake asked. Mr. Bowers pushed open a small door in the side of the barn.

"Still on her morning deliveries," he said.

Sunlight filtered into the dark building from two small windows where the hay loft would have been had the barn been built for hay. Instead of a loft, there was a great empty vaulted space above the building's lone occupant, Jake's Standard J-1 biplane.

"Shouldn't she be back by now?" he asked. He showed no interest in the wrecked plane having done his best to forget about it over the months.

"She had some errands to run," the father said, and he pushed open one of the huge double doors allowing light to pour in. The broken J-1 sat like an actress in the final scene of a tragic play lighted by a wide beam, the sole player on the empty set.

"This was a waste," Jake said.

"What?"

"Building this; for that wreck," Jake said and pointed at the Standard. "Money could have been better spent."

"This is a great barn! It's not built like others, you know." Mr. Bowers strode across the floor sharing the spotlight with the Standard stealing the scene.

"There are no center poles like your ordinary barn," he said. The roof's trussed, like a bridge. That way we could fit your airplane in here without taking off the wings. You like it?"

"I've seen it before," Jake said.

"Yes, but you've never done anything with it. You've put a lot of money into this."

"Consider it a gift."

"But what about your airplane?"

Jake glanced around at the biplane with its shattered gear and torn fabric. The smell of mice was strong in the air, so he knew they were in the wings feasting on the rib stitching.

409

Chunks of dried mud, on there since June, still clung to the underside. He shook his head.

"I'll never fix it," he said. "You should take it out and burn it; put the barn to good use. Consider it my gift to you and Susan."

The two men looked at each other without talking. Finally, Jake walked away. "I need a ride to the train station."

"When's your train leave?"

"I don't know. There's bound to be one along sometime. I'll take whatever they got, so long as it heads south, today." Jake left.

Susan's father closed the large door, plunging the biplane back into the shadows.

Jake was the last pilot to leave the Bowers' place. He waited in the front seat of Mr. Bowers' Buick and listened to Lucky Lundy's motor fade beyond the trees to the west.

"Goin' Hollywood, Jake!" Lundy had said to him after he started the Robin's motor. "I hear they run through pilots out there pretty quick; figure they could use one with experience. Maybe you should give it a try."

"You be careful with all those starlets out there, Lundy. I hear they're all sex starved!" Jake had said over the motor.

"Sure you don't want to come along; can't handle them all myself. Well, maybe I could, but I don't want to be greedy."

Jake sat in the Buick, smiling to himself, remembering Lundy. His motor had faded into the wind. Mr. Bowers walked heavily down the porch stairs and slid into the car. They were down the driveway and onto the dirt road when Jake gave a quick glance over his shoulder at the house under the tree and the new barn beside the pasture. He was surprised to feel the dull pang of loneliness that seemed to follow through his life, strike at his insides.

"Tell Susan I'm sorry I missed her," he said. "But I have a job waiting in Florida. She'll understand. I'll get back when I can." He knew it was a lie. He knew the father knew. He hoped Susan would know.

410

"If anyone wants the plane, they can have it. There's no title or anything. Just give it to them. I can get another. They're cheap."

Susan sat on the back of the truck, alone, parked on a short road behind a stand of leafless oak trees. The road was on a slight hill from where she could see the town of Luther and the railroad tracks leading in both directions. The day was warm for the first of December, and a slight breeze dried the tears on her cheeks.

The Buick had passed her without stopping. There was no way they could have seen her, she thought. She recognized Jake's face and his unkempt hair. He was looking off to his right at the empty fields while her father drove.

Now, the Buick passed her again headed back home. Without looking, she knew only her father would be inside.

She sat alone on the hill for over an hour. She tried to make time stop long enough for Jake to appear somehow, maybe walking alone on the road below, away from the train station, his small bag in one hand the other deep inside his coat pocket. She would jump in the truck and race down the hill, then pull alongside him and, leaning casually over the door, call, "Hey, stranger, need a lift?" Then she would see his thin smile and those blue eyes looking at her, his expression contrite. She would say nothing for a moment, then forgiving him, she would push open the door and...

Gradually, the wind shifted from southwest to west, then as the clouds moved in and started to block the sun, the wind shifted to northwest, and the temperature dropped. Susan sat on the truck alone on the hill. The train appeared on the western horizon, first as a puff of gray smoke, then as a long snake made up of boxcars and two passenger cars behind the locomotive.

She stood and watched the train glide behind the few buildings in town and stop. Only the very end of the caboose showed. Less than ten minutes later, a cloud of black smoke and a distant shrill of the whistle announced its departure. The whistle bounced across the cold landscape and faded.

411

Susan climbed onto the truck's railings and watched until the last car vanished into the endless rolling farms and woods. Jake Hollow was gone.

"Good morning, I'm Mr. Dwyer. I am pleased to finally meet you."

Art stepped around the low mahogany gate held open by the bank representative. For a small town like Boone, Iowa, the First Community Bank was an impressive affair.

"When my secretary said Wendall Grieve was here I had to drop what I was doing. I have been anxious to meet you." He showed Art to a padded leather chair beside the railing across from a polished oak desk. The clatter of adding machines and muffled voices of a dozen clerks made a pleasant background in the cavernous stone building. A steam radiator hissed gently against the wall.

"Would you care for some coffee?" he asked.

"No thank-you," Art said and was about to add something when Dwyer spoke.

"You know, opening an account via the telephone is, ahh, somewhat unusual, so I am pleased you are here to sign the papers now. Are you relocating to Iowa?"

"Mr. Dayer..."

"Dwyer."

"Whatever..."

A crash and a shout interrupted Art, and his head turned toward the front door where a bevy of clerks ran to assist someone who had slipped on the polished marble floor.

"We get at least one a day taking a good spill out there." The clerks helped the fallen man to his feet.

"Henry!" Art called.

"Hello, Art!" Henry answered and waved. He picked his leather flying helmet off the floor and walked carefully to the gate, fiddled briefly with the latch and stepped over.

"That floor is slicker than snot!" he said in a booming voice and took Art's hand.

"Mr. Dwyer," Art said, "I'd like you to meet Wendall Grieve." Henry grasped Dwyer's hand and shook it several

412

times. The banker only stared with a polite smile at the huge
stranger dressed in a brown leather suit spotted with grease.

"You are Wendall Grieve?"

"Oh, yes, sir!" Henry undid the front of his suit and
dropped the gauntlets and helmet onto Dwyer's desk.

"But he called you, Henry."

"Yes, so he did...Henry's my nick name!"

"Old family nickname," Art added quickly. "It's a custom
in Utah...very Mormon."

Dwyer nodded his head slowly at the two men. "Yes, well,
Mr..."

"Grieve," Henry said. "Call me, Mr. Grieve. I like that!"
He smiled at Art

"You do have some form of personal identification with
you? A driver's license or something?"

"Certainly!" The new Wendall Grieve reached deep inside
his suit.

"Why are you dressed like that?" Dwyer asked. "The suit."

Henry looked puzzled for a moment then said, "'Cause if
I wasn't dressed like this I'd freeze my fanny off! Here."
Henry pulled the real and well-deceased Wendall Grieve's
wallet out and dropped the contents including his acquired
mother's obituary on the desk.

Dwyer looked at Art who merely smiled, then carefully
picked the items from the desk. Convinced the world was full
of surprises and Wendall's identity was valid, he slid a form
across the desk for Henry to sign.

"Where I've placed the X, please."

Henry took the pen and with a glance at the birth certificate
to check the spelling he carefully penned: Wendall Grieve.
The transformation was made; Henry was Wendall.

"Well, Mr. Grieve, I'd like to be the first to welcome you
to Boone. I'm sure you'll find this town has everything one
could expect to find in any of the bigger cities, say Des Moines
or Waterloo."

Dwyer held the gate open for Art and Wendall and latched
it carefully before leading them to the front door.

"Will you be living in town?" he asked.

413

"I don't know yet," Wendall answered. "I've some business down south I need to look in on first. Say, do you give away these calenders?"

Outside, Art leaned against the Packard and looked at Wendall. The tall man in the bearskin flying suit drew several stares from pedestrians.

"Where's your airplane?" Art asked.

Wendall pointed. "Down the street and around behind the grain elevator. Town doesn't go much beyond that. A few folks came out to see me after I landed, so I paid a kid a dollar to keep them off."

"Are you certain you want to live here, Henry?"

"Wendall!"

Art smiled and knew the answer.

"Andrew's going to take me to the station. I'll catch a train to Des Moines, then south out of there."

"Texas?"

"No, Florida," Art said.

"Florida! You and Jake both got warm weather on the brain."

"And I'm not so sure what's on your brain to keep you here," Art said and spread his arms indicating the surrounding town. "I don't know about Jake Hollow, but I've got a few ideas I'd like to try out."

They looked uneasily at each other, and Wendall shifted his weight from foot to foot.

"I could use someone like you, Hen...Wendall."

"I'll think about it, but I wanna stay right here for now...you know."

"I met her," Art said.

"Susan?"

"Early this morning, at her house. She was about to leave. Delivers milk or something."

"Yeah."

"She seems nice." Art grasped Wendall's arm and made him look at him. "Is that what you want?"

"More than anything, Art. More than anything!"

414

"Andrew," Art called through the open door, "let's get to the station. The sooner I get outa' here, the sooner you get to go home."

The Packard started with a deep rumble. Wendall glanced at the long hood and ran his finger along the chrome trim.

"Andrew gets to keep the car when I leave," Art said. "Part of his pay off."

Wendall leaned his head through the door after Art stepped inside. "Andrew, you take care of this machine, ya here?"

"Bettah 'an you did, Henry!" The black man laughed and extended his hand. Wendall shook it vigorously.

"Get along, that train might be waitin' for you right now!" He stepped back, and the long Packard moved away leaving Henry-turned-Wendall alone with his helmet and gloves in his hand.

John Bowers was in his second floor bedroom when he heard the airplane motor overhead. From the window he could see the entire landing strip including the three wrecked Standard LS-5's left behind by the now vanished bootleg airline. The barn was near the end of the field and to his right, close enough that he could see his daughter step out the small side door and look skyward.

The biplane was no more than 300 feet above her, a slow moving arrangement of wires and sticks wrapped in cloth, somehow producing flight. Except for the pleasant bark of the OX-5 engine, the world was silent. The clouds had thickened all that day, and the temperature, as afternoon faded, dipped below freezing.

The now infrequent breaks in the overcast allowed brief shafts of sunlight to dart at the ground then vanish. The biplane circled the barn, and sunlight broke through, again catching the machine descending to the muddy pasture where it touched with a slight bounce.

Mr. Bowers saw the tall pilot, dressed entirely in brown leather, climb down from the cockpit. He saw his daughter cautiously approach the man, and from the puffs of steam at their faces, knew they were speaking.

The flyer held out his hand still in its glove. It hung motionless for a long while, then she took it. They walked slowly toward the barn with hands still locked. Susan slid her arm around the pilot's waist, and they disappeared through the small door.

Chapter
XXVIII

1979

"I just want to see what's out there," Terry Marcin said.

"But it's not your property. You just can't go and land an airplane on anyone's land." Meredith laughed as she spoke, because Terry tickled her behind whenever she took the not-in-this-library-you-don't tone with him. "Stop that!"

The air was warm, and cumulus clouds piled up to the west.

"Look, if we stand around arguing all evening, it'll just get dark, and it'll rain, then tomorrow the field will be way too muddy to land in. Just climb into the front seat."

"Oh, front seat is it? How privileged I feel! What have I done to deserve the honor? Normally, you make me sit in the baggage area."

"Oooyyah, why you!" Terry screwed his face up in his best Wallace Beery imitation while rubbing his hand across his nose and mouth.

"Oh, now do Jimmy Stewart!"

"Get in." They both giggled the silly way only people in love can.

"Switch is off; throttle's back, and the brakes are on!" she called, and Terry felt immense pride in the woman who had taken so well to the sky. He pulled the propeller through until gas dripped onto the pavement. The only sound was a distant hiss of car tires on the highway and the clacking of the impulse coupler on the left magneto each time the propeller snapped through half a turn.

"Okay, switch ON! Throttle BACK! And brakes ON!" he called.

"Switch ON; throttle BACK; brakes ON; you're CUTE!" she replied. Her smile beamed through the plexiglass windshield.

"Women!" he muttered and spun the propeller. The engine barked to life and idled softly. He stepped back from the

Champ admiring the silver propeller and the way the sunlight fluttered in a wavy arc on the spinning disk.

"Oil pressure come up okay?" he asked climbing into the back seat.

"Came right up," she answered, shouting slightly to be heard above the motor.

He closed the door and buckled himself in while she taxied to the runway, scanning the pattern for other planes. They saw no one, although two miles out, Wendall Grieve's Beechraft Baron entered a downwind to land. He had announced his position on the unicom frequency, but Terry carried no radio—none required—none needed.

Meredith held the brakes and ran the engine up. She checked carburetor heat then each magneto and moved the joystick in a full square checking aileron and elevator movement.

"Sounds good," Terry shouted. "Let's take the runway," he said, and they both checked the sky again. "Now remember, hold a little right rudder when that tail starts to come up, not much..."

"How much?"

"Whatever it takes."

"Big help," she mumbled under her breath. The throttle came forward, and the engine howled. The tail slowly lifted as they rolled down the pavement. She held direction well, although oversteering slightly with her feet. The airspeed nudged fifty, she pulled back on the stick, and they lifted away.

"Good! Very good!" Terry settled back in the rear seat and let her fly. He saw Wendall's Baron enter the traffic pattern and turn base leg. The rest of the sky was vacant except for the scattered clouds building slowly into thunderstorms.

He tapped Meredith's left shoulder indicating she should turn toward the barn. She knew the route. Terry had shown it to her before, and the two had speculated a number of times about its contents. Wendall's interest in the building, and his reluctance to speak of it, further cloaked it in mystery.

418

They followed the river, then picked up a long stretch of gravel road that led over the field where the barn was located. Terry had thought many times about landing on the road, but that would have meant leaving the plane while he walked the hundred or so yards to the barn. He had decided to wait until the field was harvested and simply taxi up to it.

"Look!" he yelled and again tapped Meredith's shoulder. She nodded.

"They've cut the hay."

"Let me take it," he said and took the controls. From the back seat he descended the Champ and made a pass over the alfalfa field. The hay was recently cut and baled into huge round bales scattered haphazardly across the field.

"This'll be fun," he said.

"You going to land?" she asked. "Is there room? It's so hilly."

"No problem," he said. "I'll just touch down at the very edge of the hill, miss those two bales, then turn slightly and go around that other one in the middle there." He sounded self-assured, but to Meredith it looked as though they were to land in the middle of a chess game, with any one of the pieces able to rip a wing off.

The Aeronca Champ descended slowly over the bushy stand of hedge trees bordering the ravine at the end of the field. Terry eased the stick forward, and they skimmed the grass.

"You're going to hit that..."

With the tail still up, Terry skirted past the first two bales.

"No faith!" he called.

They rolled safely past another bale and stopped well before even a third of the field was used.

"Not bad," she said.

"Next time I'll try it with my eyes open."

The low rumble of distant thunder met them stepping out of the taildragger. The wind was dead calm, and the air thick with the smell of fresh hay. Terry looked quickly over his shoulder to make sure Meredith was following. They approached the barn staring up at the huge doors and the tiny eye-like windows high above them. The sky rumbled again to the west. She took his hand.

"Well, I've seen enough; let's go home!" Terry quipped and pulled her toward the Champ.

"What? No!"

"It's just an old barn, for Chrissakes," he kidded.

"And it might be full of...of..."

"Old barn stuff."

"Then why is your boss, Mr. Grieve, so interested?"

"Maybe he wants to buy it. Maybe it's the one thing in the county he doesn't own."

"Oh, he doesn't own the whole county," Terry said.

"Damn near."

They walked hand-in-hand toward the large double doors below the small windows. The building seemed to stare down on them like a great monster, dormant but awaiting victims. Terry pressed his right eye to the crack between the two door halves. Inside, all was dark.

"What do you see?"

"Oh, just old barn stuff—pots of gold, naked virgins doing the dance of the seven veils..."

"Oh, get out of the way," she said and gently pushed his head lower so she could peek. "It's dark."

"Keep looking, there's light from those windows up there." His eyes adjusted slowly. "There is something in there; something big."

"Is that a propeller?" she asked.

"You know, I think it is!"

Suddenly, a roar of engines passed over their heads, and they both jumped back. Off to their right Wendall's Baron banked low over the road and circled back toward them.

"What's he want?" Terry asked himself.

The twin, its landing gear retracted, pointed straight at them and made another low pass. Terry ducked while Meredith watched the plane speed over. The Baron banked again, and in a steep turn, dropped its gear and lined up on the empty gravel road.

"He's landing," Terry said.

420

The Baron touched the gravel road sending up a cloud of gray dust making Terry think of all the nicks he would have to file out of the propellers the next day at work. It rolled past the barn and turned onto a grassy area several hundred feet away.

"I bet he owns this," Meredith said and pressed her face against the door to look inside. "What is in there?"

Grieve killed both engines, the door popped open, and he hopped out adjusting his baseball cap. He climbed over the fence and headed toward them.

"That is a propeller," Meredith said. "There's some kind of airplane in there."

Terry took her hand and walked to meet Grieve. Halfway to him he called, "Hello, Mr. Grieve."

"Kinda figured you'd be out here. Hello, Miss Waters."

"Hello, Mr. Grieve. Do you own this?" Terry squeezed her hand. "Terry!"

"Quit crushing the lady's fingers, Terry. For Chrissake, she has a valid question." He held his arm out for Meredith, and she took it with a smirk aimed at Terry. "Come along, my dear, I'll show you."

Terry followed Meredith and his employer to the small door at the side of the barn.

"There used to be a house over there." Wendall pointed toward the road where a Datsun pick-up inched its way around the Baron. The occupants stared at the plane and drove off. "You can still see parts of the foundation, right next to the oak tree."

"Whose house was it?" Meredith asked.

"My wife's. Well, actually her father's, then I bought it from him in 1933. He lived in it until he died in '37."

"Did you live there?"

"No," Wendall said. He gazed at the oak tree and spoke softly, his voice constricted. "My wife, her name was Susan, always wanted to move in there, but we never did. Too far from town. I had too many business dealings in town. We share-cropped the place out, and one of the tenants burned it down after the war, somewhere around '46 or '47, I think."

421

"I didn't know you were married," Terry spoke.

"You didn't ask." Wendall and Meredith both shot Terry looks of extreme reproach.

"Is your wife still alive?" The question was awkward, but Meredith was curious.

"No," he said. "No, she died some years ago." He waved his hand vaguely to indicate the distant nature of his wife's death. Meredith caught the hurt in his expression.

Thunder rumbled to the west, and Wendall said, "Well, if you're still curious about the damn barn, we'd better take a look before it rains on us. Come on."

He rummaged inside his pocket and pulled out a small key. This he inserted into a rusty lock and with a little jiggling it popped open. The door was slightly jammed, but he put his weight to it and it opened.

"Ohh," Meredith said.

"What is it?" Terry asked. "Doesn't really look like a Jenny."

"It ain't," Wendall said and lead them inside the dank building lit only by the two windows high above the numerous cracks in the walls. Sparrows flew between the overhead trusses. He pulled a dusty tarpaulin off the fuselage.

"It's a Standard, a 1918 Standard J-1. Built in Elizabeth, New Jersey, wherever the hell that is."

The biplane sat as it had been deposited in 1929, upon the sawhorses. The gear was smashed, and the fabric virtually gone from years of mice and birds nesting and feeding. Strips hung from the wings and fuselage like Spanish moss.

"Susan always said she wanted to rebuild it and learn to fly someday."

"Why didn't she?" Terry asked.

"It didn't actually belong to us. Belonged to a friend of hers, fella named Jake something..." Again, he waved his hand dismissing the past. "I guess she was just waiting for him to come back...I guess."

"And he never did?" Meredith asked.

"No," Wendall said. "I've got no need for the damned thing."

422

"It's beautiful," Terry said and ran his finger along the filthy longerons.

"You want it?"

Terry looked, his mouth opening in slight movements.

"Take it! I got no use for it." Wendall looked the biplane over from one end to the other and said, "I got no use for it." His voice cracked, and he turned, pushing his baseball cap back on his head. "Lock the door when you leave." He was gone before Terry could thank him.

Meredith and Terry were still in the barn when they heard Wendall's Baron take off. The sound of the two engines faded, and the rumble of the approaching thunderstorms grew louder.

"We'd better go," Terry said.

Meredith slipped her arms around his waist and pressed herself to him. Looking up into his eyes she kissed his chin with a light peck and said, "Teach me to fly?"

THE END

Epilogue

James and Herbert sold their trucks and moved both families to Brownsville, Texas where they opened a bait shop called Captain Jimmy's. During the war, the shop caught on and they soon had a small diner attached. In 1978, Herbert was shot during a robbery; his assailants never found. James, now almost blind, spends most of the day in a wheelchair on the front porch. His grand nephew and his wife run the bait shop and tavern.

Randall bought back his filling station from Sam's mother. It never was listed in Sam's name with the county. Randall paid less to rebuy the station then he sold it to Sam for. During the war, Randall made a small fortune selling black market tires, gas and oil—all rationed items. He died in 1946 in a hunting accident. Someone mistook him for a duck. Clara sold her diners one by one, until all she had was one road house outside of town in 1937. That year she died from influenza. Her obituary took four lines and made no mention of the incidents of 1929. Mayor Twimms died in 1935, and his obituary took 24 lines to extol his career's accomplishments. There was also no mention of the events of 1929, although everyone knew.

Mr. Howard left the Standard Aircraft Company when the reorganized New Standard Company refused to hire him. He and his wife wandered over to Moline, Illinois where he worked briefly for the Mono Aircraft Corporation before it folded. In 1937, he was offered a job in Wichita at the Cessna Factory as factory representative but saw no future with them so took a job with Culver promoting the Culver Cadet before that firm went belly up. In 1948, he moved to Hackensack, New Jersey and became a salesman for a Pontiac dealership where he stayed until his death in 1957. His wife worked for J.C. Penney Co. and retired in 1963 after which she returned to Lincoln, Nebraska where she still lives. When asked by a writer doing research on the old Standard factory, what became of her husband's books and papers from that era she replied, "I don't know; I think it was all thrown out when we left. He didn't make much at that job. I don't think he liked it."

Nick Barchek spent two weeks in the Lincoln hospital during which time he became quite attached to a young nurse named Kathleen. They had met briefly during his ordeal with the mayor and the reporters. After Doc had left, there was no need for Nick to remain in Lincoln, not that he really wanted to, either. He returned to New York after first swapping addresses with Kathleen. Upon returning to his office in New York, his supervisor, Patrick Fitzpatrick, showed him a letter received several days before his return. It was postmarked from Houston, Texas and read: THANKS FOR A WONDERFUL TIME. WE SHALL HAVE TO DO IT AGAIN SOMEDAY. YOUR SERVANT, THE DOCTOR.

Nick Barchek wrote regularly to Kathleen, and in February, 1930, in the middle of a blizzard he took the subway to Pennsylvania Station where she arrived on the train from Chicago. Together, they shared his apartment until June of that year when they were married. She found work as a nurse in Manhattan, and he stayed with the Treasury Department, taking over Fitzpatrick's position in the reorganized post-Prohibition office (see Fitzpatrick). He left government service in 1938 and moved with Kathleen to Fairlawn, New Jersey where he worked for a small publishing firm. In 1942, he joined the U.S. Army and worked as an I.G. auditor in the E.T.O. until February, 1945, when he was discharged after slipping on an icy sidewalk in France and injuring his back. He returned to New Jersey, and while confined to his bed for six months with back pain, wrote a best-selling novel about barnstormers and bootleggers. He now lives in seclusion with his wife Kathleen on his estate in Alpine, New Jersey.

Patrick Fitzpatrick suffered severe ridicule from his superiors in Washington for wasting funds chasing after Doc.

Although his career appeared stuck forever in neutral in 1929, things improved greatly when Franklin D. Roosevelt became president in 1933. One of Fitz's old Treasury pals (he still had a few) called him one day from Washington offering him a position there. Fitz jumped at the opportunity and found himself moving rapidly up the promotion ladder until he settled in as Assistant Undersecretary to the Treasurer in 1939. In 1949, he retired and took a position with an accounting firm in Arlington, Virginia, eventually buying in as a partner. In 1964, the firm was purchased by Dow Jones Inc., and Fitz received a handsome seven figure settlement. He now lives with his wife in Arlington where he raises Bernese Mountain dogs.

Wendell Grieve died in Boone, Iowa in 1984. The bulk of his estate was left to charity. The entire airport operation, however, he willed to Terry Marcin who runs it to this day along with his wife and chief flight instructor, Meredith. Terry devotes much of his spare time to rebuilding the J-1.

No one knows what became of Phil Moyer. Canadian National Railroad records show no mention of him after 1929.

And Jake? Jake Hollow found another flying job..., but that's another story.